GENETIC TURNING POINTS

The Ethics of Human Genetic Intervention

JAMES C. PETERSON

WILLIAM B. EERDMANS PUBLISHING COMPANY
GRAND RAPIDS, MICHIGAN / CAMBRIDGE, U.K.

© 2001 James C. Peterson

Published 2001 by
Wm. B. Eerdmans Publishing Co.
255 Jefferson Ave. S.E., Grand Rapids, Michigan 49503 /
P.O. Box 163, Cambridge CB3 9PU U.K.

Printed in the United States of America

06 05 04 03 02 01   7 6 5 4 3 2

Library of Congress Cataloging-in-Publication Data

Peterson, James C., 1957-
Genetic turning points: the ethics of human genetic intervention /
James C. Peterson.
p.   cm.
Includes bibliographical references and indexes.
ISBN 0-8028-4920-2 (pbk.: alk. paper)
1. Human genetics — Religious aspects — Christianity.
2. Medical genetics — Religious aspects — Christianity.
3. Genetic engineering — Religious aspects — Christianity.
I. Title.

QH438.7.P485   2001
174'.28 — dc21
                                00-067687

www.eerdmans.com

For Laurie

EPHESIANS 3:20-21

Contents

vii

Contents

PART I
GENETIC RESEARCH

PART II
GENETIC TESTING

Contents

PART III
GENETIC DRUGS:
ADDING GENE PRODUCTS TO THE BODY

CONTENTS

PART IV
GENETIC SURGERY:
CHANGING GENES IN THE HUMAN BODY

Contents

Acknowledgments

Human genetics raises fundamental questions across a wide range of disciplines. I am grateful for the scholars who have read portions of this manuscript in their areas of expertise. Of course it is my responsibility where mistakes or infelicities remain. Readers have included: geneticists V. Elving Anderson, Hessel Bouma, Caroline Freund, Don Munro, and Henry Tillinghast, clinical cytologist Frank Grass, physician Jose Bufill, professor of chemistry Chris Dahm, insurance executive James A. Peterson, clergy Rick Hughes and Merridee Peterson, philosophers Frank Beckwith, Dave Leal, and Robert Prevost, theologians John Jefferson Davis, Dennis Hollinger, and Randy Maddox, and bioethicists John Kilner and Robert Weir. Particular thanks go to James F. Childress, who not only read part of this manuscript, but bore with me as I first began to explore some of these ideas in my Ph.D. dissertation at the University of Virginia. French Anderson, John Fletcher, Thaddeus Kelley, and Daniel Westberg also advised that endeavor.

The Charlotte area has been a rich one to develop this study. Back in 1994, at the prompting of Professor of Nursing Lynda Opdyke, Sister Jerome sent me a letter summoning me to serve on the ethics committee of the Mercy Hospitals which are now part of the Carolina Medical Center Hospital System. Work there and on the Board of the Bioethics Resource Group has kept me in close touch with practitioners thoughtfully seeking the best for their patients in often conflicting circumstances. Teaching as an adjunct professor at Gordon-Conwell Theological Seminary has also been a source of insightful dialog with students and faculty seeking to best serve parishioners in often difficult times.

Outside of Charlotte, Robert Weir, Jeff Murray, and Ray Crowe welcomed me as a Research Fellow in Molecular and Clinical Genetics at the University of Iowa. It was an invaluable experience of firsthand work in genetics laboratories and clinics. Another point of ongoing genetics discussion and reference has been the American Scientific Affiliation. That association of several thousand scientists is an invaluable source of expertise and thoughtful fellowship.

At my professional home, Wingate University, Dick and Carol Dickson generously endowed a chair in ethics, the first endowed chair in the university's hundred-year history. President Jerry McGee and the board of trustees stunned me with that post early in my career. Also from the university a Spivey Instructorship halved my teaching load for one year and granted travel funds to a superb genetic counseling conference at St. Jude Children's Research Hospital. Staff can make or break a place. My division's secretary, Pam Merrill, has always been quick to meet secretarial needs expertly. Susan Sganga has so expeditiously fulfilled inter-library loan requests that Wingate's library has functioned much as the Bodleian does for me when at Oxford University.

Three foundations have supported this project. The Henry R. Luce Foundation did so indirectly. Some years ago they sent me to the far east and then on around the world as a Luce Scholar. While there are no direct references to that in the text, the experience has been pervasive in my work ever since. The John M. Templeton Foundation gave me an award for course work in science and theology and has funded a faculty research group that has continued that work at Oxford University. John Roche and Alister McGrath have hosted the program of rich discussion. The essential gift of concentrated time to write came from the Louisville Institute, a Lilly Endowment program for the Study of American Religion. Led by its executive director James Lewis and associate director David Wood, the Institute covered the cost of a teaching replacement for a full year. My department chair, Dr. Edwin Bagley, and the Academic Vice President, Dr. Robert Shaw, could not have been more gracious in the administration of the resulting leave.

Finally, heartfelt thanks to my wife Laurie who was always the first to be subjected to each new page of this book, and to my daughters Laura and Noelle who were usually patient when the study door was closed for Daddy to write.

Introduction

For years now, every day or two people have been asking me to recommend a book that, as far as I had been able to tell, did not yet exist. Whether we met in genetic labs or medical clinics, classrooms or pews, what they wanted was a book that would help them to recognize and start thinking through the ethical challenges of human genetic intervention. They realized that the rapidly expanding technology was offering new choices and demanding decisions. But the sheer number and importance of the questions and decisions were daunting. Should a cardiologist reveal to a patient that the genetic test she ordered for a heart condition also indicates that the patient is likely to face the dementia of Alzheimer's disease? Should companies be able to patent human genes? Would changing some of our children's genes make them mere objects of manufacture? Should genetic intervention be limited to cure of disease? What do philosophical and theological traditions say about genetic intervention? Is it ever appropriate to clone human beings? Thoughtful texts were available that addressed one aspect here or another there. What was not available was a book that organized and addressed the full range of interconnected questions; questions, and the way one answers them, that affect the specific decisions one increasingly has to make.

This book is an attempt to meet that need. Its first intended contribution is to sort through clinical decisions and myriad implications in order to analyze the key turning points in the discussion and their practical outworking. The ethical questions are addressed in the order in which the particular technologies that raise them are becoming available. That means beginning with genetic research and then moving on to genetic testing, genetically created pharmaceuticals, and finally genetic surgery that directly alters a person's

1

genes. This order allows a building process of addressing each question in relation both to familiar categories and to what has already been worked through within the book. At each level there are both points of continuity with earlier genetic and nongenetic interventions and aspects that are quite new.

Prior knowledge of genetics or ethics is not assumed. The demands of the subject call for both an army of specialists and an active interdisciplinary discussion. This text is intended to contribute to the ongoing scholarly discussion while remaining accessible across disciplines. For example, medical practitioners will be quickly brought up to speed on the essentials of the involved philosophical ethics and pastors on the needed genetics. Actually these two professions are already beginning to feel the most direct need for thinking through the implications of genetic technology. New genetic tests are multiplying rapidly and are increasingly ordered and interpreted by family physicians.[1] Most people in the United States who seek professional help for thinking through life choices turn first to clergy. Clergy need to teach thoughtfully ahead in regard to genetic choices and be ready for premarital and other counseling as it arises. This book is intended for these professions as well as more specialized clinical geneticists, theologians, and bioethicists. It will also be of use to the well-educated lay person who might be highly motivated by impending decisions. Such decisions are probably near or already present for anyone holding this book.

Part of the value of covering the range of issues together is to see the extensive interrelationships. For example, how one resolves one ethical question about genetic research may be directly connected to thinking through another concerning changing people's inheritance. Probably most of what is asserted in this book is assumed by some readers and contested by others. If the book belabors a point that the reader has already worked out, the reader should feel free to skip ahead. If the book touches too lightly on a point that the reader considers contentious, that may be an occasion for another book. This book builds its argument progressively but is organized so that the reader can use the table of contents to go directly to an issue of particular interest. Cross-references within each chapter refer to other relevant passages in the text. Each of the fifteen chapters is short enough to quickly locate a specific discussion.

Fifteen chapters is also an ideal framework for a standard semester course. It is increasingly important that all college students and certainly graduate students in a number of professions have the kind of orientation of-

---

1. Francis S. Collins, "Shattuck Lecture — Medical and Societal Consequences of the Human Genome Project," *New England Journal of Medicine* 341, no. 1 (1999): 35.

fered here. Much of the content of this text has already been well received in continuing medical education and in courses for both college and seminary students. Genetics is so central to human life, and increasingly to medical care, that it raises almost all the basic issues usually addressed in a standard bioethics course, from informed consent to the goals of medicine. It can be a fresh and compelling point of entry.

No author can avoid writing from a specific understanding of the world and what matters. This is constantly evident when thinking through the implications of a technology that is so formative for human beings. Whether painstakingly building from a starting foundation or seeking what John Rawls called a "wide reflective equilibrium," one's evaluation of human genetic intervention would do well to be coherent with one's convictions about the nature and purpose of life. I have found the classic Christian tradition the most convincing description of reality and so will include resources from that tradition throughout the following discussion of multiple questions and views.[2]

LeRoy Walters describes the last forty years of theological reflection on human genetics as divided between "cosmic theology" that thinks about the grand scheme of God's plan for humanity and "casuistic analysis" that addresses precise questions of practice.[3] One of my intentions here is to bring these two streams together. If one is consistent, one's understanding of the purpose of life will shape the concrete decisions one makes in life. Ideally, Christians will be conscious and open about how their cosmic theology shapes their practice and how their practice questions their cosmic theology. An integration of these two will be the second intended contribution of this work and will be of great interest to Christians who are making these decisions. Drawing from the Christian tradition will also help those who are not Christians to better understand some of the people with whom they work. For people in intellectual, social, or religious communities other than the varied Christian ones, the tradition might still be recognized as a rich resource. It has been developing questions and insights over two thousand years of reflection on our common human condition: a condition that is in many cases touched by the possibilities of human genetic intervention.

---

2. Dena S. Davis and Laurie Zoloth have brought together a thoughtful series on professional bioethics and religion. *Notes from a Narrow Ridge: Religion and Bioethics* (Hagerstown, Md.: University Publishing Group, 1999).

3. LeRoy Walters, "Human Genetic Intervention and the Theologians: Cosmic Theology and Casuistic Analysis," in *Christian Ethics: Problems and Prospects,* ed. Lisa Sowle Cahill and James F. Childress (Cleveland: Pilgrim, 1996), pp. 235-49.

Why Focus on Human Genetic Intervention?

Genes store the directions for the physical form of all life on earth, from the cold virus to orchids to human beings. All depend on the same basic system of genetics for their structure and operating instructions. Being able to read and shape that code is transforming agriculture, animal husbandry, and other industries as well. As important as these uses are, this study will focus on *human* genetics.

How genes work was largely unknown to us until a few decades ago. Human beings have long recognized the reality of inheritance and in some cases had already made rather sophisticated observations. Rabbinic law, for example, was aware that hemophilia, which only occurs in men, is always transmitted from the mother's side. But how inheritance occurred was a mystery. In about 1865 an Augustinian monk, Gregor Mendel, posited units of inheritance from his study of pea plants in his monastery's garden, but it was not until 1953 that Watson and Crick recognized the physical structure of these units of inheritance (DNA).[4] Since that step, knowledge about DNA has quickly accumulated from the efforts of countless dedicated people. The foundation for this development was laid by long-term investment in basic research with particular interest in curing cancer. Such study was drawn to what became the most promising level, molecular genetics. Government and university research has continued. Combined with the coordinated and independent efforts of industry, genetics has received an unusual infusion of talent, capital, and time.[5] The Human Genome Project alone, funded primarily by the Department of Energy and the National Institutes of Health, has been spending millions of dollars to record the sequence of the chemicals that form the human genetic code. The media loves a horse race and followed an ongoing wrangle between the private corporation Celera Genomics (CG) and the publicly funded Human Genome Project (HGP) over the ownership and release of new sequencing data. By 1999 allusions were being made to a working draft of the entire human genome.[6]

On June 26, 2000, the completion of a working draft was officially announced in a joint press conference by Francis Collins (HGP) and Craig Venter (CG). Reading the sequence is far from being able to understand or use it, but the basic scientific advance of making the sequence known is ex-

---

4. James Watson and Francis Crick, *Nature* 171 (30 May 1953): 964.

5. As a case in point, in an address at the dedication of Monsanto's new Life Sciences Research Center, Chesterfield, Missouri, 24 Oct. 1984, C.E.O. Frank Press stated that the facility was "the largest research investment in the company's history."

6. Collins, "Shattuck Lecture," p. 28.

tremely useful to better understand and begin intervention in the human genome. Labs have also completed most of the consensus sequence of some bacteria, yeast, the fruit fly *Drosophila melanogaster,* and the nematode *Caenorhabditis elegans.* Sequencing these organisms is particularly important because it gives us models to study how the human code functions.

The knowledge gained has already begun accelerating practical applications. What was considered distant science fiction not long ago, such as engineering the genes of bacteria to make *human* insulin, has already been done on a commercially successful scale for over a decade. The scientific advance in genetics is playing a growing role in how we predict, diagnose, prevent, and treat disease in medical care. Almost all disease has a genetic component. Genetics offers new therapies and a chance to further the movement from repairing damage done by disease to preventing the damage from occurring in the first place. Vaccination is a far better response to polio than the iron lung. Human genetic intervention might eventually offer increase in physical capacity such as resistance to disease. It is not possible to predict exactly when certain capabilities will be attained, but despite setbacks many developments have come much more quickly than generally expected. It may be that at this pace of development the next one hundred years will be the century of biology the way the last century was characterized by developments in electronics.

Lee Silver, a professor of molecular biology at Princeton University, writes, "For better and worse, a new age is upon us — an age in which we as humans will gain the ability to change the nature of our species."[7] We have long been able to transform ourselves culturally; rather than just getting around by walking, we can fly through space. But now we are developing the possibility of changing our physical nature as well. Such a capability has been called a "revolution,"[8] "the most awesome and powerful skill acquired by man since the splitting of the atom,"[9] "the threshold of self transfiguration."[10] Projections and metaphors can easily move ahead of what is actually possible for the near future. The "threshold" appears to be a wide one, but beginning pur-

7. Lee M. Silver, *Remaking Eden: How Genetic Engineering and Cloning Will Transform the American Family* (New York: Avon, 1998), p. 13.

8. Craig W. Ellison, "The Ethics of Human Engineering," in *Modifying Man: Implications and Ethics,* ed. Craig Ellison (Washington, D.C.: University Press of America, 1978), p. 3.

9. John Naisbitt, *Megatrends: Ten New Directions Transforming Our Lives* (New York: Warner, 1984), p. 74.

10. Victor C. Ferkiss, *Technological Man: The Myth and the Reality* (New York: George Braziller, New American Library, 1969), p. 28.

posefully to alter in some degree human physical endowment is no longer inconceivable. It has already begun.

Research in human genetics has been rapidly expanding for two decades now. This is where techniques are created and information gathered that makes all the other interventions possible. A second wave of work in genetics is genetic testing. Physicians have long done primitive genetic testing by taking a family history. This gives some idea of the gene pool from which the present patient has received genetic heritage. New techniques developed from the Human Genome Project make such genetic testing more precise and readily available. There are now hundreds of genetic tests, with more being added at an accelerating pace.

A marriage of these tests with the manufacturing process for computer chips is making them increasingly economical. The oligo-nucleotide probe array or "gene chip" is a small wafer with precisely ordered layers of DNA on its surface. The DNA on the chip fluoresces when it comes into contact with matching sequences of DNA. This provides a quick and inexpensive way to read a person's genetic code for the presence of genetic variations associated with various diseases. The greatest expense is in developing the templates for manufacturing such a chip. Once that is in place, individual chips cost about the same to make and process whether they have on board one test or five hundred. Economics is likely to drive toward bundled tests on a limited number of chip types. Currently there are about nine hundred genetic tests in use that could be placed on such a chip. The number of available tests has been roughly doubling each year. Who should have access to the resulting information?

The third wave of genetic impact in clinics is likely to be in pharmaceuticals. The human body knows how to make untold numbers of exquisitely detailed chemical compounds. Those recipes reside in our DNA. With the code in hand it will often be possible to make the needed compounds outside the body. This has already been done with great success in the case of *Humulin*. Diabetics used to inject porcine insulin when their bodies did not make enough insulin. While sustaining, it was not a perfect match for human insulin and had attendant risks from its animal source. With the discovery of the human body's code for insulin, it became possible to implant the directions for human insulin into bacteria. The commandeered machinery of the bacteria then follows the directions and dutifully manufactures human insulin. This is human insulin which has never been inside a human body. It is the treatment of choice. Other genetically based pharmaceuticals will probably be of great use as well.

It is much easier to read the genetic code than it is to repair or otherwise change it, yet the fourth wave of genetic technology may well be a kind of surgery that directly alters a person's genes. It may be widely used to treat genetic defects that can be resolved by the introduction of the missing genetic material. The first such treatment to show some probable success began September 14, 1990. It was intended to address a four-year-old girl's adenosine deaminase deficiency.[11] Other potential uses and incremental improvements in techniques have been pursued in the decade since.[12] For example, William Schwartz catalogs progress toward a gene-therapy contribution in the treatment of ALS, cancer, coronary artery disease, cystic fibrosis, epilepsy, glaucoma, Huntington disease, organ transplantation, stroke, and viral infection.[13] Yet there was not yet one clear case of cure by human gene therapy.

That appears to have changed in April 2000. Marina Cavazzana-Calvo and Alain Fischer announced then in *Science* that their team had saved the lives of two babies afflicted with SCID-X1 (severe combined immunodeficiency-X1).[14] The bone marrow of the babies lacked part of the genetic instructions needed for a working immune system. The physicians were able to insert the needed genetic material into marrow cells which then multiplied and displaced cells with the defective gene. At the time of the announcement the babies were continuing to sustain their newly functioning immune systems ten months after treatment. While the gene-therapy community has been chastened by the plethora of unfulfilled promises that have swirled around grant proposals and news stories, it appears that here at last is confirmation that in some cases gene therapy might actually cure a condition.

Of course techniques developed for one use can often be put to another. A genetic intervention to increase the muscle mass of a patient with degener-

11. ADA deficiency was chosen because it "is one of the few known diseases where gene corrected cells should have a selective growth advantage in the patient's body." W. French Anderson, editorial, "The Beginning," *Human Gene Therapy* 1 (1990): 372.

12. More than a decade ago W. French Anderson was already suggesting that gene transfer might be developed to serve as a minutely specific drug delivery system: "Whither Goest Thou, Gene Therapy?" *Human Gene Therapy* 1 (1990): 227, or as preventative enhancement in altering blood lipid metabolism to help protect from some kinds of cardiovascular disease: "Human Gene Therapy: Scientific and Ethical Considerations," *The Journal of Medicine and Philosophy* 10 (1985): 288. Other early examples include Kenneth W. Culver, W. French Anderson, and R. Michael Blaese, "Lymphocyte Gene Therapy," *Human Gene Therapy* 2 (1991): 107-9; Donald B. Kohn et al., "Toward Gene Therapy for Gaucher Disease," *Human Gene Therapy* 2 (1991): 101-5; and Douglas J. Jolly, "HIV Infection and Gene Transfer Therapy," *Human Gene Transfer* 2 (1991): 111.

13. William B. Schwartz, *Life Without Disease: The Pursuit of Medical Utopia* (Berkeley: University of California Press, 1998), pp. 131-47.

14. Marina Cavazzana-Calvo et al., "Gene Therapy of Human Severe Combined Immunodeficiency (SCID)-X1 Disease," *Science* 288, 5466 (28 Apr. 2000): 669-72.

ative muscle disease might be able to bring about the increased muscle mass desired by a competitive weight lifter.[15] While somatic therapy by definition treats only the initial recipient, germline intervention would eliminate the disease from both the patient and the patient's descendants. Such an intervention could take place once yet produce dramatic help for many descendants. That efficiency of minimal intervention for maximum service is one of the principal rationales for germline intervention.[16] Germline changes could also begin as an unintended consequence of relieving suffering for a particular individual[17] or as a deliberate step in preventing disease that due to physiology could not be treated by genetic therapy affecting only one person's body. Lipid storage diseases such as Tay-Sachs or syndromes such as Lesch-Nyhan would be examples.

Why Ethics?

The Need for Clarity

Some choices are better than others. The focus of this work is to think through ethically the goals of human genetic intervention and how they are pursued. The intent is to be immediately practical, yet emphasize underlying attitudes and goals that will not become outdated as new questions arise. Such analysis will still be relevant as new techniques develop. Part of that analysis will be to set a clear agenda of ethical questions that need attention. Presenting the issues in a fair and manageable form would be a major contribution to clearing up the cacophony that has long attended this subject. The Enquete Commission in West Germany declares that a genetic inheritance free from any form of genetic engineering is a basic human right,[18] while Joseph Fletcher writes in *The New England Journal of Medicine* that every child

15. Example from Erik Parens, "Is Better Always Good? The Enhancement Project," in *Enhancing Human Traits: Ethical and Social Implications*, ed. Erik Parens (Washington, D.C.: Georgetown University Press, 1998), p. 2.

16. LeRoy Walters, "The Ethics of Human Gene Therapy," in *Contemporary Issues in Bioethics*, ed. Tom L. Beauchamp and LeRoy Walters, 3rd ed. (Belmont, Calif.: Wadsworth, 1989), p. 523.

17. Robert Mullan Cook-Deegan, "Human Gene Therapy and Congress," *Human Gene Therapy* 1 (1990): 168-69; Marc Lappe, "Ethical Issues in Manipulating the Human Germ Line," *The Journal of Medicine and Philosophy* 16 (1991): 621-39.

18. Enquete Commission, "Prospects and Risks of Gene Technology: The Report of the Enquete Commission to the Bundestag of the Federal Republic of Germany," *Bioethics* 2, no. 3 (1988): 254-63; Jasper Becker, "Rights on DNA," *Nature* 295 (18 Feb. 1982): 545.

has a right to genetic engineering for a sound physical and mental constitution.[19] For Paul Ramsey, human germline intervention is "playing God."[20] For Donald MacKay, it is fulfilling the role of human beings as God's obedient children.[21]

Care in terminology will be crucial. The language used to state questions and distinctions often reveals already set conclusions. For example, originally the most frequent term used to refer to intervention in human genetics was *engineering*. Scholars used the term in quite different ways but tended to focus on the mass effects of genetic intervention. Proponents called for action on behalf of distant future societies by redressing the progressive damage being done to the human gene pool. Opponents tended to speak in terms of a societal danger of totalitarian oppression through biological means and the potential loss of humanity as we know it. With the advance of possible clinical applications, the current discussion has centered instead on whether to treat particular individuals in need. Such choices, only now entering the realm of possibility, are immediate and personal and so tend to come to the fore. The dominant descriptor that comes with them is not *engineering* but *therapy*.

These two terms contrast in referent model, connotation, and focus. While *engineering* is associated with an industrial construction paradigm, *therapy* refers to the medical model of person-centered care. Engineering shapes malleable material, while therapy heals a person. The shift in focus from engineering to therapy makes intervention much more attractive. Yet both individual and societal effects, near and future, still need to be addressed. Rather than assume at the start one or the other of these models and sets of connotations and expectations, I am using the more generic term *intervention*. That will encourage openness to see more clearly without prejudgment.

As for the term *human genetics*, it does not refer in this study to a disembodied monolith with some ontological reality of its own. The human genome is always manifested in particular persons, one at a time. Cumulative effects over the sum total of human beings are important because they are affecting many particular human beings. It is in the sense of the genetic code as found in individual human beings that I refer to intervention in the human genome.

19. Joseph Fletcher, "Ethical Aspects of Genetic Controls," *The New England Journal of Medicine* 285, no. 14 (1971): 776.

20. Paul Ramsey, *Fabricated Man: The Ethics of Genetic Control* (New Haven: Yale University Press, 1978), p. 151.

21. Donald M. MacKay, *Human Science and Human Dignity* (Downers Grove, Ill.: InterVarsity Press, 1979), p. 59.

Even if some of the effects of intervention are society wide, the genetic intervention is directly in specific individuals and there has its ethical import for each person or through them on other persons and wider communities.

The Need for Widespread Reflection

Careful consideration from across society is warranted for human genetic intervention. Since new knowledge and applications may be put to harmful use or carry damaging side effects, discernment and choice should not be left solely to those who develop the technologies.[22] In the past scientists have not been able to assure the beneficial use of the knowledge or technologies they produce. Yet there is some resistance to societal involvement in the control and shaping of technology. These objections can be summarized as four concerns. Each one follows, along with a response.

First, John Robertson argues that in the United States the study and dissemination of technology is protected under the first amendment as a right of free speech. It can only be curtailed if it is a "clear and present danger."[23] In response, there is a long-honored distinction between ideas and actions. While freedom of ideas and expression in research is protected, all actions are not.[24] Individuals, families, and communities can and should decide how to best use genetic technology.

Second, some have claimed that only the scientists directly involved have the expertise and awareness to discern the weight of ethical concerns.[25] This view is voiced more often in private than in public, for political and funding reasons. It is the case that scientists are often in a unique position to understand what is currently possible. That leads to a special responsibility. They would do well to be the first ones to raise ethical questions due to their earlier awareness of what can be done. On the other hand, no one can completely comprehend

22. Eric T. Juengst, "The NIH 'Points to Consider' and the Limits of Human Gene Therapy," *Human Gene Therapy* 1 (1990): 429; Daniel Callahan, "Ethical Issues in the Control of Science," in *Genetics and the Law,* ed. Aubrey Milunsky and George J. Annas, 3rd ed. (New York: Plenum, 1984), pp. 23-24.

23. John A. Robertson, "The Scientist's Right to Research: A Constitutional Analysis," *Southern California Law Review* 51 (1978): 1203-79.

24. Sheldon Krimsky, "Science Perverted: Can It Happen Here?" *Hastings Center Report* 13, no. 6 (1983): 42.

25. For example, Christian Barnard at "Hearings Before the Subcommittee on Government Research, United States Senate, S.J. Res. 145" (Washington, D.C.: U.S. Government Printing Office, 1968), p. 79.

the full range of what is technically possible in regard to intervention in human genetics, let alone the societal implications.[26] In such complex research, scientists are generally focused on specific fragments of the work. When they do address larger social implications, they have no unique expertise and in fact may be blinded by their personal involvement. Not everyone involved in the laudable task of biomedical research is always above the temptation and pressures of personal advancement and prestige found in most occupations.[27] There is considerable pressure in grant writing to make ambitious promises and to provide dramatic results.[28] While scientific study has built-in safeguards against fraud and misrepresentation, the sheer volume of material makes it difficult to assure complete accuracy even at purely technical levels.[29] The halo of expertise from the scientific endeavor does not guarantee authority for ethical questions.

A third concern is that in the competitive world of research, short periods of time separate the prestige and copyright rewards of being the first to make a new discovery from the comparative anonymity of being second. In *Invisible Frontiers*, Stephen Hall traces competition between genetic labs.[30] A six-month wait for U.S. government approval of plasmid BR322 was a quick response for the intricate and massive bureaucracies that granted it but was agonizingly slow for the involved scientists. Basic research in genetics is needed for economic development in multiple applications from new drugs to agriculture. To inhibit such development and attendant copyrights could be economically crippling and a significant burden to a research group or country. However, this concern that research not be slowed by regulation is more a call for timely oversight than a rationale for no accountability.

Fourth, no one government has the power to limit genetic research worldwide, nor is the consensus available to coordinate such an effort, especially when one considers that potentially hostile countries have made genetic engineering a central part of military research efforts. William

26. Von Ulrich Eibach, "Leben Schopfung aus Menschenhand? Ethische Aspekte genetischer Forschung und Technik," *Zeitschrift für Evangelische Ethik* 24 (April 1980): 129-30.

27. Alexander Kohn, *False Prophets* (London: Basil Blackwell, 1987).

28. Jon Beckwith, "Social and Political Uses of Genetics in the United States: Past and Present," in *Annals of the New York Academy of Sciences, 1976* (New York: New York Academy of the Sciences, 1976), p. 54.

29. R. L. Engler et al., "Misrepresentation and Responsibility in Medical Research," *New England Journal of Medicine* 317 (26 Nov. 1987): 1383-89; W. W. Stewart and N. Feder, "The Integrity of Scientific Literature," *Nature* 325 (1987): 207-14; William Broad and Nicholas Wade, *Betrayers of the Truth: Fraud and Deceit in the Halls of Science* (New York: Simon and Schuster, 1982).

30. Stephen S. Hall, *Invisible Frontiers* (New York: Atlantic Monthly, 1987), p. 169.

Kucewicz and John H. Birkner have each argued this case in regard to the former Soviet Union.[31] With the breakup of the USSR, threats seem less likely from that quarter but have possibly increased from other directions due to the dispersal of experts to smaller belligerents. Countries have found it prudent to pursue research at least to understand the potential threat and possible defense. Further, short of a pervasive police state, it is difficult to detect genetic work, which is often a relatively unassuming low-capital activity, especially when so much immediately applicable information has already been disseminated.

Control then is problematic in varying degrees and raises at least three distinct questions: (1) What applications of technology are appropriate? (2) What should government restrain or support? (3) What means should government use to carry out its policy? While government funding within the United States would be the most easily guided as a matter of public policy, privately funded development or application in other countries would be the most difficult to control. Recognizing such difficulties at the second and third questions does not absolve the need to address all three challenges, beginning with the first one. The first question of what technology is appropriate is not merely a question for governments. Most decisions will be made by individuals, families, and communities.

The Need to Think Ahead

Bernard Davis writes that "useful public discussion must be built, as far as possible, on recognition of what is likely to become technically feasible soon. Preoccupation with very remote dangers will dilute discussion of issues of real concern, and can develop diffuse anxieties that will distort the issues."[32] Davis is concerned here that to be effective in shaping choice, reflection needs to focus on proximate issues. The closer at hand the choice is, the more accessible the actual issue and context. While some potential genetic interventions are probably distant if attainable, others, such as genetic testing, have already begun. As discussed above, genetic intervention is progressing at remarkable

31. William Kucewicz, "The Threat of Soviet Genetic Engineering," series in *The Wall Street Journal* beginning 23 Apr. 1984; John H. Birkner, "Biotechnology Transfer — National Security Implications," in *Biotechnology in Society,* ed. Joseph G. Perpich (New York: Pergamon, 1986), pp. 205-6.

32. Bernard D. Davis, "Ethical and Technical Aspects of Genetic Intervention," *The New England Journal of Medicine* 285, no. 14 (1971): 800; also Council for Science and Society, *Human Procreation* (Oxford: Oxford University Press, 1984), p. 7.

speed. The history of scientific investigation has been filled with surprises. There have been unforeseen setbacks, while at other times advances have occurred that even experts did not anticipate. Orville Wright wrote in 1921 that the limits of flight had been reached, and Albert Einstein in 1932 thought nuclear fission an impossibility.[33] Commercial fusion has been "imminent" for decades, while genetic intervention has advanced much more quickly than often predicted. No one actually knows how quickly various capabilities will be at hand. Considering the unpredictable timing and importance of the involved decisions, it is only prudent to address what is possible now *and* to think ahead.

Elisabeth Beck-Gernsheim observes that "a pattern is starting to form: new biomedical aids are introduced in order to prevent or relieve suffering within a narrowly defined range of unambiguous 'problem cases.' Then a transitional or habituation phase sets in, during which the field of application is continually expanded."[34] While intervention should be incremental in an area of such implications, even incremental change should take place in the light of considered long-range goals and concerns. Since long-range goals tend to reflect deeply held values and worldviews, a working consensus across society is even more difficult to obtain than immediate cooperative choices, yet the most effective time to shape the application of a technology toward thoughtful goals is before it is entrenched.

"To assess the impacts of technology too late in the decision cycle is likely to create strong polarization of attitudes and blunt the potential effectiveness of the study for the decision process. One cannot put too high an importance on assessing the state of the development of a technology so that the study may be timely in the decision cycle."[35] Error is hard to detect, but much easier to correct, at the beginning of such an endeavor. Policy is more malleable before the technological imperative gains momentum.[36] Once in place, a technological innovation gains vested interests and shapes perception. *Stare decisis* is not felt only in the courtroom. Technology shapes a society's life and values as much as it is a function of them. The piecemeal automatic application of technology as it becomes available can over time lead to results no one

---

33. George Will, "Man-Made Evolution," *Washington Post*, 3 Aug. 1978, p. A23.

34. Elisabeth Beck-Gernsheim, *The Social Implications of Bioengineering*, trans. Laimdota Mazzarins (Atlantic Highlands, N.J.: Humanities, 1995).

35. Joseph Coates, "The Identification and Selection of Candidates and Priorities for Technology Assessment," in *Technology Assessment*, vol. 2 (London: Gordon and Breach Science Publishers, 1974), p. 81.

36. Jacques Ellul, *The Technological Society*, trans. John Wilkinson (New York: Alfred Knopf, Vintage Books, 1964), p. 143.

would have deliberately chosen[37] or to costs that would not have originally been accepted.[38] Hans Jonas writes that

> familiar, traditionally aspired to goals may be better satisfied by means of new technologies whose emergence they have inspired. But the reverse may also happen and is becoming more typical: new technologies may inspire or create, or even force upon us, new goals that nobody has ever thought of before, simply by offering us their feasibility. Who ever wished to have presented to him, in his living room, great operas or open heart surgery or the recovery of corpses from a plane crash (not to mention the ads for soap, refrigerators, and sanitary napkins that go with them)?[39]

Jonas is quite right that techniques now seen as necessities were often undreamed of not long before their development. That does not mean, however, that they have created desires. People have long enjoyed music and opera. Having it available in a person's own home used to be possible only for the emperor of China or others of such rank or means. The technology of television makes that appreciated experience more widely available. Whereas before only a surgeon could see the wonder of open-heart surgery, now others can better understand it. Technology in that case is not creating interest but is bringing resources within reach of people who otherwise could not have even considered it. Access is increased but not always new interests.

The interests, magnitude, and implications of genetic intervention warrant discussion well in advance of full development and application. While appraisal will have to be refined as methods of intervention are developed and new consequences come to light, techniques of such impact as those found in genetic intervention call for the beginning of careful evaluation before they are fully in practice. The underlying issues are ones which need to be addressed for a wide area of possible development and will not be irrelevant even after marked technical advance.

Finally, even if many interventions occur only in the future, evaluation of their appropriateness already affects acceptance of projects now. One of the primary arguments against further research or intervention in some directions is concern about a slippery slope, that even if current applications

37. Jonathan Glover, *What Sort of People Should There Be?* (New York: Penguin, 1984), p. 18.

38. Richard McCormick, testimony at a hearing of the Human Genetic Engineering Committee of the United States Congress, 1983, Committee Print 170, 311.

39. Hans Jonas, *Technik, Medizin und Ethik* (Frankfurt, 1985), p. 19, as quoted by Beck-Gernsheim, *Social Implications of Bioengineering,* p. 28.

are acceptable they will lead to unacceptable ones. This slide can be a *conceptual* slippery slope due to no clear distinction between desired and disastrous steps, or a *social* one if there is a good clear distinction, but it will not be honored in practice. Either slippery-slope argument is concerned that intervention for one acceptable use will be extended to ones which are not. A perception of future applications can impinge on acceptance of current efforts. Future possibilities affect present use.

When society is faced with issues of such import and complexity, there are often several stages of moral reflection.[40] First, cases come to the fore in which different views often conflict and responses polarize. Second, ethical debate works toward more carefully defined and tested lines between acceptable and unacceptable choices. Third, public policies are shaped to reflect the consensus distinctions and lead further in the shaping and strengthening of the demarcations. For genetic intervention, we are well into the first stage and need to carefully think through the second before the third is set.

Why Include Christian Resources?

One cannot write from a completely neutral perspective. The very fact that one has chosen a particular topic and way of addressing it is already revealing. Rather than claiming the impossibility of indifferent neutrality, I will treat a wide variety of views as fairly and respectfully as I can to see what we can learn from each. My own conclusion is that the classic Christian tradition makes the best sense of a wide range of concerns, so its contribution will be featured prominently in the dialogue among alternative perspectives. Theological vocabulary and concepts will be explained at their first use. This should make the ideas accessible without driving the richness of the tradition out of the discussion.[41] Of course essential noetic experiences of the tradition cannot be captured discursively.[42] A poem about something as simple as a Saturday afternoon in a backyard can be informative and even evocative, but it is not of itself the same as catching the scent of cut grass or the aroma of a baking apple pie. Much of the Christian tradition will not communicate in

40. John C. Fletcher, "Evolution of Ethical Debate about Human Gene Therapy," *Human Gene Therapy* 1, no. 1 (1990): 56-57.

41. Courtney S. Campbell, "Religion and Moral Meaning in Bioethics," in *Theology, Religious Traditions, and Bioethics: A Special Supplement to the Hastings Center Report* 20 (July/Aug. 1990), ed. Daniel Callahan and Courtney S. Campbell, pp. 4-10.

42. H. Tristram Engelhardt Jr., *The Foundations of Christian Bioethics* (Amsterdam: Swets & Zeitlinger, 2000), p. 180.

words alone either. The Christian tradition is also not always monolithic, yet some questions have been largely settled for the classic mainstream, such as the description of the person of Jesus Christ at the universal Councils of Nicaea and Chalcedon. Other ideas, such as the description of God as immutable, are currently challenged.[43] Such consensus and adjustment are typical of science as well. It has undergone countless refinements and even paradigm shifts, as seen in the move from Newtonian physics to quantum mechanics.

It has been argued that "in a secular society such as ours, if philosophy is to discharge one of the traditional roles of the humanities (i.e., to assess the core ideas and images of the ambient culture critically), then either the theological concerns will need to be exorcized from secular discussions or a way must be sought to articulate them in general secular terms."[44] Such reduction of theological language to general secular terms would be a great loss. Theological language without intending actual reference to God has lost its point. Further, stripping public dialogue of theological language would silence the voice, concern, and contribution of many citizens. While Craig Gay may be right that currently "the central spheres of modern social and cultural life are, for the most part, both theoretically and practically atheist,"[45] religion and religious institutions still play a formative role for many people.

As to whether our government and public policy should be secular, one could argue that our constitutional safeguards include the protection of the freedom of religious thought and practice, not their exclusion from our common life. An extensive literature has developed addressing how we can best relate moral and religious belief, politics and law, in our pluralistic society.[46] At

43. John Sanders, *The God Who Risks: A Theology of Providence* (Downers Grove, Ill.: InterVarsity Press, 1998).

44. H. Tristram Engelhardt Jr., "Human Nature Technologically Revisited," in *Ethics, Politics, and Human Nature,* ed. Ellen Frankel Paul, Fred D. Miller Jr., and Jeffrey Paul (Oxford: Basil Blackwell, 1991), p. 182.

45. Craig Gay, *The Way of the (Modern) World or, Why It's Tempting to Live As If God Doesn't Exist* (Grand Rapids: Eerdmans, 1998), p. 11.

46. For example, Robert Audi and Nicholas Wolterstorff, *Religion in the Public Square: The Place of Religious Convictions in Political Debate* (Lanham, Md.: Rowman & Littlefield, 1997); Ronald F. Thiemann, *Religion in Public Life: A Dilemma for Democracy* (Washington, D.C.: Georgetown University Press, 1996); Stephen L. Carter, *The Culture of Disbelief: How American Law and Politics Trivializes Religious Devotion* (New York: Doubleday, 1994); Michael J. Perry, *Love and Power: The Role of Religion and Morality in American Politics* (New York: Oxford University Press, 1991) and *Morality, Politics, and Law* (New York: Oxford University Press, 1988). For a more sociological emphasis, see James Davison Hunter, *Culture Wars: The Struggle to Define America* (New York: Basic Books, 1991). For a specific example in bioethics, see James F. Childress, "Religion, Morality, and Public Policy," in *Notes from a Narrow Ridge,* ed. Davis and Zoloth, pp. 65-85.

times the conclusions of one community can seem incomprehensible to another, yet helpful questions and insight can be offered from one tradition to another, religious or philosophical.[47] Topics, perception, evaluation, if self-consistent, always arise from a worldview of what is and what matters. Why privilege the plurality of "nonreligious" views? They offer no more assured or agreed consensus than religious ones.[48] Lisa Cahill is quite right that public discussion is most effective when it embodies "a commitment to civil exchanges among traditions, many of which have an overlapping membership, and which meet on the basis of common concern."[49] Our use of genetics is such a common concern. Genetics matters to all who have genes or care about someone who does.

Jacques Ellul has argued throughout his work that we have become so enamored with technology that we no longer recognize what we are doing or why.[50] The two millennia of experience and reflection in the Christian tradition can offer some needed perspective in the dialogue that follows.

Book Structure

In Broad Outline

This book will build its case sequentially. It would be helpful to read it from the beginning in presented order, but it can be accessed at any point with supporting cross-references. The table of contents is sufficiently detailed to allow readers to go directly to the sections that are most helpful to their interests. This serves the increasingly common learning style of searching for particular information as on a CD-ROM or web site. In that case the cross-references may be seen as web page links.

The first three chapters establish context that will be helpful throughout. These are followed by four parts that address specific types of genetic intervention. The types of genetic intervention will be addressed in the order of

47. See, for example, the meta-ethical arguments of Alasdair MacIntyre in *Whose Justice? Which Rationality?* (Notre Dame: Notre Dame Press, 1988) and in *Three Rival Versions of Moral Inquiry: Encyclopedia, Genealogy, and Tradition* (Notre Dame: Notre Dame Press, 1989).

48. Engelhardt, *The Foundations of Christian Bioethics*, pp. 20-97.

49. Lisa Sowle Cahill, "Can Theology Have a Role in Public Bioethical Discourse?" in *Theology, Religious Traditions, and Bioethics*, p. 10.

50. See, for example, Jacques Ellul, *The Technological Society*, and *The Technological Bluff*, trans. Geoffrey W. Bromiley (Grand Rapids: Eerdmans, 1990).

how soon they will be widely implemented in medical care and the degree of their distinctiveness from presently common practice. Each type of intervention will be the focus of one part. Part I will consider genetic research. Part II will consider genetic testing. Part III discusses genetic pharmaceuticals, and part IV the direct change of a person's genes. Commonly these last two parts are discussed together under the one title of gene transfer. While gene transfer is often technically important to achieving each, in practice gene products can be given to a person without altering that person's own genes. Since this book is focused on the best use of genetic technology rather than the description of laboratory techniques, I have divided the common category of gene transfer into the two recipient-centered categories of genetic drugs and genetic surgery. Genetic surgery does not include the genetic pharmaceuticals often addressed under gene transfer, yet it is a broader category than gene transfer in that it does include gene removal, inactivation, or duplication (such as in cloning). Again the key practical difference is that genetic pharmaceuticals involve no direct alteration of a person's genes. Instead the human being is receiving a genetically synthesized product. In contrast, genetic surgery to some degree establishes or changes a person's genetic endowment.

Considering genetic pharmaceuticals before genetic surgery has at least three advantages. First, our discussion proceeds in the order of widespread availability of the intervention. Genetic drugs already have a history of effective use. Genetic surgery is just beginning. Second, addressing genetic drugs as a separate category helps to keep our complicated task of analysis more manageable by raising issues of cure and enhancement before adding the further complication of directly changing a person's genes. Third, these two categories, as the categories of research and testing, each emphasize the continuity of genetics with previous interventions, problems, and discussions. We do not have to face genetic intervention as if nothing like it had ever been considered before. We can learn from the insights of related past discussions and better recognize new challenges.

Within each of the four parts there will be three chapters. The first chapter of each part will begin with the decisions that need to be addressed by individuals, then the second in regard to families, and the third for communities. This is the order from relatively simple to more complex as more people and institutions become involved. It is also the order from actual direct intervention to its rippling side effects. There are then four levels of intervention, cross-linked with three levels of who is affected. This delineation into twelve chapters will help the reader to find information and to keep track of an immense and intricately interrelated body of material.

Chapter by Chapter

More specifically, this introduction is presenting the goals and structure of the book. Chapter 1 begins by considering whether or not a prominent outlook addressed in this book, a Christian approach, is necessarily at odds with a scientific perspective. The chapter seeks to defuse the deep mutual suspicion too often found between science and Christianity. Actually the two have been allies for much of their histories and still have much in common. The second part of the chapter briefly overviews the role of genetics in human body structure and behavior. This is crucial material for anyone who is not familiar with it. Ethical evaluation depends on the facts of the case. Explanation there will clear up some popular misconceptions of what genes can and cannot do. The third part of the first chapter considers the illuminating power of some aspects of genetics for several attitudes that will aid the rest of the study.

Chapter 2 describes the powerful shaping influence of technology on our lives. Most people do not realize how deeply technology forms us, from where we live to who our friends are. Since technology is the sum of our tools to shape the world to what we want, it raises the most basic question. What are we trying to do? What is our purpose? Chapter 3 seeks to stimulate the reader's thinking about such foundational matters by addressing that question from the perspective of the most prominent voice in the topical dialogues to follow: the Christian tradition. The first three chapters as a whole then work toward a clearer picture of science in general, genetic science in particular, technology, and human purpose.

With the above in hand we will investigate four levels of genetic intervention. Part I, "Genetic Research," consists of chapters 4, 5, and 6. Chapter 4, "Searching for Genes and the Individual," reflects on worthy goals for research and respectfully treating the individual as a partner. This raises the importance of the participant's competence and understanding. Chapter 5, "Searching for Genes and Family," highlights the familial nature of genes. While genes are quintessentially personal, they are shared with relatives. One person's choice to participate in genetic research often affects others. This is most acutely felt in cases of surrogacy, when one person decides on behalf of another such as a child. The status of developing life in the womb is addressed in this chapter. Chapter 6, "Searching for Genes and Community," considers group consent, gene patenting, and social investment.

Part II, "Genetic Testing," includes chapters 7, 8, and 9. Chapter 7, "Testing Genes and the Individual," examines how genetic testing can provide an individual with information to understand, plan, prevent, or treat a condition. There are, however, difficulties in establishing and conveying genetic information, which usually comes in terms of probabilities. Chapter 8, "Testing Genes and Family," considers family effects of new genetic information both for untested relatives and for a couple's procreation. Changing the natural course and zygote selection are among the themes to be addressed there. Chapter 9, "Testing Genes and Community," discusses the role of insurance companies, employers, government, and churches.

Part III, "Genetic Drugs: Adding Gene Products to the Body," includes chapters 10, 11, and 12. Chapter 10, "Genetic Drugs and the Individual," tests the often proffered boundary between correction and enhancement. Such a distinction depends on the definitions of disease and health. Chapter 11, "Genetic Drugs and a Family's Children," offers an alternative of four standards for appropriate intervention: safety, genuine improvement for the recipient, supporting an open future, and best use of available resources. Chapter 12, "Genetic Drugs and Community," thinks through professional standards and government interest in equality of opportunity.

Part IV, "Genetic Surgery: Changing Genes in the Human Body," consists of chapters 13, 14, and 15. Chapter 13, "Changing Genes and the Individual," considers the case for and against altering our physical form, attitudes encouraged by genetic change, and the case of human cloning. Chapter 14, "Changing Genes and the Family Line," examines reasons for changing genetic heritage and what concerns we should have for descendants. Chapter 15, "Changing Genes and Community," discusses the historic dangers of coercion, racism, and eugenics, how diversity might be welcomed, and who should make genetic choices for future generations.

Since each chapter ends with a brief summation easily accessible from the table of contents, the conclusion will not try to repeat all findings. It will pursue two themes that have run throughout the book. Genes are an *important* part of human life, yet only a *part* of human life. What role should shaping them play?

CONTEXT

CHAPTER 1

Science: To Better Understand
Part of the World and Ourselves

Science and the Christian Tradition

Since Christian perspectives will play a substantive role in the conversation that follows, it is important to consider at the outset whether or not scientific and Christian views are inherently incompatible. There are people who are convinced that science and the Christian tradition are engaged in a fight to the death. In the eyes of some, eventually one and only one, Christianity, will be left standing as the truth. There are others who return the favor with a different expectation of who will win. The mortal-combat picture is constantly pushed by the media system that depends on conflict to draw attention to itself so that it can then sell that attention to advertisers. Before we can focus on specifically genetic science, we need to dispel that distortion.

In all their variety, the scientific and Christian traditions have often been allied. In fact to this day they have significant common cause. They share a great deal in history, method, and understanding of reality. The story of the dialogue between scientific and Christian traditions, with all its highways and dead-ends, and importantly its future prospects, is far better detailed and advanced in other books than there is time to do in this one.[1] But a brief description begins this chapter to help set needed context for our task of evaluating the best use of a particular science, human genetics. If you are already familiar with the connections between the scientific and Christian tra-

---

1. Oxford professor Alister E. McGrath has recently written a judicious introduction to these important themes. The book includes a key bibliography and is entitled *The Foundations of Dialogue in Science and Religion* (Oxford: Blackwell, 1998).

ditions, it would be most efficient to skip ahead to the second part of this chapter. If not, this overview of the conversation between science and the Christian tradition offers important context. The second part of the chapter introduces the role of genes in human physical form and behavior. That is important to clarify in that there are many popular misconceptions of what genes can and cannot do. The third section of this chapter will introduce several attitudes that have long been lauded by the Christian tradition and are helpful in thinking through the issues in this book. Each of those virtues is illuminated in part by an aspect of genetics.

Part of the confusion about how science and the Christian tradition relate to each other is in the definitions of both "science" and "Christian tradition." As the term "science" first developed from the Latin *scientia,* it simply meant any rigorous and assured system of belief. By this definition, theology was often referred to as the "queen of the sciences." Over time science has become associated with particular bodies of knowledge such as those found in physics. While this is currently common usage, it does have the disadvantage of not recognizing how much such content has varied over the years. Newtonian physics, revolutionary and compelling in its time, is not the physics taught at a university level today. The most effective definition of science is probably one based on method. "Science" is a way of carefully observing the physical world to improve continually the accuracy of hypotheses concerning causal connections. In other words, science tries to understand how the physical world works. Technology, which attempts not so much to understand the world as to change it, will be addressed in chapter 2.

As for the "Christian tradition," even twenty years ago Lonnie Kliever's book on Christian theology was entitled not just *The Spectrum: A Survey of Contemporary Theology,* but *The **Shattered** Spectrum* [bold is mine].[2] The range of scholars influenced by and claiming the inspiration of the Christian tradition is startling. While insights will be gleaned and valued from many of these sources, this work will focus on the classic core of Christian teaching that is still central for many and at least in some sense historically influential for all in the tradition. That core of ideas and commitments may be seen in Christian Scripture,[3] the Apostles' Creed that the early church used to describe itself to

---

2. Lonnie D. Kliever, *The Shattered Spectrum: A Survey of Contemporary Theology* (Atlanta: John Knox, 1981).

3. Allen Verhey succinctly describes the ongoing discussion of how Scripture should be understood and how it should shape Christian life in "Scripture and Ethics: Practices, Performance, and Prescriptions," in *Christian Ethics: Problems and Prospects,* ed. Lisa Sowle Cahill and James F. Childress (Cleveland: Pilgrim, 1996), pp. 18-44.

new adherents,[4] and the more detailed exposition of the early universal councils at Nicaea (325) and Chalcedon (451). This would include then what is held in common by at least the historic Eastern Orthodox, Roman Catholic, and Protestant communities on through their contemporary heirs.

Continuing and refining the opening question, have science and historic Christian theology always been antithetical? Actually in its late medieval roots, science was nurtured by the church.[5] The first scientists, scholars who focused on observing and understanding the physical world, called themselves "natural philosophers." Their housing and salary were provided by the church at the world's first universities, Bologna (1150), Paris (1200), and Oxford (1226). These universities were founded by the church to prepare scholars to serve the church. Natural philosophy was a recognized way of discovering more about God and God's world. It was in fact required of all students as part of their basic undergraduate education, before specializing in either law, medicine, or theology. All professionals, including educated clergy, were expected to be conversant with the best natural philosophy of their day as part of their basic education.

The University of London geneticist, R. J. Berry, points out that this connection between the Christian tradition and the study of nature continued for the professors of natural philosophy that followed through succeeding centuries, even as their academic posts began to receive funding from outside the church. Most often these natural philosophers "were explicit and often devout believers" in the Christian tradition, or at least theism, the belief that there is a Creator-God.[6]

Johannes Kepler (1571-1630) wrote of "thinking God's thoughts after him" as he contemplated the stars; Blaise Pascal (1623-62) experienced a pro-

---

4. The Apostles' Creed is quite short for ease in memorization. It is simply:

"I believe in God the Father Almighty, Maker of heaven and earth, and in Jesus Christ His only Son Our Lord, Who was conceived by the Holy Spirit, born of the virgin Mary, suffered under Pontius Pilate, was crucified, dead, and buried. On the third day He rose from the dead and ascended into heaven and is seated at the right hand of God the Father Almighty; from there He shall come to judge the quick and the dead. I believe in the Holy Spirit, the holy catholic Church, the communion of the saints, the forgiveness of sins, the resurrection of the body, and the life everlasting."

5. In the early Middle Ages, Islamic scholars played an important role. See Michael Robert Negus in *God, Humanity and the Cosmos*, ed. C. Southgate (Edinburgh: T & T Clark, 1999).

6. R. J. Berry, *God and the Biologist: Faith at the Frontiers of Science* (Leicester, U.K.: Apollos, 1996), p. 11.

found religious awakening and became a committed Jansenist (these were a group believing in the controlling effects of divine grace); Robert Boyle (1627-91) gave money to make the New Testament available to the native peoples of North America; Isaac Newton (1642-77) devoted himself as much to theology as to physics.[7]

Of particular interest to this study, the first to recognize the units of inheritance that we now call genes, Gregor Mendel (1822-84), was an Augustinian monk who did his experiments with the pea plants he grew in the monastery garden.

It has been argued that this historic connection between theism and the development of science is not accidental. Stanley Jaki goes so far as to propose that science might only arise in such a theistic context.[8] At the least, the intellectual milieu of Christian theology did provide three perspectives that welcomed the development of the scientific method. First, because the Creator could make the world in any way he wished, the only way to know what God actually chose was to carefully look at it. Speculation about how things should be is inconclusive without going out and looking at what God actually did. Science strives for this careful and honest observation of what is actually present. Second, because God is consistent, one can expect that the creation God made is probably consistent as well. Observations of how the material world works at one point can be generalized to other similar cases. Without this expectation, observation would always be limited to spot-checking local minutiae. There would be no patterns to discover or apply. Third, because the world was a creation and not of itself divine, one did not risk offending greater powers by manipulating it for observation. In sum, Christian theology suggested that the material world was contingent, regular, and available for experiment. The lack of any one of these foundational expectations would have forestalled the development of the scientific method in the Christian context where it thrived.

As the scientific method built a long record of success these expectations became widely accepted outside theistic circles, yet theology has remained important to many scientists. Edward J. Larson and Larry Witham recently repeated a survey of scientists done in 1916. They were meticulous in following the original design. The random selection of one thousand scientists came from the same source, the respectively current editions of *American Men and Women of Science*. The apportionment of scientific disciplines was retained at

7. Berry, *God and the Biologist*, p. 11.
8. Stanley Jaki, *The Road of Science and the Ways to God* (Chicago: University of Chicago Press, 1978).

one-half biologists, one-quarter mathematicians, and one-quarter physicists or astronomers. Also, identical questions were asked in the original order. The percentage of recognized scientists who affirmed that they "believe in a God to whom one may pray in expectation of receiving an answer" was about 40 percent for both 1916 and the present.[9] The expectation of some that theism would gradually wither among scientists was simply not met.

As to specifically Christian theists, an example of continued presence would be the American Scientific Affiliation. It currently has about two thousand members, all of whom affirm the Apostles' Creed as part of joining the association,[10] and most of whom hold Ph.D.s in the natural sciences.[11] The organization exists to facilitate dialogue between science and the classic Christian tradition. Their active journal is *Perspectives on Science and Christian Faith*. Across the Atlantic, the Society of Ordained Scientists and Christians in Science are similar affiliations in Great Britain. The latter has its own journal, *Science and Christian Belief*. Further examples of ongoing interaction in Britain include a volume of essays on Christian faith by fourteen well-known scientists, edited by R. J. Berry and entitled *Real Science, Real Faith*.[12] Nevill Mott edited a similar but more broadly religious volume called *Can Scientists Believe? Some Examples of the Attitude of Scientists to Religion*.[13] Ted Peters has recently edited a volume with scientists and theologians from both sides of the Atlantic contributing: *Science and Theology: The New Consonance*; and there have been other more focused collaborations.[14] An example is that of the University of Minnesota geneticist V. Elving Anderson writing a joint study with the philosopher Bruce Reichenbach on a Christian ethic for the biological sciences.[15]

There is a long and continued history of connection between science

---

9. Edward J. Larson and Larry Witham, "Scientists Are Still Keeping the Faith," *Nature* 386 (3 Apr. 1997): 435-36.

10. Quoted in note 4.

11. The American Scientific Affiliation home office can be reached at asa@asa3.org (web page: www.asa3.org), or Post Office Box 668, Ipswich, MA 01983.

12. R. J. Berry, ed., *Real Science, Real Faith* (Eastbourne, U.K.: Monarch, 1991). Berry has written recently with Malcolm Jeeves, *Science, Life and Christian Belief: A Survey and Assessment* (Leicester, U.K.: Apollos, 1998). Another anthology by scientists and theologians in dialogue is edited by Fraser Watts, *Science Meets Faith* (London: SPCK, 1998).

13. Nevill Mott, ed., *Can Scientists Believe? Some Examples of the Attitude of Scientists to Religion* (London: James and James, 1991).

14. Ted Peters, ed., *Science and Theology: The New Consonance* (Boulder, Colo.: Westview, 1998).

15. Bruce R. Reichenbach and V. Elving Anderson, *On Behalf of God: A Christian Ethic for Biology* (Grand Rapids: Eerdmans, 1995).

and theists, and specifically for our study, members of the Christian tradition. What has spawned the idea that they are incompatible? The perceived conflict has been precisely where they have the most in common: method and intertwined history. This is not surprising, in that it is where the two disciplines overlap that they have the most potential to help or abrade each other.

Method

Adjusting to Reality

The first point of important methodological agreement is that there is one reality. Human interpretations and descriptions vary, but reality itself is not infinitely malleable to human taste or desire. There is a there, there. If all human beings ceased to exist, there would still be much that would remain. Both scientific theory and traditional Christian theology affirm that central point. Trying to recognize what actually is cannot be assumed for some academics today. For example, members of the deconstructionist literary movement so emphasize the centrality of human interpretation that human interpretation seems to be all that is left.[16] One can believe whatever one wants so long as one does not believe that it is independently true. Truth is always specific to truth for me or truth for you. In contrast, science and the Christian tradition both recognize that human interpretations vary. But this is attributed to how limited human knowledge is and to mistakes in the interpretation of the knowledge that is available, not that there is no reality to recognize. Because human beings do err, interpretations should always be open to reform.[17] The point is not thereby to create reality but to become more aware of what actually is reality. On this central point, science and the Christian tradition are allies.

Interpretations of available data need to adjust as more is learned. Yet there are core facts and theories that seem quite assured. Part of the power of science has been its willingness to change an accepted interpretation of data to explain better new data, hence more accurately recognizing reality. Thomas Kuhn called this sometimes-wrenching change a "paradigm shift."[18]

---

16. Noretta Koertge, ed., *A House Built on Sand: Exposing Postmodern Myths about Science* (New York: Oxford University Press, 1998). Terry Eagleton, *The Illusions of Postmodernism* (Oxford: Blackwell, 1996).

17. R. C. Lewontin, "Facts and the Factitious in Natural Sciences," *Critical Inquiry* 18:140-53.

18. Thomas Kuhn, *The Structure of Scientific Revolutions* (Chicago: University of Chicago Press, 1970) and *The Essential Tension* (Chicago: University of Chicago Press, 1977).

When new data are contrary to the current explanation and there is an alternative that better accounts for it, most of the scientific community eventually, not always smoothly, switches to the new description. Both theology and scientific method seek to recognize what actually is and are willing to adjust past understandings to get closer to it. One of the themes of the Protestant Reformation was "reformed and always reforming." The Roman Catholic Church continues a careful *aggiornamento,* a bringing up-to-date, that can develop considerable movement when seen from the perspective of passing centuries. This reforming or updating is not to surrender or flee the truth. It is to better approach it. Understanding can increase.

If in this process science observes something that is an aspect of reality, it will not be a surprise to God. All truth is God's truth. Now the new information may conflict with a human interpretation of God's revelation, in which case it offers an opportunity to better hear what God actually revealed or to revisit the surety of the scientific claim.[19] This has often been called the two-books approach. One can expect the book of God's word and the book of God's works to fit together.[20] Part of the learning process is to see better how they do agree.

Aspiring to Be Comprehensive

To the theologian in a sense everything is a theological question, including the interpretations of scientific observation, because everything is related to God. Theology attempts to be comprehensive because of the nature of its subject, God who encompasses everything. Science also attempts to be comprehensive, but only of a smaller subset, the vast material universe. In pursuing comprehensive explanation, science has a smaller data base. Both science and the Christian tradition welcome observation of the physical world and seek to be comprehensive, coherent, and fruitful in how they interpret it. Classic Christian theology is also convinced that God exists and has chosen to partially reveal himself. It would take yet another book to describe the justification of such confidence in God's presence and self-revelation.[21] The revela-

19. Nicholas Wolterstorff, *Reason Within the Bounds of Religion,* 2nd ed. (Grand Rapids: Eerdmans, 1984).

20. Francis Bacon, *Advancement of Learning* (1605).

21. See, for example, Alister McGrath, *A Passion for Truth: The Intellectual Coherence of Evangelicalism* (Downers Grove, Ill.: InterVarsity Press, 1996). William Lane Craig, Gary Habermas, John Hare, Basil Mitchell, Alvin Plantinga, Richard Swinburne, and Nicholas Wolterstorff, each with a unique perspective, are examples of others worth reading on this point.

tion of God in Scripture, reason, personal experience, and tradition provides information and insights to better understand more of reality than what is directly available to scientific observation. Such sources are not available to the usual experimental manipulation of the scientific method. Nancey Murphy has suggested the scientific method as described by Imre Lakatos[22] as a way to evaluate whether something is a revelation from God.[23] But even if successful in that regard, the scientific method would have a role in verification of revelation, not replacing it.

The extra data available to theology, such as revelation from God, raises a key distinction between science and "scientism." Science is primarily a method of seeing how much one can explain by physical cause. This has been a powerful heuristic. One is more likely to find physical connections if one keeps looking for them. This is a methodological focus, not a metaphysical claim that only the physical exists. The commitment to doggedly search for physical causation has been quite successful in its limited scope. It is welcomed by many theists.[24] It is quite another claim, indeed a metaphysical one, to claim that physical causation is all that is possible. That is "scientism," a worldview claim that the only real things are those one can physically measure.

This philosophy had some influence in the past under the label "logical positivism."[25] Although some gifted science popularizers still seem to assume it, it has fallen into disrepute among philosophers of science because it leaves no room for many statements that are widely recognized as meaningful, such as descriptions of precise historical events. That Abraham Lincoln was shot by John Wilkes Booth is an understandable and communicative statement, yet nonsense for strict logical positivism, because there is no presently available method to observe it. Most scholars of the logical-positivist school even-

22. Imre Lakatos, "Falsification and the Methodology of Scientific Research Programmes," in *Criticism and the Growth of Knowledge,* ed. Imre Lakatos and Alan Musgrave (Cambridge: Cambridge University Press, 1970).

23. Nancey Murphy, *Theology in the Age of Scientific Reasoning* (Ithaca, N.Y.: Cornell University Press, 1990).

24. The intelligent-design movement has emphasized that looking for physical causation is not the only possible way to do science. On the contrary, it believes that recognizing design in instnces of "irreducible complexity" has greater consistency and explanatory power. See Michael J. Behe, *Darwin's Black Box: The Biochemical Challenge to Evolution* (San Francisco: Touchstone, 1998); also William A. Dembski, *The Design Inference: Eliminating Chance Through Small Probabilities* (Cambridge: Cambridge University Press, 1998) and *Intelligent Design: The Bridge Between Science and Theology* (Downers Grove, Ill: InterVarsity Press, 1999). Intelligent-design theorists also maintain a web site at www.discovery.org.

25. Founded by the Vienna Circle and known first in English through the work of A. J. Ayer, such as his 1936 book, *Language, Truth, and Logic.*

tually adjusted to the acceptance of reasonable inference from observational evidence. This would allow then for history and also for theoretical physics and other sciences describing events that cannot be directly observed. Much of what we seem to know occurs at points that are not physically measurable. There is more to reality than what can be directly observed. The insistence of scientism that only the directly observed is real is after all rather immodest. One thinks of Carl Sagan profoundly expounding in his television series that the physical cosmos is all that ever was, all that is, and all that ever will be. The esteemed astronomer probably did not know what was on the other side of Cleveland. How could he claim to know what had or had not been, what is or will be for the entire universe?

Christian theology, and for that matter any theism, is in direct conflict with such a religion of scientism. Norman and Lucia Hall wrote, "Science and religion are diametrically opposed at their deepest philosophical levels. And, because their two world views make claims to the same intellectual territory — that of origin of the universe and humankind's relationship to it — conflict is inevitable."[26] As Alan Padgett points out in response, the authors are quite right that their worldview of philosophical materialism is contrary to religious viewpoints.[27] If the broad range of religions agrees on anything, it is that there is more to reality than just the physical.[28] The mistake is in equating the Halls' own worldview with science. Science is usually pursued as a method that looks for physical causation,[29] not a metaphysical claim that the physical is all that exists. Science is not "scientism."

Scientism can also fall into what might be called "reductionism." A *New York Times* article on the successful decoding of the genetic instructions of a worm, the nematode *Caenorhabditis elegans*, declared, "In the last ten years we have come to realize humans are more like worms than we ever imagined."[30] The warranted enthusiasm of the quoted scientist no doubt extends from the powerful service such an animal model can provide for better un-

---

26. Norman F. and Lucia Hall, "Is the War Between Science and Religion Over?" *The Humanist* 46, no. 3 (1986): 26, as quoted by Alan G. Padgett, "The Mutuality of Theology and Science: An Example from Time and Thermodynamics," *Christian Scholars Review* 26, no. 1 (1996): 13.

27. Padgett, "The Mutuality of Theology and Science," p. 13.

28. Edmund F. Perry, *World Theology* (Cambridge: Cambridge University Press, 1991).

29. As referenced in footnote 24, the intelligent-design movement is seeking to make the case that good science can establish that something other than the material universe is an important cause within the material universe.

30. *New York Times*, 11 Dec. 1998.

derstanding some mechanisms found in both nematodes and humans. At the cellular level, nematodes and human beings have much in common. Since nematodes are relatively simple and multiply quickly, they are readily accessible for studying how their genes work. That can give insight into how human genes work. In contrast, reductionism is the abuse of this powerful method. It is present when the helpful use of the animal model for describing one aspect of human existence is extended to claim that all human existence can be explained adequately as variations of this baseline commonality. Daisies and human beings use the same DNA system to encode life-bearing structure, but that does not mean that daisies live at all the levels and ways characteristic of human beings.

Human life cannot be fully described by what happens at the level of physics or genetics. Because one can describe an instance of molecular movement does not mean that one has understood all that is actually happening in an instance of glucose converting to lactic acid to generate ATP. The event might be further described as part of the contraction and release of a particular muscle. Reality might include that this muscle is increasing the tension within a human vocal cord. In fact more of what is happening is that a deliberate tone is being produced. The tone is part of someone singing. She is singing a selection from an oratorio called *The Messiah*. Her singing may be for pay, aesthetic enjoyment, camaraderie with her fellow chorus members, worship, or some combination of all the above.

Describing the human world at the simplest level of physical phenomenon can be insightful but so often remains incomplete. Human action and motivation is too ambiguous and complex for what Donald MacKay castigated as "nothing buttery."[31] "Nothing buttery" is when an action is described as "nothing but" a manifestation of genetic drives or some other elemental part. Phenomena often have multiple levels that cannot be reduced to the most elemental.[32] Examining only one aspect at a time to better understand its contribution can be a helpful exercise, but that is quite different from declaring the whole merely the sum of its smallest parts. The more complex the phenomenon, often the more important the emergent properties that cannot be atomized. Context is crucial to meaning in texts and physical systems. Christian theology can welcome science as a fruitful method within its area of expertise. Both do well to be modest and tentative, ready to learn.

31. Donald MacKay, *Human Science and Human Dignity* (London: Hodder & Stoughton, 1979). See more currently Malcolm A. Jeeves, *Human Nature at the Millennium: Reflections on the Integration of Psychology and Christianity* (Leicester, U.K.: Apollos, 1997).

32. Michael Polanyi, *Knowing and Being* (London: Routledge & Kegan Paul, 1969).

A History of Conflict?

At the beginning of this chapter, the long and continued interaction of science and the Christian tradition was noted. The quality of that dialogue has varied as much as science and theology vary within themselves. A number of historians have superbly traced some of the variations and subtleties of the interaction over the centuries.[33] Here there is only time to briefly cite two notable examples. The first is far enough in the past to offer clear insight with less acrimony. What happened in the Galileo incident?

Galileo Galilei

Both Galileo Galilei (1564-1642) and the teaching authorities of the Roman Catholic Church expected that observation of the physical world and the teachings of Scripture would agree. Galileo stated his expectation repeatedly and explicitly that "Holy Scripture and nature proceed alike from the divine Word, the former as a dictate of the Holy Spirit and the later as the faithful executrix of God's commands."[34] Where Galileo disagreed with some church authorities and most of the scientists of his day was on the best available explanation of the scriptural and observational data.[35] Their debate was not one between science and religion. It was an intramural debate among Christians as to what was the best way to understand the relevant Scripture and observational data.[36]

It did not feel as if the earth was moving. There was no strong breeze from one direction, no rumbling vibration, and a ball tossed in the air did not land at a new location as the earth moved beneath it. Also the plainest and simplest reading of many scriptural texts seemed to say that the earth did not move. Cardinal Bellarmine quoted passages such as Psalm 104:5, "He set the

33. For example, John Hedley Brooke, *Science and Religion: Some Historical Perspectives* (Cambridge: Cambridge University Press, 1993); David C. Lindberg, *The Beginnings of Western Science: The European Scientific Tradition in Philosophical, Religious, and Institutional Context 600 B.C. to A.D. 1450* (Chicago: University of Chicago Press, 1992); and David C. Lindberg and Ronald L. Numbers, eds., *God and Nature: Historical Essays on the Encounter Between Christianity and Science* (Berkeley: University of California Press, 1986).

34. William R. Shea, "Galileo and the Church," in *God and Nature*, p. 126.

35. Shea, "Galileo and the Church," p. 119.

36. The concern with a central text is characteristic of the extended conversation in any tradition, whether of science or moral inquiry. See Nancey Murphy, "Using MacIntyre's Method in Christian Ethics," in *Virtues and Practices in the Christian Tradition*, ed. Nancey Murphy, Brad J. Kallenberg, and Mark Thiessen Nation (Harrisburg, Pa.: Trinity Press International, 1997), pp. 31-33.

earth on its foundations; it can never be moved," and Ecclesiastes 1:5, "The sun rises and the sun sets, and hurries back to where it rises." Galileo had observed with his telescope that Jupiter had moons, which meant that the earth was not the center of all orbits. This and other observations could be explained by the sun being at the center of the planetary system. But Tycho Brahe, another eminent astronomer and mathematician of the time, harmonized Galileo's observations with an unmoving earth. Brahe argued that the planets do circle the sun, and then the sun with its retinue of planets circles around the earth. That theory accounted for what could be physically observed and did not contradict the most common interpretation of Scripture. Galileo, however, remained convinced that the Catholic priest Nicholas Copernicus had the best theory. All the planets, including the earth, circled the sun.

Galileo was anxious for papal endorsement of Copernicus's theory and met with the pope many times to convince him. In the meantime the discussion was complicated by personal and political motivations.[37] Some of the forces involved include that the Roman Catholic Church had been charged by Protestants with not taking Scripture seriously enough. Here was an opportunity to show that Scripture was receiving its full due. Further, Galileo was closely associated with Paola Sarpi, who had led a Venetian revolt against the Roman Curia. Incredibly, in the midst of these discussions, Galileo also published a book entitled *The Dialogues* that had a dim-witted and ridiculed character named "Simplico" (the Simpleton). Simplico was the foil character who voiced foolish opinions, including, in at least one passage, an argument personally stated by the pope to Galileo.[38] Whatever the motivations, *The Dialogues* was officially banned and not dropped from the index of prohibited books until 1832.

The famous case reminds us that both science and theology interpret data. Brahe's geocentric model accounted for the observational data available at the time as well as Galileo's did. The majority of astronomers of the day sided with Brahe. As more observations became available, scientific consensus eventually followed a laudable tradition of self-correction to improve that interpretation and moved toward the Copernican hypothesis that Galileo supported. The transition was neither smooth nor quick. It seems difficult for human nature, whether in scientist or theologian, to change perspective. The practice of Christian theology has been uneven in such self-correction as well. There is certainly precedent for change. References have already been made in

---

37. David Lindberg nicely uncovers the complex motivations in "Galileo, the Church, and the Cosmos," a chapter of *Science and the Christian Tradition: Twelve Case Histories,* ed. David C. Lindberg and Ronald L. Numbers (Chicago: University of Chicago Press, 2000).

38. Brooke, *Science and Religion,* pp. 98-109.

this chapter to the Reformation and *aggiornamento*. The trick for science and theology is to hold a stable system for investigation while being willing to adjust it to account for new data.

Theologians in the classic Christian tradition hold to the revelation of Jesus Christ as always central to the tradition, yet need to be willing to better understand that revelation. Too often theologians have confused a trustworthy Scripture with the trustworthiness of their first reading. New data from the sciences can sometimes help readers to see more clearly what a text is actually affirming. That is not to give scientific observation a trump card over revelation. One can err in observation, and new data can change interpretations of what the earlier data meant. But especially core scientific data that has been repeatedly confirmed is a worthy dialogue partner.

The medieval world knew that the earth was a globe but resisted the counterintuitive idea that it was moving through space. That observation is now assured. With that information in hand one can better understand the point of the Scripture passages quoted earlier. Some are speaking from the perspective of the viewer. As God accommodates revelation to limited human language, the text is also accommodated to other aspects of inconsequential human perspective. When Ecclesiastes 1:5 states that "the sun rises and the sun sets, and hurries back to where it rises," the sentence refers to the cyclical appearance of the sun as an illustration. It is not teaching astronomy, that the earth is still and the sun is moving around it. We still use the same language today of sunrise and sunset with no intent of an astronomical claim. When Psalm 104:5 says, "He set the earth on its foundations; it can never be moved," it may be a metaphor for God's power and our weakness. The earth is where God puts it. We cannot move it. It is not intended as a lesson in astronomy. The Scripture has not been circumvented in clarifying what it is affirming. The texts are becoming better understood.

For the theologian in the classic Christian tradition that should be welcome. If the data or interpretation of a phenomenon is contrary to an understanding of revealed truth, it is time for a more careful look at the data and theory from both the scientific observation and the theological perspectives, to find out where the error lies and more of what the actual reality is. Apparent conflict is an opportunity to better apprehend the truth.

The Scopes Trial

A second case that is often cited for the incontrovertible conflict of science and Christian theology is that of the Scopes trial in 1925. Edward Larson has compiled a meticulous and fair record of that show trial in his Pulitzer Prize-winning book *Summer for the Gods: The Scopes Trial and America's Continuing Debate over Science and Religion.*[39] There William Jennings Bryan dramatically clashed with Clarence Darrow over whether the community should choose school curriculum for their children or whether a teacher's freedom of speech should prevail. Stump speeches and rhetoric shifted the debate to a conflict between Bryan's understanding of the book of Genesis and Darrow's advocacy of one theory of evolution.

Part of the confusion of the debate was over which theory of evolution and which Christian interpretation of Scripture. Simon and Schuster recently published a book called *The Darwin Wars.* It is not about a conflict between evolution and religion, but about the struggle between different Darwinian theories.[40] The Christian tradition has varied in response to theories of evolution as well. Henry M. Morris has taken a lead in the movement that propounds that if the theory of evolution is correct, there was no Adam, and if there was no Adam, no fall. If there was no fall, there is no need for redemption. If there is no need for redemption, there is no cross, and if no cross, no Christ. The faith stands or falls on how God created the world.[41] In contrast, the Roman Catholic priest Teilhard de Chardin raptures that evolution is the best way of summarizing God's plan for the world. Creation is designed to evolve physically and spiritually, continually becoming more complex yet more united, until all is one in Christ.[42] It seems that most of the possible responses in between have been articulated as well. It is another book to sort out that discussion.[43] This study will have to continue without attempting to resolve it. It will suffice for our task here to proceed on the basis of what the Apostles' Creed affirms, that God is the maker of heaven and earth, without trying to establish here how God chose to do it.

Ian Barbour has described the possible relationships between science and religion as falling into four types. These interactions can be ones of "conflict, independence, dialogue, or integration."[44] To use his elegant typology to

---

39. Edward J. Larson, *Summer for the Gods: The Scopes Trial and America's Continuing Debate over Science and Religion* (Cambridge, Mass.: Harvard University Press, 1997).

40. Andrew Brown, *The Darwin Wars* (New York: Simon and Schuster, 1999).

41. Henry M. Morris and John D. Morris, *The Modern Creation Trilogy: Scripture and Creation, Science and Creation, Society and Creation* (Master Books, 1997).

42. Pierre Teilhard de Chardin, *Christianity and Evolution* (New York: Harcourt, Brace, Jovanovich, 1969).

43. A starting point for an interested reader might be J. P. Moreland and John Mark Reynolds, *Three Views on Creation and Evolution* (Grand Rapids: Zondervan, 1999).

44. Ian Barbour, *Religion in an Age of Science* (San Francisco: HarperSanFrancisco, 1990), pp. 3-30.

express my own conclusions, often heartfelt *conflict* between science and classic Christian theology is actually apparent and temporary in that there is only one reality which is being gradually uncovered. Different data available to theology and to science means that there will be areas of *independent* investigation. Both disciplines need to keep in active *dialogue* with each other for each to learn from the other at those points where they overlap, which should lead toward *integration* in those cases.

The Science of Human Genetics in Particular

With an understanding of the relationship between science and the Christian tradition in hand, we can go on to listen to some key points of genetic science in particular. An important foundation for ethical analysis is to get the facts straight. This is true because if the facts of a case are changed, the evaluation will be as well. That Noelle took ten thousand dollars from someone else's bank account sounds like larceny, until the facts are clarified that Noelle was duly given power of attorney to manage the account. If the further fact is introduced that she is planning to use the money for her own benefit, ethical evaluation raises concern, until it becomes clear that the account owner while still competent left instructions for her to do precisely that. Ethics is not inconsistent in these varying evaluations. It is taking into account new data. Scientific description as well changes as new data becomes available. But there is a fast-developing core of relatively assured understanding about our genetic heritage and its role. The following will briefly introduce some of the most essential information for our task of considering human genetic intervention.

Genetics and the Human Body

The human eye is marvelously complex. Could you design one in all its intricacy? One would have to know the exact chemical formula for the translucent cornea and how to vary its thickness to focus the light on the retina. Rod and cone light receptors would need exquisite array and connections to the optic nerve. Then brain tissue would have to be in place to interpret optic-nerve impulses into understandable pictures. It would be a task of vast intricacy. Since you are reading this book, it is likely that you have successfully assembled precisely this design already, probably twice. How did you do that?

All of the necessary information for design, assembly, and basic operation of your eye and the rest of you is contained in your DNA. DNA is an abbreviation for deoxyribonucleic acid. It consists of a varied sequential pattern of just four chemicals: adenosine (A), cytosine (C), guanine (G), and thymine (T). They are set in a double helix that looks something like a spiraling ladder and in human beings is organized into two sets of twenty-three chromosomes. Each rung of the ladder is a pair of A-T, T-A, G-C, or C-G. Each sequence of three rungs codes for a particular amino acid. It is the assembly of these amino acids into proteins that builds the incredibly intricate and interactive system called the human body.

By their sequence the four constituent parts of DNA encode a vast amount of varied information, just as the twenty-six letters of the English alphabet, by varying their order, can record the plays of Shakespeare, love letters, or a VCR manual. Another parallel is found in computers that run word processing, spread sheets, and vivid simulations from a simple binary on/off code. DNA consists of only four distinct units that by their order carry all the information needed to structure the particular form of life that bears it. In fact, except for a few viruses that use RNA, it appears that DNA is the chemical code for the structure of all the rest of earthly life. Yeast, daisies, mice, and human beings all record and duplicate their instructions in the same system. At the cellular level human and other mammalian cells are almost interchangeable in many respects. That is why medical research for human medicine can be advanced by studies with mice and other animals. Chimpanzees are the most genetically similar to human beings, with about 98 percent of their DNA sequence held in common with us. The 2 percent that is different is extremely important, but human DNA does reflect considerable physical continuity with other physical beings.

DNA also has a strong community dimension within humanity. Each person is genetically unique,[45] yet an individual's genes have come from just two other people. Except for an infinitesimal percentage of random mutations an individual has no genes that were not present in one of her two parents. Each parent received genes from just two people as well. On the one hand one is closely tied genetically to just a few people, yet on the other one's genes have been passed on for untold years from people who lived before, and future people may carry on those genes present now. Our genes combine uniquely and intimately in individuals, yet are held in common by families, ancestors, and descendants. The system of human genetics is one of diversity but also unity; variation and continuity. Genes tie together individuals, humanity, and life.

45. Unless through twinning there might be one or two other persons sharing the same genetic combination.

These associations and the rest of the environment are as influential to the individual's form as genetic heritage. "Genotype" is the individual's genetic instructions. "Phenotype" is how those genetic instructions are expressed by the organism. Genotype is not phenotype. There are myriad steps in between. Some physical characteristics are shaped more by genetics and some more by environment. No amount of nutrition, stretching, or other environmental influence is able to add substantially to one's height beyond a certain range of genetic predisposition, yet lack of optimum nutrition can severely limit height.[46] Genes can predispose one to build muscle mass, but only exercise, diet, and hormonal activity will actuate them. Our physical form and health are the result of an intricate interaction between genes and their surroundings within and outside the body. Even in the most limited variations of water crystallizing to form a snowflake, when multiplied countless times in reaction to a changing environment, the snowflake that results may not be precisely like any other. Small variations multiplied in an interactive environment can result in large differences. If one twin of genetically identical twins has diabetes mellitus, the other twin has a one-in-three chance of eventually developing the disease. That is a much higher rate than that of the general population, yet two out of three identical twins never develop the diabetes mellitus that affects their twin with identical genes.[47] Genes are important, but not by themselves physical destiny.

Genetics and Human Behavior

Within the animal kingdom it is clear that genes guide not only structure but also certain behaviors. Seek water, hide by day, flee the unusual, and fly south for the winter would be examples. Genes provide both structure and operating instructions. In a computer such would be called hardware and software. Human beings have genetic tendencies too.[48] For instance, an inherited aver-

---

46. Jim Fix, the long-distance runner, probably added years to his life by his conditioning yet still died of a heart attack at 52. His father had died of one at 43.

47. Daniel Drell, "FAQs," *Human Genome News* 9, nos. 1-2 (1998): 4.

48. Edward O. Wilson has sought to explain these gene-based behavioral tendencies as the result of evolution. *Sociobiology: The New Synthesis* (Cambridge: Harvard University Press, 1975), *On Human Nature* (Toronto: Bantam, 1982), and Charles J. Lumsden and E. O. Wilson, *Genes, Mind and Culture: The Coevolutionary Process* (Cambridge: Harvard University Press, 1981). On this theme applied directly to ethics and theology see Arthur Peacocke, ed., "The Challenge of Sociobiology to Ethics and Theology," *Zygon* 19, no. 2 (1984): 115-232. For a critique of sociobiology see John C. Avise, *The Genetic Gods: Evolu-*

sion to certain smells may protect people from eating rotten foods that are often poisonous. Other studies have shown that infants with no experience of falling or training to avoid falling, resist moving onto a transparent surface that appears to be a drop-off. They seem to have an inherent antipathy to unsupported heights. A team led by R. P. Epstein has found genetic variation associated with novelty seeking, correlating forms of a particular gene with personality test scores.[49] The K. P. Lesch team has found a correlation between a particular polymorphism and degrees of anxiety.[50] Mary Midgley asks in response to those who argue that all human behavior is learned from the environment, "How do all the children of eighteen months pass the word along the grapevine that now is the time to join the subculture, to start climbing furniture, toddling out of the house, playing with fire, breaking windows, taking things to pieces, messing with mud, and chasing the ducks? For these are perfectly specific things which all healthy children can be depended on to do, not only unconditioned, but in the face of all deterrents."[51]

Genes may also contribute to far worse behaviors than messing with mud. A 1993 Netherlands study found an extended family mutation on the X chromosome that interferes with production of monoamine oxidase A. Lacking this enzyme affects levels of serotonin and noradrenalin in the brain. This seems to predispose to impulsive and aggressive behavior associated with criminal activity.[52] Of the family members with the mutation, most have been caught committing crimes such as arson. Of those without the mutation, none.

It should be noted that the debate about whether the human soul is a separate entity intimately linked to the body or an emergent phenomenon of the body (dualism versus monism) is not central to this discussion. Both monists and dualists agree that there is at least a deep mutual connection and influence between soul and body. The body's physical states, including ones

tion and Belief in Human Affairs (Cambridge, Mass.: Harvard University Press, 1998), pp. 155-56.

49. R. P. Epstein et al., "Dopamine D4 Receptor (D4DR) Exon III Polymorphism Associated with the Human Personality Trait of Novelty Seeking," Nature 12 (1996): 78-80.

50. K. P. Lesch et al., "Association of Anxiety-Related Traits with a Polymorphism in the Serotonin Transporter Gene Regulatory Region," Science 274 (1996): 1527-31.

51. Mary Midgley, Beast and Man: The Roots of Human Nature (New York: Meridian, 1978), p. 56.

52. H. G. Brunner et al., "Abnormal Behavior Associated with a Point Mutation in the Structural Gene for Monoamine Oxidase A," Science 262, no. 5133 (22 Oct. 1993): 578-80.

connected to its genes, make a difference to the person/soul. When the body is tired, the soul will find it difficult to pray.

Of course, not all behavior is genetically determined. Such a reductionist description would ignore the documented influence of environment and our choices.[53] Not only do experiences shape our ongoing choices as persons, but they can sometimes affect the activity level of the genes that we have. It appears, for example, that seizures can turn on the expression of certain genes involved in the brain's response to injury that would have otherwise remained quiescent.[54] As with physical characteristics, genotype is not phenotype. Having a gene that has an influence on behavior does not tell to what degree it will be expressed, nor is it usually determinative even if expressed to the maximum possible influence. Human beings develop in a complex interaction that includes far more than just genes. Over the years behavioral geneticists have fairly consistently concluded inheritability rates for a number of behaviors at rates ranging usually from roughly 40 to 60 percent.[55] Identical twins raised apart have more similarities in personality traits and social attitudes than fraternal twins raised together.[56] Genetic heritage is influential but not determinative. Inheritance seems to have as much, not more, influence as environment for a number of personality differences.[57]

This is why it is not possible to clone an individual's personality. Only a person's genes could be cloned. The resulting baby would be a younger genetic twin. The child would have much in common with the older sibling but would develop as a unique person. Duplicating someone's genes would lead to another human being much younger than the first, with strong physical resemblance. The genetic twin would experience the womb, infancy, and childhood in a markedly different environment and make his or her own choices. A different person results. If you have ever met identical twins, you have met naturally occurring clones, two people who have the same genetic code and who despite sharing an environment and identical genes are still unique individuals, different people. Our genes are formative for our physical structure

---

53. Holmes Rolston III, *Genes, Genesis and God: Values and Their Origins in Natural and Human History* (Cambridge: Cambridge University Press, 1999).

54. The geneticist V. Elving Anderson notes this point in his 1995 book with Reichenbach, *On Behalf of God: A Christian Ethic for Biology*, p. 253.

55. For example, Thomas J. Bouchard Jr., "Genes, Environment, and Personality," *Science* 264 (17 June 1994): 1700-1701.

56. Auke Tellegen et al., "Personality Similarity in Twins Reared Apart and Together," *Journal of Personality and Social Psychology* 54 (1988): 1031-35.

57. Reichenbach and Anderson, *On Behalf of God: A Christian Ethic for Biology*, p. 268.

and some temperamental inclinations, but not finally determinative of the complete person.[58]

For an organism such as a worker ant, genes are destiny. The ant is born with instinct, genetic instructions to behave one certain way. The behaviors may be intricate but they are rote and unchangeable. A more complex animal such as a Labrador retriever may still have strong genetic instructions. Present a Labrador retriever puppy with a tub of water and he will be more likely to jump into it than drink from it. Yet at this level the instinct is not destiny. The puppy can be trained to avoid water or only to jump in on command. Environment is beginning to play the most formative role. By the time we reach human beings, we too have genetic drives such as to protect our young, but we can choose to channel or resist them. They are not destiny but rather different degrees of tendency. For example, the same temperament could be channeled to be constructively steadfast or harmfully stubborn; another to be energetic or impulsive. But it is true that some genetically related conditions such as schizophrenia can be so powerfully destructive that they are difficult to channel positively.

Glenn McGee gives a mirthful caricature of the idea of genetic tendency always controlling us. According to that view, not his own, "we think that we choose, while we are really determined, pushed, or persuaded by the genetic call of our innards. It is as if the overeater hears a voice from the pantry, which is really projected from within the genome: 'eat more guacamole.' "[59] Of course one's genes have no voice box or telepathic skills. However, they do "communicate." What actually may be happening at the pantry door is that fat is a high-energy food of particular value when food is scarce. Since for most of human history food has been in short supply, physical pleasure in eating fat would encourage one to seek it out, a successful strategy. When food is in short supply, the natural (genetic) tendency to enjoy the taste and texture of fat and begin with the premium ice cream is helpful. When one faces the pantry in times of plenty and remembers how good that rich smooth (fat) taste is, one tends to reach for the Haagen Dazs before the celery. The natural desire for that taste can be overridden by one's awareness and conscious choice that under conditions of plenty, the celery would be better

58. V. Elving Anderson succinctly explains the links between genes and behavior in "Genes, Behavior, and Responsibility: Research Perspectives," in *The Genetic Frontier: Ethics, Law, and Policy,* ed. Mark S. Frankel and Albert H. Teich (Washington, D.C.: AAAS, 1994), pp. 105-30. See also Theodosius Dobzhansky, "The Myth of Genetic Predestination and of Tabula Rasa," *Perspectives in Biology and Medicine* 19 (1976): 156-70.

59. Glenn McGee, *The Perfect Baby: A Pragmatic Approach to Genetics* (Lanham, Md.: Rowman & Littlefield, 1997), p. 61.

for health after all. Genes can encode rewards such as physical pleasure for valued behaviors, but what they reward does not have to be pursued. Genetic tendency does not have to be obeyed. It does for worker ants, but not for human beings. Humans have a sufficient level of cognitive ability to decide whether to encourage or enable a particular tendency. These propensities can vary in their intensity, yet are usually not destiny.

An apt metaphor might be that of a sailboat. The size, shape, and condition of the hull and sails would be genetically endowed. The strength and direction of the wind and water currents would be the environment, and the way one sets the sails and rudder would correspond to one's will. Some people hit the jackpot in the genetic lottery. They are born with countless genetic advantages of strong hull and sails. The wind of environment is at their back, and yet they just sail in circles. There are other people who have poor initial resources and a contrary wind, yet manage to achieve great things. Genetics reminds us that everyone starts with a different set of givens. Genes and environment deal varied hands. Our will plays a decisive role in what we do with them and hence what we become.

Genes Illuminate Needed Attitudes

Seeking to understand human genetics can be a fascinating exploration in and of itself. It can also illuminate aspects of long-praised attitudes in the Christian tradition. Of course simple observation of nature in general or genetics in particular cannot serve as a standard of good actions or attitudes. There are natural evils such as malaria. The standard one uses to sort out natural goods from natural evils may be one's actual standard. Attitudes of humility, respect, community, responsibility, and compassion have long been affirmed in the Christian tradition. Why, is another book. For here my judgment is that these are helpful for thinking through ethical decisions in general and needed genetic choices in particular. If that is not immediately clear, just think of the effect of their opposites on genetic choices. What if our intervention comes to be characterized by arrogance, belittling of self and others, egoism, irresponsibility, and insensitivity to the needs of others? As I introduce these attitudes, it is interesting to see that aspects of human genetics can in part illuminate each one. This part of the chapter also exemplifies one aspect of the constructive dialogue possible between science and the Christian tradition. As mentioned in the beginning of this chapter, the two are often described as antagonists. It might be helpful to see a less combative interaction.

Humility

Wonder can refer to being puzzled at something before us. It is an apt description for our current view of human genetics. Even as we now complete the initial sequencing of the human genome, the new data has revealed how much we do not know. Only about 10 percent of our DNA seems to be active (transcribed at some time into mRNA for making proteins). The remaining 90 percent is often called "junk DNA," but that term is more a reflection of our ignorance than knowledge. While we can now read the sequence of nucleotides (ACTG), we often do not know what the sequence means. "Junk DNA" could turn out to be essential for gene expression, copying, structure, some other function, or no function at all.[60] One can learn a great deal about a new country, including some of its language, history, geography, and culture, yet such knowledge reveals whole areas of investigation that one might have not even considered before. In most human endeavors the more we actually do learn, the more we become aware of what we do not know. That is certainly the case in regard to human genetics. Humility, a cardinal virtue of scientific and theological investigation, is well in order. Humility is not self-deprecation or degradation. It is simple recognition of one's own limitations and the importance of others. "Jesus knew that the Father had put all things under his power, and that he had come from God and was returning to God; so he got up from the meal, wrapped a towel around his waist, poured water into a basin and began to wash his disciples' feet."[61] Jesus was fully aware of who he was, yet was so other-focused as to gladly meet an evident need. Jesus did not depreciate himself. He simply thought of others first. He was even unselfconscious enough to use the humble and freeing words "I do not know."[62] We do well to speak honestly the same when it is the case.

Respect

We can wonder at what we *do not* know and at what we *do* know. The intricacy and beauty of human genetics is a wonder. One example is the 3,000,000,000 base pairs of DNA in each cell of the human body, which

---

60. James C. Peterson, "Ethical Standards for Genetic Intervention," in *On Moral Medicine: Theological Perspectives in Medical Ethics,* ed. Stephen E. Lammers and Allen Verhey, 2nd ed. (Grand Rapids: Eerdmans, 1998), p. 579.

61. John 13:3-5.

62. Matthew 24:36.

could uncoil and stretch out to about two meters in a straight line. That is the proportional equivalent of winding thirty miles of fine wire into a tight enough form to fit inside a walnut shell, yet in such a way that the wire could be unwound, copied, and returned repeatedly to its small package at will. That is an engineering feat worthy of approbation. Recognition of such, repeated in each of all 3,000,000,000 nucleated cells that compose one human body, should evoke respect for each human being as a marvel of vast intricacy.

For one who is convinced of the truth of the Christian tradition, praise would also be due to the designer. Granted, this observation is fraught with difficulties worked through time and again in the history of natural theology. Does the existence of painful genetic conditions at birth mean that God delights in suffering or punishes to the third and fourth generation? No, for if one takes seriously the data of scriptural revelation both of those extrapolations are contravened. Scripture offers other explanations for such suffering, which we will examine more closely in chapter 3. What use then might genetic data have for knowledge of God? Genetics may powerfully illustrate or suggest implications of what is revealed elsewhere.

For example, the Christian tradition has always taught that God is the Creator of all. As our awareness of the size of the universe has increased, our appreciation for that claim should as well. In our galaxy alone there are about 100,000,000,000 stars. It appears that our Milky Way Galaxy is of average size among about 10,000,000,000 galaxies. Those figures expand the sense of power, indeed magnitude, that is involved in describing God as Creator. At the genetic level, each human being has roughly 3,000,000,000 DNA base pairs encoding our genetic information in every nucleated cell. A mistake at just one point can cause a potentially lethal condition such as cystic fibrosis. Our bodies have a level of detail that can also expand one's sense of what it means to describe God as Creator. If God is the ultimate Creator, God is as vast in subtle intricacy as in astronomical space. Thanksgiving would be appropriate both for what God has done and for who God is to be able to do it. That is compounded by the fact that we are contemplating not some distant object but rather our own bodies. The body is a wonder we know intimately and depend on. An appropriate quote from the Psalms is carved in the entrance gates of the Cavendish Physics Laboratory at the University of Cambridge, "Great are the works of the Lord; they are pondered by all who delight in them."[63]

---

63. Psalm 111:2. Berry, *God and the Biologist*, p. 57.

Community

Most of us are genetic individuals.[64] Our precise combination of genes has never been seen before and will never be again. We are each genetically unique. Yet those genes came from only two other people, who received their genes from their parents. There is remarkable variety and yet continuity in our genetic system. Two parents become one in physical intimacy. A new unique individual begins, so that now three people are present. That individual may well become one with another, yielding another new individual. It is an interesting parallel, built into our very nature, to the nature of God as trinity. God, of course, has no genetically encoded body, nor is God a family, but God is described in the Christian tradition as one nature in three persons: Father, Son, and Holy Spirit. Unity and distinction are part of the essence of God and in a different way of us, God's creation.

This genetic link with others also reminds us of our interdependence with fellow human beings. Our very genes are part of a community of human beings and the wider community of life. Whether we seek or dread the responsibility, our choices have effects beyond ourselves. Our actions affect what genes will be present in future people, as the actions of others have deeply shaped what genes we inherited. While our genetic heritage is quintessentially intimate to each of us, it is shared with family and a whole history of ancestors in the past and descendants to come. As genetic individuals, we are inextricably part of a community of family, humanity, and life.[65]

Responsibility

The degree of influence varies, but we all have genetic tendencies that are helpful to carry out and others that are not. In mice there is a gene called fos B that is crucial to nurturing behavior. If the gene is knocked out, mice mothers ignore their pups and the pups die. Humans carry the fos B gene. At the human level that one gene does not determine parental care, but having a natural desire to nurture could contribute to that positive behavior. On the other

64. As noted earlier, a few of us have one or two other people who share our genetic code as identical twins or triplets.

65. Berry suggests that this community extends to all creation. In the words of Romans 8:19-21, all the rest of creation waits in eager anticipation for human beings to fulfill their rightful part as the crown of creation. The earth is like a concerto waiting for and anticipating the featured performer who has not yet fully played the featured part. See *God and the Biologist*, p. 104.

hand, because one has a tendency to eat until one has built up a substantial layer of fat, a useful drive in a context of food scarcity, does not mean that such is the most healthy behavior in the current context of abundant food and typically longer lives in industrialized countries. A tendency to respond with violence when angered is usually not the best response in our highly interdependent society. The desire for multiple sexual partners could have advantages for spreading one's genes but does not automatically mean that such behavior is best for all involved. Genetic tendency does not have to be mechanically obeyed. It does for worker ants, but not for human beings. Human beings have a sufficient level of cognitive ability to decide whether to encourage or enable a particular tendency. These propensities can vary in their intensity, yet are not destiny.

Genetic tendency to behave in a particular way is not an ethical trump card. Simply having a desire or tendency does not answer the question of whether it is appropriate to carry it out. While the Christian tradition states that God created the physical world and declared it good, the tradition also speaks of human beings having a "fallen" nature.[66] The belief that "God created me this way, so it must be what I should do" ignores that part of the scriptural tradition and would require the acceptance of greed, pedophilia, sadomasochism, and many other destructive behaviors to which people are inclined. Desire cannot be automatically equated with God's will. Also desires which can be good to act upon in one context might be harmful in another. Being sexually aroused by one's spouse is to be welcomed, but sexual desire toward someone else is not. Now one is not responsible for what one desires, but one is responsible for which desires one entertains and enables and when one does so. Discernment of what is worthwhile, right, and good requires more than the presence of personal inclination.

Genetically inheriting sexual attraction toward one's own sex has been the subject of much speculation of late. Several studies have concluded that one might inherit such a drive.[67] So far under peer review the studies have not been found conclusive by the scientific community.[68] Bailey and Pillard drew

66. Ted Peters offers an explication of genetic influence, sin, and human responsibility in *Playing God? Genetic Determinism and Human Freedom* (New York: Routledge, 1997), particularly pp. 63-94.

67. For example, Dean H. Hamer et al., "A Linkage Between DNA Markers on the X Chromosome and Male Sexual Orientation," *Science* 261 (16 July 1993): 321-27.

68. Neil Risch, Elizabeth Squires-Wheeler, and Bronya J. B. Keats, "Male Sexual Orientation and Genetic Evidence," *Science* 262 (24 Dec. 1993): 2063-64, with reply from Hamer, p. 2065; Eliot Marshall, "NIH's Gay Gene Study Questioned," *Science* 268 (30 June 1995): 1841.

attention when they originally claimed that when an identical twin reports homosexual inclination about half of their genetic twins report the same. For fraternal twins raised together who share an environment and family genes but not identical genes, they found a concordance rate of about 22 percent.[69] These figures have been contested, and Bailey has halved the original estimate to a concordance rate of about one-quarter for identical twins. By this measure the percentage difference between those with identical genes and those with similar genes does not appear to be statistically significant. The difference between 22 percent and 25 percent could be a random variation. Even if one interpreted the study as showing a slight increase when a greater number of genes are held in common, in both cases the concordance rate is far less than 100 percent. The strongest implication one could draw is that a genetic component may be present but is not determinative by itself. It would not be surprising if eventually varying degrees of genetic propensity toward same-sex attraction are found. The important point for ethics and genetics is that having a genetic inclination or even a strong physical drive to do something does not even begin to address whether that is a good, right, or helpful thing to do. That would have to be addressed by effects on health and persons,[70] as well as, for Christians in the classic tradition, the insights and direction of Scripture.[71]

For the classic Christian tradition, the ultimate standard for all behavior is God's will recognized through Scripture, tradition, reason, and experience such as the direct leading of God's Holy Spirit.[72] The resulting life is exemplified and taught by Jesus Christ and his apostles. It can only be enjoyed by grace from beyond ourselves and is developed through a process of learn-

---

69. J. Michael Bailey, Richard C. Pillard, Michael C. Neale, and Yvonne Agyei, "Heritable Factors Influence Sexual Orientation in Women," *Archives of General Psychology* 50 (1993): 217-23, and J. Michael Bailey and Richard C. Pillard, "A Genetic Study of Male Sexual Orientation," *Archives of General Psychology* 48 (1991): 1089-1096.

70. Thomas E. Schmidt gently weighs health issues and causes in *Straight and Narrow? Compassion and Clarity in the Homosexuality Debate* (Downers Grove, Ill.: InterVarsity Press, 1995). Jeffrey Satinover focuses on the psychiatric literature in *Homosexuality and the Politics of Truth* (Grand Rapids: Baker, 1996).

71. Richard B. Hayes listens carefully to the relevant Scripture in *The Moral Vision of the New Testament: A Contemporary Introduction to New Testament Ethics* (San Francisco: HarperCollins, 1996).

72. Each branch within Christianity usually appeals to all four ways of recognizing God's will, but with an emphasis on one way of knowing more than the other three. For example, tradition is particularly shaping for an Eastern Orthodox perspective, reason through natural law for Roman Catholicism, and sensitivity to the direct experience of the Holy Spirit for Pentecostals or the Society of Friends (Quakers).

ing over time. One is not only welcomed into God's family by grace but is also called to live by grace. The love, joy, peace, patience, kindness, goodness, faithfulness, gentleness, and self-control that are to be characteristic of the Christian are described as the fruit of God's present Spirit.[73] The distinctive life that can be developed over time for those who belong to Jesus Christ is to stand out as a clear sign that God is at work in the lives of God's people. They could not achieve it on their own.[74] The quality of their lives is to reveal God's presence and intervention. One is not necessarily limited to the tendencies or even capabilities of one's genes.

C. S. Lewis suggested that in the next life the people who will be recognized as having been the greatest heroes in this one will be people whom no one has heard of before. They will be recognized for what they achieved with what they had. This is not to lower the standard of what we should be and do, but rather to recognize that we begin at different points. God judges by the heart, as God did in choosing David as king. When Samuel was sent to anoint a new king from the family of Jesse, he immediately expected that Eliab would be the one, but God instructed him, "Do not consider his appearance or height, for the Lord does not look at the things that human beings look at. Human beings look at the outward appearance, but the Lord looks at the heart."[75] Jesus told the parable of the stewards who were responsible for multiplying whatever amount of talents they had, and said explicitly, "To whom much has been given, much shall be required."[76] In regard to our genes and environment, God knows where we start, offers power to grow, and holds us responsible for what we do with what we have. Chapters 3, 11, and 13 will speak in more detail about the connection between genetics, choice, and behavior.

Compassion

It is widely recognized that we are not all that we could be. Genetics reveals where some of this physical brokenness is embodied. Our roughly fifty thousand genes are central to human development. When any one or a combination is randomly mutated, it is occasionally an improvement but usually detrimental. The harm can range from unnoticeable to strongly felt. More than

73. Galatians 5:22.
74. John 15:5.
75. 1 Samuel 16:7.
76. Luke 12:48.

six thousand diseases have been cataloged so far as directly attributable to a defect in a single gene.[77] The number recognized continues to increase. Cystic fibrosis is an example of a disease that can be triggered by a mistake at just one rung of the DNA ladder (a point mutation) in a large number of different locations around the transmembrane conductance regulator gene. Even complex behaviors can be traced back to defects in a single gene. Lesch-Nyhan disease, with symptoms of retardation, cerebral palsy, and compulsive self-mutilation, seems to be one. Many other diseases as well can be traced to a deleterious combination of several genes. Yet others involve chromosomal abnormalities such as Trisomy 21 (Down syndrome) that result in varying degrees of mental retardation and physical difficulties.

John C. Fletcher cites some of the statistics that convey the extent of genetic disease. "About 22% of newborn deaths in developed nations are due to congenital malformations or genetic disorders. About one-third of children admitted to pediatric units in Western nations need treatment for the complications of genetic disorders, congenital defects, or mental retardation. . . . Nearly half of all naturally implanted embryos die in spontaneous abortions in the first few weeks after fertilization, largely due to abnormalities that increase with maternal age . . . contributing to the infertility that affects 2.4 million United States married couples of reproductive age. . . ."[78] The World Health Organization lists genetics as involved in 50 percent of the cases of blindness and more than 50 percent of the cases of deafness occurring in developed countries.[79]

Genetic endowment also varies one's susceptibility to disease.[80] Some people can smoke tobacco all their lives without lung cancer, while others are afflicted with lung cancer without any clear exposure to an outside carcinogen. The degree of the body's ability to cope with carcinogens varies as widely as environmental exposure to them. It is already established that Alzheimer's, insulin-dependent diabetes mellitus, and certain cancers such as retinoblas-

77. D. N. Cooper and J. Schmidtke, "Diagnosis of Genetic Disease Using Recombinant DNA," *Human Genetics* 73 (1986): 1-11. For an authoritative and continuously updated list see the latest edition of Victor A. McKusick, *Mendelian Inheritance in Man* (Baltimore: Johns Hopkins University Press). This information is also web accessible at http:/www.ncbi.nlm.gov/Omim.

78. John C. Fletcher, "Controversies in Research Ethics Affecting the Future of Human Gene Therapy," *Human Gene Therapy* 1 (1990): 308.

79. WHO Scientific Group on the Control of Hereditary Diseases, *Hereditary Diseases Report* (Geneva: World Health Organization, 1996), p. 19.

80. Dr. Theodore Friedmann in testimony before a congressional hearing, in *Human Genetic Engineering* (Washington, D.C.: United States Congress, Committee Print 170, 1983), pp. 275-83.

toma show strong family inheritance patterns. If one has the genotype for cystic fibrosis, one will be stricken with the disease. Cancer, on the other hand, has both genetic and environmental components, with some types more genetic and others less. Having a predisposition does not guarantee that one will develop the disease. Still other afflictions such as AIDS appear at this time to have very little genetic component at all.[81]

Our genes do not decide everything, yet they are foundational for much of our physical form and function. That can include shaping mental states as well. As physical beings, human beings are substantially dependent on the proper function of their genetic endowment. When the body lacks needed genetic information or carries a defective version of a gene, the cost in suffering, medical care, and loss of productivity and lives is often staggering. When one's neighbor suffers, one should compassionately do what one can to help. There are many times when genetic science may make such service possible.

Chapter Summary

Science and the Christian tradition have had a long and varied history of interaction. They have often been allies. Each affirms that there is one reality and is willing to adjust scientific and theological theories to better describe it. Both have found it difficult at times to give up long-held paradigms, whether Newtonian physics or the unmoving earth. Where they most differ is in sources of data about reality. The classic Christian tradition is convinced that God has revealed a great deal through the Christian Scripture, reason, tradition, and experience. Science limits its investigation to observations of the material world and usually to hypotheses about material causation within that world.[82] This focus on physical causation does not try to address other sources of information, so scientists who are Christians have felt free to use the scientific method as a useful tool without any threat to their larger worldview. When scholars extend science to exclude anything nonmaterial from reality, they are advocating not science but "scientism." Scientism is a metaphysical claim that the material is all that exists. Such a metaphysical claim is simply not inherent to the scientific method.

From the perspective of the Christian tradition, any conflict between

81. These examples are cited by Francis Collins, director of the Human Genome Project, in "The Impact of the Human Genome Project on Cancer Diagnosis and Treatment," presentation to Ethical Boundaries in Cancer Genetics conference, Memphis, 26 May 1999.

82. See footnote 24 on the intelligent-design movement.

current scientific hypotheses and the Christian tradition is only apparent and temporary in that there is only one reality both are seeking to recognize. All truth is God's truth. In any given instance of disagreement, probably both or at least one will eventually have to adjust to better approximate the truth. Such differences are a call to better apprehend reality, a goal they both share. Different data availability does lead to areas of independent investigation, but both disciplines do well to maintain active dialogue with each other. That offers the opportunity to learn from the insights and challenges of the other where they overlap. Such dialogue can ideally lead toward integration at those points to the degree that each one is accurate.

As to genetic science in particular, rapid advance in human genetics has shown the important role of genetics in human structure and behavior. It has become clear that genes are deeply influential but usually not determinative. Better understanding of them cannot require but can illuminate at least five attitudes helpful to the ethical evaluation of their use:

1. *Humility* is elicited by recognition of how much we do not know.
2. *Respect* is warranted for human beings in all their vast intricacy and for what it entails to know God as Creator.
3. *Community* is unavoidably embodied in the unique yet intertwined nature of how we receive and pass on our genes.
4. *Responsibility* remains for what we do with what we have, since genes are influential but not determinative.
5. *Compassion* is elicited by the staggering suffering and loss we can now trace directly to genetic cause or susceptibility.

"Perfecting" our genes would not provide a perfect life. But granted the tragic losses involved in our current genetic state, should we intervene? Life is more than the physical and the physical is more than our genes, but should we attempt to change our genes at those points where they do make a difference? This is a question of technological intervention, and that is where chapter 2 begins. We would do well to take with us to the questions of application a due humility toward self and respect for self and others. That will be essential to building a community that recognizes personal and group responsibility to act with compassion.

What Technology Can Do for Us and to Us

About ten thousand people stood together in the dark. They were crowded around the courthouse in Wabash, Indiana, waiting expectantly. The courthouse bell began to ring. Suddenly the town square was as light as day. The gathered throng had expected it. Many had traveled miles to see it happen. But when it did, there was no roar of appreciation, no applause, just stunned silence. In the blinding light some people gasped, and others fell on their knees, overwhelmed. All their lives, if there was any outdoor light at night, it was the slight illumination of the moon and stars. All one could see was at best indistinct grey outlines in the darkness. Indoors there might be the faint flame of a candle, fire, or gas lamp that flickered and smoked. Here suddenly was intense light everywhere, as if the sun had been turned on at will. All was color and detail, steady, unblinking. At the throw of a switch, the town fathers had lit four three-thousand-candle arc lights on top of the courthouse, and the Wabash of 1880 would never again be seen in quite the same way. At least at night.[1]

The dramatic lighting of the town square held the media's attention, but the transformation came elsewhere. Downtown lighting was spectacular, but that was not where electric lighting made the greatest difference. What most changed the way people lived was in the rapid but incremental steps of extending electricity to one house at a time. Wiring a house was not the stuff of headlines, but it was a transforming moment for each household it

---

1. David E. Nye, *Electrifying America: Social Meanings of a New Technology, 1880-1940* (Cambridge, Mass.: MIT Press, 1991), pp. 2-3. Firsthand accounts: *Wabash Plain Dealer,* 7, 14, 21, 28 Feb. 1880.

reached. Electric light was so valued that by 1934, 96 percent of the households in Muncie, Indiana, had electric lighting at the same time that more than half had no central heating, 34 percent had no running hot water, and 18 percent were still using outdoor privies. By 1945, half of Muncie's households were still without a telephone, but almost all used electric light.[2] Electric lighting had many valuable uses, but it was in one home at a time that it most changed the way people lived, not only in bringing light but in opening the way for all the other electric machines that so shape our time, from television to the personal computer.

Much as the spotlights on top of the Wabash courthouse suddenly dazzled onlookers in the square, cloning now rivets public attention. Speculation about trying to duplicate people has dazzled and maybe to some degree blinded us in a brilliant flash of media focus. Actually, there are already more widely transforming events in genetics that have received less notice. For example, it is the rapid incremental steps of new genetic tests that are already changing what we see and how we live, one person at a time. In coming chapters we will examine the choices involved in the whole range of new genetic technologies. We will approach them in the order of their widespread implementation. But first in this chapter we will establish needed context by briefly looking at the nature and impact of technology more broadly.

Technology shapes us far more than most people realize. We often fear it as it first arrives and then come to take it for granted as necessary to our lives. What can technology actually do for us, and just as important, what will it not be able to do for us? If we can guide it toward particular ends, which ones should we prefer? The goal of this chapter is to avoid mindless boosterism of every new thing and just as much the stunned inaction of a deer caught in the glare of headlights. What should we see and do by this new genetic light, without letting it blind us? We need to start by better understanding the momentum of the tools we use to shape our world; the tools which also end up so deeply shaping us.

Technology Is Necessary

I have described science as a method of careful observation of the physical world. In contrast, technology is the sum of the tools that we use not just to understand our physical world, but to shape it. Our tools can be as simple and solid as a hammer or as intricate and conceptual as a hospital system. Ian

2. Nye, *Electrifying America*, p. 22.

Barbour refines the definition this way. Technology is "the application of organized knowledge to practical tasks by ordered systems of people and machines."[3] In some ways the distinction between science and technology is quite artificial. The two drive each other. Copernicus's solar-system theory intrigued Galileo to study the planets with his new tool, the telescope. It was the telescope that enabled him to make observations of Jupiter's moons. Those observations were then substantial evidence for the solar-system theory of Copernicus. While science and technology are intertwined, it may be that "science is concerned more with what is, not what ought to be."[4] In contrast, technology is all about changing the physical world to make it more what we want.

Adapting the physical world to our needs is necessary for most of us to survive. That does not refer only to those depending on respirators, kidney dialysis, or daily doses of blood thinner. Very few of us live where it is always near 20 degrees centigrade (72 degrees Fahrenheit), with ample food available that can be caught with bare hands and eaten raw. The simple fish hook is a tool, as is the fire and a pan for cooking, as are the clothes on our backs. The Amish, famous for maintaining older ways, depend on technology too. Horse-pulled reapers, hand pumps for obtaining water, and barn design are all technologies. They are just the continued technologies of an earlier era. Making and using tools is characteristic of human beings, who would otherwise be hard-pressed to live out the week. Philip Hefner has described human beings as "created co-creators" with God.[5] The emphasis on the creative role that God has given us is welcome, yet even paired with "created," the noun "co-creator" might be misunderstood as claiming human beings as relative equals with God. Maybe the description "creative creatures" may better capture human beings and our mandate. According to the Christian tradition and others, this needed capacity to create tools is not contrary to God's provision. It is God's provision. We need technology to survive.

Technological Change Is Accelerating

While technology is necessary to our survival, it has developed far beyond that minimum. Its development is also accelerating. Certain inventions have

3. Ian Barbour, *Ethics in an Age of Technology: The Gifford Lectures*, vol. 2 (San Francisco: HarperSanFrancisco, 1993), p. 3.

4. R. J. Berry, *God and the Biologist: Faith at the Frontiers of Science* (Leicester, U.K.: Apollo, 1996), p. 27.

5. Philip Hefner, *The Human Factor: Evolution, Culture, and Religion* (Minneapolis: Fortress, 1993).

increased the pace of technological development. The printing press made information about what others had discovered more widely available. Telecommunications has dramatically quickened the spread of information, and the worldwide web has already compounded that. From one country to another different people receive credit for inventing the telephone. That is not just national chauvinism. Different people did invent the telephone independently of each other within a period of a few years. Without rapid communications they did not know what the others were doing. With modern communications new discoveries are quickly heralded, so that people are working on advancing the next step rather than duplicating what has already been worked out elsewhere. The pace of invention quickens.

Not only are the people who do research more efficient, but there are more of them. It has been suggested that most of the full-time scientists who have ever lived are working today. New information is discovered and quickly put to use discovering more. Product cycles are short and becoming shorter. From invention to widespread use takes less time. The television was invented in the 1930s and took thirty years to become ubiquitous in our homes. Personal computers took less than fifteen years from inception to widespread home use. Louise Brown, the first in-vitro baby, was born in 1978. Within ten years, thousands of babies had been born from the same method all over the world.

Technology Is Becoming More Intricate

Robert Pool describes how a Turkish Airlines DC-10 crashed shortly after takeoff from the Paris airport in 1974. A cargo door blew open when the plane reached twelve thousand feet. The sudden loss of pressure collapsed the passenger compartment floor, which severed the control lines to the tail control surfaces and brought the plane down. No one person designs the more than five million individual parts in a commercial airliner. Different teams of engineers had made design choices that interacted in a lethal way. The doors opened out rather than in. That made it easier to load the plane but also meant that air pressure at flying altitudes would be trying to open the doors rather than pushing them securely in place. The team that designed the door handles allowed the handles to close without fully engaging the pins that were to keep the door secure. The group that chose to route the control cables under the passenger floor had chosen a path that was well protected from outside injury but vulnerable to passenger-floor movement. The builders of the passenger floor chose lightweight materials that were more than adequate to carry foot traffic and airline seats, but not to stay rigid if there was a loss of cabin pressure below the deck. It was the interaction of all these choices that was lethal.[6]

The science of ecology has helped us to realize how interrelated the natural world is. Lose one species here and the effects can be drastic at quite a distance. The ecology of modern technical systems is also vulnerable to seemingly irrelevant changes. When Hurricane Andrew knocked out power in Miami, cars started to run out of gas. Gasoline was abundant but stored in tanks only accessed by electric pumps. Changing one part has ripple effects far afield. It is like pulling on one thread in a tapestry and discovering that it is connected to many other threads. Such intricacy means both that it is increasingly difficult to foresee the results of new technology introduction and to withdraw damaging technology that has already been integrated. The complexity of modern technology means that we are more dependent on experts to maintain it and that whole systems depend on single parts. The experts are dependent on other experts to manage the subsystems fundamental to their systems. This makes it difficult to predict all eventual implications when introducing or removing a technology. Effects and the expertise needed to predict and manage them are dispersed. The discipline of technology assessment has been developed with particular emphasis on this problem of expecting the "unintended, indirect, and delayed" effects of present and new technology.[7] Surprising results can be positive or negative.[8] We need to choose, develop, and use our tools wisely, for much is at stake.

Technology Is Formative

Technology affects how we live and who we are. It has been suggested that if one has a hammer in hand, everything looks like a nail. Our tools do influence our perception and choices. Think of how deeply shaped we have been by the introduction of electricity described earlier. It does not just make cer-

6. Robert Pool, *Beyond Engineering: How Society Shapes Technology* (New York: Oxford University Press, 1997), pp. 133-34.

7. Joseph Coates, "The Identification and Selection of Candidates and Priorities for Technology Assessment," in *Technology Assessment,* vol. 2 (London: Gordon and Breach Science Publishers, 1974), p. 77.

8. Alan Kay wrote that Daniel Bricklin and Robert Frankston designed computer spreadsheet software as a convenient editor for accounting past expenditure. VisiCalc instead was enormously popular as a simulation tool for future business decisions. An unintended use was its most significant contribution. "Computer Software," *The Scientific American* 251 (Sept. 1984): 52.

tain tasks easier with dishwashers and vacuum cleaners, but it changes our environment with air conditioning, entertains us with music and images, and even extends the hours of "the day" available to us for work and play. It makes possible whole new tool systems such as the computer. It is so ubiquitous that it is its absence in a power outage that we notice, not its presence.

Think of another invention, the automobile. Contrast life in the rural/suburban college town of Wingate, North Carolina, before and after the advent of the automobile system. Friends would be drawn from the few family groups within walking distance. With the automobile, one's socially closest friends may live tens of miles away. People are able to spend more time with others who share special interests. While there is only one family with twins in Wingate, one can drive a short distance to Charlotte and join the "Mothers of Twins Club" with hundreds of members. One can also associate with a particular religious group. Before the automobile, one's choice for worship within walking distance was First Baptist Church or the United Methodist Church. Now within a half-hour's drive are several hundred different congregations and tens of different religions.

Since the advent of the automobile one can associate with more specialized groups that are similar in some way, yet one is regularly exposed to more people with whom one might not have as much in common. To reach a particular place of worship one may have driven past tens of places gathering people with quite different commitments. Meeting a stranger used to be a noteworthy event, yet now one is exposed to countless strangers. On a trip to the grocery store, one sees hundreds of fellow shoppers, clerks, and cashiers. One might be surprised to recognize any of them. Exposure to more people can enrich one's life, but it also makes it easier to enrich one's life at the expense of another. Not only can one flee a crime scene in the get-away car, but one can use the stolen goods in relative anonymity a few minutes away by car from their original owner.

Even the flavors of our food and when we eat it have been affected by the automobile. It makes possible in suburban or rural housing the Arcadian dream of living in a quiet garden setting far from industry. This means that family members are often dispersed during the work day and gather again at its end. The main meal of the day used to be at noon, but now for most it is in the evening when family members have returned from work places that can be quite distant from where they live. The food at dinner has probably come from a giant supermarket accessible only by car or from a restaurant drive-up window.

A major portion of the work force is involved in making possible the system of high personal mobility offered by the automobile. There are jobs involved in making basic materials such as glass and steel for cars, then design, transport, parts, assembly, sales, and finance to buy the machine. To operate it, there are insurers, appraisers, traffic police, courts, driving instructors, and maintenance and repair mechanics. Further, the automobile depends on an intricate and pervasive system of road construction and maintenance, parking lots, garages, parking decks, petroleum discovery, refining, transport, service stations, toll booths, map making . . . that absorbs a substantial part of our personal and national economy.

Time is affected as well. The average American drives about twelve thousand miles a year. If one includes stop lights, drive-up windows, finding parking places, traffic tie-ups, and other limitations on speed, an average of thirty miles an hour for these twelve thousand miles would be generous. That places most Americans in their steel boxes for at least four hundred hours a year. That is the equivalent of a full-time job from January first to mid March. The automobile is just *one* technology system that affects one's friends, religion, crime, food, education, land use, personal and social economy, place of work, and for many their very livelihood. Technology shapes our lives.

When the automobile was first introduced, it was available only to the wealthy. With mass production and the government's commitment to build roads, it became an option to more people. Today much of the United States is structured so that one cannot work, obtain food, or worship in a community without it. Residential neighborhoods are deliberately zoned away from industrial areas. Grocery stores have abundant variety in centralized locations. What begins as optional, if widely adopted, tends to become necessary to participate at all. In some parts of the world a shiny new car would be useless, except to polish the finish and listen to the radio until the first tank of gas ran out and the battery died. Technology often comes in complex and integrated systems that need and then enforce widespread adoption.

This integrated nature means that it is difficult to remove one part of the system without affecting others. Technology becomes entrenched over time and often builds its own momentum. In part II we will examine in detail the impact of new genetic diagnostic tests. Here, it might simply be noted as an example of technological momentum that when such tests exist there are substantial pressures to use them. The conscientious physician would want an available aid to patient care. Also would any physician wish to be placed on the witness stand and admit that there was an inexpensive accurate test for the patient's condition that was not used? The existence of the technology is often pushed forward into use by multiple societal dynamics.

One of these dynamics is the widespread assumption that new is better. Products can receive a noticeable bump-up in sales simply by adding to the packaging the word *new*. There is an optimism that technology will always produce a better way, as if human beings could resolve any question with enough technological insight and application.[9] Further, innovation is craved. Last year's product or method is not good enough. Surely there is more power, more space, more speed, more something to add to the experience. People who lived quite well a few years ago without any internet access at all feel the need for more and more download speed. Ironically, while innovation is desired, the media has become so adept at projecting future developments to sell magazines, newspapers, and science-fiction films, that when new technologies become available people often feel that the capability has already been around for some time. Expectations are high and often fulfilled that new capabilities and the refinement of already present ones will occur frequently.

There is some resistance to this trend of constant innovation. England was traumatized by the "mad cow disease" disaster.[10] English beef producers fed ground-up beef by-products such as cow brain to their cattle to speed growth. This practice made possible the spread of a brain disease lethal to cattle and apparently deadly for some humans who ate their meat. This understandably sensitized people to possible danger from the agricultural industry that provides their food. With the advent of genetically modified (GM) plants, the reaction has been not so much any specific objection but rather a horror that changes are again being made in the food system. Virtually all our food is genetically modified and has been for centuries. Standard plant cross-breeding that has produced our seed for millennia is a clumsy and slow type of genetic modification but GM nonetheless. Domestic sheep have been intentionally bred, genetically modified, to grow high-quality wool and docilely accept human handling to regularly shave it off. For the most part GM is just a more precise and quick way of cross-breeding, though admittedly genes can be introduced that have not been in plants before. The main impetus for concern may be that after the mad cow disease fiasco, people have reason not to trust the food industry. In practice the public debates have been more over whether civil disobedience or private property should prevail when a group masses to trample a farmer's crop, than concerning the actual effects of the technology.

Can we thoughtfully resist or direct the new technologies that so deeply shape us, or are they unstoppable juggernauts? Are we so quickly dependent on their accelerating and interlocked development that there really is no time or leverage for choice? Is technology like the Star Trek aliens, the

---

9. Craig M. Gay, "The Technological Ethos and the Spirit of (Post)Modern Nihilism," *Christian Scholars Review* 28, no. 1 (1998): 90-110.

10. Bovine spongiform encephelopathy.

Borg, who always announce their presence with "We will assimilate you. Resistance is futile"?

Technology Is Malleable

Refrigerators are ubiquitous, and almost every one of them hums. The hum is the sound of the electric compressor. In contrast, gas refrigerators are utterly silent. With no moving parts they do not wear out. They are also more energy efficient and no more expensive to mass produce. Why do we not all have gas refrigerators? It is the more efficient technology and was invented first. The commercial triumph of the electric refrigerator is an example that the most efficient technology is not always the one that is adopted. In this case the key reason electricity prevailed was the timely and concerted effort of a relatively small group, the leaders of an electric power generating corporation.[11] What could be more beneficial to the sale of their product, in this case electricity, than an appliance in each person's home that uses their product twenty-four hours a day and could only be turned off at substantial financial loss to the owner? The corporation was willing to invest great sums of financial resources to mass produce the electric refrigerator so that they could drive the per unit cost down to competitive levels with gas ones. They then opened stores across the land to make their refrigerators more accessible and sustained a massive marketing campaign to introduce people to their product. The resulting economics of scale and development of widely available electric power not only gave electric refrigerators the lead but also provided the structural supports for their dominance to become complete. Electric power lines became more accessible than gas ones. Gas refrigerators still exist, but usually in commercial buildings that have the size and independence to have a custom system. Electric refrigerators own the domestic market.

This story does not reveal that economics determines all. That is not the case.[12] It is an example that the most efficient technology is not always the one widely adopted. The disc operating system that currently dominates computer software is often cited as a similar case, as is the triumph of VHS video tape over another type of video tape called Beta. Generally, "Americans

---

11. Ruth Schwartz Cowan, "How the Refrigerator Got Its Hum," in *The Social Shaping of Technology,* ed. Donald MacKenzie and Judy Wajcman (Philadelphia: Open University Press, 1985), pp. 202-18.

12. Arnold Pacey nicely contravenes that conclusion in his book *The Maze of Ingenuity: Ideas and Idealism in the Development of Technology,* 2nd ed. (Cambridge, Mass.: MIT Press, 1992).

have traded away the neighborhood, local businesses, the walking city, and mass transit, for the detached suburban house, the shopping mall, and the freeway." Most American cities are "designed less to be lived in than to be moved through."[13] Whether decisions are conscious, conscientious, or not, individuals and communities make formative choices about what technology we will use.

Jacques Ellul suggests that technology has taken on an independent life of its own. "It is artificial, autonomous, self determining, and independent of all human intervention."[14] Langdon Winner sees much the same momentum but has more hope that technology can be shaped to human needs. "Human 'somnambulism,' rather than any inherent technological imperative, has allowed large technical systems to legislate the conditions of human existence."[15] We have the capability to destroy human life on the planet with nuclear missiles but have so far resisted carrying out that capability. What can be done has not always been done. Possibility is not necessity. We do have a choice.

That choice is usually greater when a new technology is just beginning. Once a technology is formed and adopted, it tends to preclude other options. A choice for is often a choice against as well. That is not always a negative. Offering a choice socially legitimizes the offered options. We no longer allow the choice of dueling. That prohibition eliminates a method of conflict resolution but also frees people from being challenged to a duel.[16] On the other hand, commitments to particular technologies can shape us in ways we later regret. Such problems are often hard to detect or predict at the beginning when they are still relatively easy to change. By the time adverse results are evident to all, the technology is often entrenched and difficult to disentangle. In the late 1990s billions of dollars were spent to rewrite computer software that was not designed to accommodate dates in the 2000s. The original shortfall in design came from an economy measure to save a computer's expensive memory space. Few considered the future confusion and expense that was being embedded in the system.

13. David E. Nye, *Consuming Power: A Social History of American Energies* (Cambridge, Mass.: MIT Press, 1998), pp. 256-57.

14. Jacques Ellul, "The Technological Order," *Technology and Culture* 3 (Fall 1962): 10.

15. Merritt Roe Smith, *Does Technology Drive History? The Dilemma of Technological Determinism* (Cambridge, Mass.: MIT Press, 1994).

16. Allen Verhey, "Luther's 'Freedom' and a Patient's Autonomy," in *Bioethics and the Future of Medicine: A Christian Appraisal,* ed. John F. Kilner, Nigel M. de S. Cameron, and David L. Schiedermayer (Grand Rapids: Eerdmans, 1995), pp. 89-90.

It is difficult to control the development of new capabilities, but one can try to guide applications thoughtfully. Technology shapes all of us. Its already pervasive presence and influence is increasing at an accelerating pace. We can choose where we want it to take us. So where do we want to go? Who do we want to be? That is the fundamental question for technology. It is certainly a key question for the developing technology of human genetic intervention.

We are back to the stunning lights that first night in Wabash, Indiana. It is essential that we let our eyes adjust to what is going on. We need to discern what to do with this new light, as well as whether to turn it up or off, or in an entirely new direction.

Chapter Summary

Technology is the sum of the tools that we use not just to understand our physical world but to shape it. Our tools can be as simple and solid as a hammer or as intricate and conceptual as a hospital system. We need technology to survive, yet we extend it far beyond that end. The pace of new technology discovery and widespread implementation is accelerating as we develop immensely complex and intricate systems. Such interdependent systems develop powerful momentum as they become embedded within the rest of society. They shape us deeply. We can shape them if we make the conscious and dedicated effort to do so, especially as they are first developed and implemented. Human genetic technology is at precisely that point as it offers us increasingly formative interventions in genetic research, testing, pharmaceuticals, and surgery. So where do we want to go? Who do we want to be?

CHAPTER 3

The Purpose of Technology from a Christian Perspective: To Sustain, Restore, and Improve

The Point of Human Life

We create and use technology to change ourselves and the world. Deciding how to develop and use technology then asks the most basic question: What are we trying to do? We need to take care lest we become very good at high-speed travel in no direction. What is our purpose? The proffered answers are myriad, from Buddhistic ceasing to desire to frenetic attempts to fulfill every desire. Since the most prominent voice in the topical dialogues to follow is a Christian perspective, it will be useful at the outset to better understand what the classic Christian tradition values.

To Love God

God is at the center of the classic Christian tradition. While this description would be obvious for most of church history, it has become more contested since the challenge of Ludwig Feuerbach (1804-1872). He claimed that God is a human creation that we project for our own use. For Feuerbach human beings are the center. "God" is a concept that some human beings find useful. Many theologians have accepted that critique. They then write of God as a flexible concept limited only by our imagination. Among other possibilities the idea of God can be used to encourage moral development, give psychological comfort, or rally community. God language is used to pursue these ends without any claim that God actually exists apart from human creation.

64

The ongoing classic Christian tradition finds this projection theory of God fundamentally mistaken. The apostle Paul wrote that an awareness of God is present in every human being who is honest with himself. The great French mathematician Blaise Pascal described each human being as having a "God-shaped vacuum" that can only be at peace if God is welcomed. But both Paul and Pascal are referring to the need to know the actual Creator, not a human projection. Chapter 1 footnoted some of the scholars who have responded to Feuerbach convincingly on this point (see note 21). The classic Christian perspective begins by recognizing God as the source of all that is, the reason we and the universe continue, and the only worthy point to which life can lead.

For those who are open to God's presence, it is a relationship with God that gives life the consistent joy and confidence that human beings were designed for. The highest, most fulfilling goal for humanity is a right and complete relationship with God. The Westminster Confession says that the purpose of human beings is to "glorify God and enjoy him forever." "To glorify God" is to live the adventure of all that God would have one to be and do to God's honor. "Enjoying God" is to welcome, appreciate, and find completion in God's presence. By God's grace both are to start now and continue without end.

This relationship is founded on what Jesus called the first and greatest command. Human beings are to "love the Lord your God with all your heart, and all your soul, and all your strength, and all your mind";[1] in other words, with everything they are and have. This love begins and continues to be grounded in a conscious choice of commitment that seeks the best for the one loved. The New Testament term for it is the Greek word *agape*. One is consciously to choose to welcome God's love and learn to respond in kind. The relationship with God is also described by *phileo*, which is also often translated as "love." It has been argued that in Koine Greek, the language of the New Testament, *phileo* and *agape* are synonyms. I see different connotations sometimes continuing from their classic forms. *Agape* is primarily an act of will and commitment to care for the other. It can be a one-sided decision. This is why Jesus could call his people to agape-love their enemies. They are not expected to generate feelings of affection toward their enemies, but rather to choose to seek what is best for others, even their enemies. In contrast, *phileo* emphasizes the sense of mutual enjoyment of the other's presence. In this friendship human beings can genuinely communicate with God, which leads to communion and increasingly union.[2] Human beings always

1. Luke 10:27-28.
2. Dallas Willard, *In Search of Guidance: Developing a Conversational Relationship*

remain the creature and God the Creator, yet human beings have received a unique potential to enter into an intimate enjoyment of God that is not possible for any other earthly creature.

Human beings have received this unique capacity to enter a relationship with God, but not to exhaust it. God is accessible, yet vastly beyond what human beings could even begin to comprehend. Rabbi Jay Holstein tells the story of walking along a Florida beach and seeing a little girl industriously running into the surf with her pail, filling it with sea water, hauling it back up the beach, and pouring the water into a dip in the sand. From a nonthreatening distance he asked her what she was doing. She proudly explained that "today I am going to empty the ocean with my pail." She knew what the brine tasted like on her tongue. She knew how the surf tugged at her legs. She could smell the salt in the air. She knew the ocean with the fullness of all her senses. Yet she knew nothing of undersea volcanoes and canyons, blue whales and barracuda, icebergs and tropical islands, water stretching from where she stood to the African coast. We too can truly know God with everything that we are, but we cannot begin to comprehend all that God is. That may be a task for eternity.

To Love Your Neighbor as Yourself

Community with God is to be extended to fellow human beings. Coupled with the goal of joyful relationship and active life with God is the goal of fellowship with God's people. Jesus said that the second greatest commandment is to love your neighbor as yourself. This begins with a commitment to seek the best for the other. That choice, as an act of will, can be extended even to one's enemies, as Jesus said that it should be. Yet love is best fulfilled when it is reciprocated in mutual care and enjoyment of each other. As God is one, yet three persons, human beings are invited to fulfill their individual potential by becoming part of a community of fellowship with God and one another. It is as if each note is most clearly distinct when uniquely contributing to a greater harmony. The christological hymn found in Colossians praises Jesus Christ in this way: "all things were created by him and for him. He is before all things and in him all things hold together. . . . For God was pleased to have all his fullness dwell in him, and through him to reconcile to himself all things."[3]

---

with God (Eugene, Ore.: Wipf & Stock, 1997), p. 164. Released under a new title by Inter-Varsity Press in 1999, *Hearing God: Developing a Conversational Relationship with God.*

3. Colossians 1:15-20.

We are not already there. To reach the end that is the real beginning requires creation, redemption, and transformation. This process is characteristic of the church and is recapitulated again and again in individuals. Creation is necessary in that one must exist to live. Redemption is the enabling of reconciliation, the at-one-ment of those formerly estranged, a place of healing. Transformation is in becoming the kind of person that can more fully love and be loved. It is natural that the goal is pursued not only in individuals but cooperatively in the church, since the process is then already beginning to fulfill part of the goal of community. The point is to become the kind of people who can and do understand, serve, and enjoy God and one another. That ultimate purpose is exemplified and encouraged in how the transformation comes about.

The Place Where We Live

If the above is the point, what role does the material world play in it? How is the physical world part of our creation, redemption, and transformation to live a life of love for God and our neighbors? The answer to that question will guide us in how to shape the material world, including our genetics.

Creation: A Place to Choose and Care

God has created a world where we can live. Indeed, it is filled with contingencies that in the slightest variance would have left no physical place for material beings. The Cambridge University physicist Sir John Polkinghorne cites as an example the exquisitely balanced forces that hold nuclei together yet can allow them to decay. Such stability yet recombination is necessary to physical life. He has joined others in describing these surprising fits as examples of "the anthropic principle." The material universe looks like it was made for us.[4]

Yet what is most remarkable about existence in this physical world is how hidden God is. After all, God is the fundamental and pervasive reality. The center, what T. S. Eliot called "the still point of a turning world," can be functionally ignored for a time. The Christian tradition sees this strange situation as a gift. Human beings are designed to live in fellowship with God, but it is an intimacy founded on invitation. If it were forced, it would not be the

---

4. John Polkinghorne, "A Potent Universe," in *Evidence of Purpose,* ed. John Marks Templeton (New York: Continuum, 1994), pp. 105-15.

same relationship. Directly confronted with God's presence as Isaiah was at his calling, we would be struck with terror. Like a shadow before a spotlight, our sheer faint existence would be overwhelmed in God's unshielded presence. Instead, God has created a physical world where we have a genuine choice as to whether we recognize and value his presence. God reveals himself more and more to those who seek him and allows those who do not to carry on for a brief time as if alone. The physical world with all its beauty and terrors, sunsets and tornadoes, directs us to seek God from admiration or fear, yet allows us to live for a time as if God were not present.

This material world also calls human beings to band together to survive and live. Yet this too is by invitation. The physical world grants sufficient physical separation that we can live in a hermit's isolation or even in urban crowds as if no other human beings existed. The physical world is a place where we can choose how we will relate to our fellow human beings. Here we can enter into life with God and others or not. We can fulfill the two great commandments, the purpose of life, or choose isolation and death. We can choose whether to care for others and the world we have received or not.

The physical world is a God-given place to do this, to choose and to care about God, others, and God's world entrusted to us. Having made this material place, God declared it "good." Contrary to Plato, Marcion, and others, there is nothing distasteful about matter. Human beings are not trapped in bodies. Human beings are bodies; God-given bodies. In the second chapter of Genesis, Adam is described as being made from the earth. The chapter goes on to say that while human beings are of the earth, they are yet uniquely in-breathed with God's Spirit; of the earth, yet distinct from it. Human beings have a unique calling embodied in the material world.

God likes the material world. God created it this way as distinct from all other possibilities. The physical is not our ultimate concern, but if we are in a right relationship with God we should care about it because it is part of our God-given form, place, and stewardship. What we do with our bodies matters. Our bodies in particular and the more general physical world as well are all the sphere of influence that we have. Small as it is, we can make genuine choices in matter, that matter. It gives us a place to learn attitudes, skills, and commitments, and to practice actually carrying them out. It is where we have the opportunity to become more of who we are meant to be. Our first assigned task with God and one another is to care for the creation in which we have been placed.

The first chapter of Genesis describes God creating human beings "in His image." There have been long debates over what "God's image" precisely entails. Three definitions have been dominant. One emphasizes that human beings have capacities that are uniquely God-like, such as the ability to reason and freedom to make choices.[5] Any conscious human being will hence be in God's image. However, Martin Luther noted that Satan also would be in God's image if capacities such as reason were sufficient.[6] A second interpretation focuses on carrying out God's delegated dominion as much as possible in the way God would. This view takes particular note that the command to rule the earth immediately follows the description of being made in God's image. A third sees the image of God in relationships with God and other human beings. As John Calvin suggests, a mirror cannot reflect what it is not oriented toward.[7]

It may be that each is correct in part. God's image includes capacity, task, and relationship. One needs certain capacities such as reason to carry out the stewardship of the world in a godly way but will only be able to achieve that way of life if in an empowering relationship with God. All three need to be present to fully bear the image and likeness of God. Reflecting God's image has to do with both the capability and the actual carrying out of choices as God would. One is only reflecting God's image if one is sufficiently close to God and oriented to God to reflect his presence. Reflecting God's image is being in right relationship with God and representing God accurately in the sphere we have been given. This would mean then that those rebelling against God by act or apathy would not be reflecting God's image. What then of Genesis 9:6, that the death penalty is required for shedding the blood of human beings since they were made in God's image? Also the New Testament letter of James warns against cursing human beings, who were made in God's image.[8] These protections may not be limited only to those reflecting God's presence well. The reference to God's image may refer to its presence at the origin of humanity or to the potential of each human being to bear that image.

God has all dominion. Some is delegated to human beings bearing his image in our limited physical sphere. Even in the material world it is not an unqualified dominion. First, it is to carry out God's direction. Second, it is limited to earth. Third, there is a limit to what human beings may do even with the physical world entrusted to them.[9] They may use part of it for food,

5. Douglas John Hall names this view as the most dominant in the church up through the time of Thomas Aquinas. *Imaging God: Dominion as Stewardship* (Grand Rapids: Eerdmans, 1986), pp. 92-93.

6. Hall, *Imaging God,* pp. 100-101.

7. Hall, *Imaging God,* p. 104.

8. James 3:9.

9. Genesis 1:27-29.

and not other parts. The granted dominion is one of ruling the creation as God's representatives. That requires that human beings care for creation as God does and remember that it is still God's creation. God had declared the creation good. After all it is his choice out of all possible contingencies. This is the world God made. It deserves respect and preservation as the work of God's hands.

Calvin DeWitt points out that the mandate to Adam includes that he should "till" the garden. When the word "till" appears in other contexts, it is most often translated as "to serve."[10] A famous example is that of Joshua challenging the people of Israel that they must choose whom they will serve: "As for me and my house, we will serve the Lord."[11] The word translated as "to serve" the Lord is what Adam is to do in the garden. Changing the context to agriculture can certainly change the meaning of the word, but DeWitt may be right that "to till" the garden continues in the sense that it is to serve the garden, helping it to reach its proper end, not to rend or destroy it. This would fit with the paired word to "keep" the garden. Adam is to "till and keep" it. The word translated as "keep" is the same word found in the classic Aaronic blessing, "The Lord bless you and keep you."[12] Both words in context and use elsewhere imply the authority and ability to care for the garden, not to destroy it. Human beings will answer to the owner for how they keep his garden.

Matter then matters as the place human beings have been given to live for a time. In the physical world human beings can make genuine choices to welcome or reject relationships. They also have the ability to affect the material world and are responsible for how they treat it. The material world is not theirs to destroy. It should be sustained as God's good creation under their rule and tilling. They are expected to learn to be like and with God and with each other in how they care for the garden and shape it to positive ends.

Redemption: A Place to Be Reconciled and to Heal

Human beings have often chosen disastrously. We have often failed to care for God, our fellows, and the world entrusted to us. Our relationships with God, one another, and the physical world have been deeply broken. Thankfully, rather than starting over God has opened a way of reconciliation and healing.

10. Calvin DeWitt, *Caring for Creation: Responsible Stewardship of God's Handiwork* (Grand Rapids: Baker, 1998), p. 45.
11. Joshua 24:15.
12. DeWitt, *Caring for Creation*, p. 44.

This redemption is centered in the person and work of Jesus of Nazareth, the Messiah, the Christ.

Jesus Christ

The relationship with God that human beings were designed for can be established through the work of Jesus Christ. The Christian tradition makes the striking claim that the one true God came to live among us in the person of Jesus of Nazareth. When one meets Jesus of Nazareth, one is meeting God face to face. This Jesus was the Messiah long promised in Hebrew prophecy. He brought about forgiveness, healing, and reconciliation with God. Through him, by accepting his gracious provision, one can become a member of God's family. This is a gift. The gift is often called "justification." Justification is making things "just-as-if" I had never lived away from God. It can also be called "redemption." One is redeemed, bought back, from trying to live a life apart from God. By God's grace, despite past brokenness, one can be declared right with God and welcomed into his family. As part of this new life and community one is to return to proper care for one another and the physical world. This includes healing damage done by sin and restoration of caring dominion for the physical world.[13] Such is exemplified in the life of Jesus.

Jesus not only achieved reconciliation for those who receive it, he also lived it. Jesus is a model for our behavior in at least three ways. First, he is described repeatedly as God among us. When one encounters Jesus Christ, one is encountering God. What God values, Jesus valued. What Jesus did was rightly reflecting what God does. If we are to reflect God's image, Jesus is the clearest presence of God that we have. Second, Jesus is described as the perfect human being. He is what human beings look like when fully reflecting God's image. In Jesus we can see what humanity is meant to be. Part of his work as the second Adam was to live a human life rightly within human constraints. If he had acted by his own divine power, he would not have been a human being who could fittingly die in our place. It is often stated in the Gospels that Jesus did miracles not by the power of his innate divinity, but by the power of the Holy Spirit working through him.[14] That is the same Holy Spirit at work in his people. Third, the New Testament states often that the church is now the body of Christ. God sovereignly chooses to work through human beings to

13. John S. Feinberg, "A Theological Basis for Genetic Intervention," in *Genetic Ethics: Do the Ends Justify the Genes?* ed. John F. Kilner, Rebecca D. Pentz, and Frank E. Young (Grand Rapids: Eerdmans, 1997), pp. 183-92. Also the works of Francis Bacon.

14. Matthew 12:28; Luke 4:14.

carry out his will. The work of Jesus Christ on earth continues under his leadership through his church. In all three senses, how Jesus treated the physical world is a model for his people now.

Jesus was deeply committed to this physical world in his incarnation, teaching, and action. The central events in his life were often substantially body events. The incarnation, birth, crucifixion, resurrection, and ascension are all physical landmarks. He cared for and sustained his body with food for some thirty years. The usual human condition of having to find and prepare food was part of his life and that of his disciples. For example, on one occasion the disciples were rebuked by locals for breaking the Sabbath by gleaning cast-off grain from a field as they passed through it. But when no food was available, Jesus recognized the need and gladly provided, even the best drink for a wedding feast.

While Jesus cared for his body, he did not place it first. He would rather help the woman at the well than make time for lunch.[15] His mission came first, even in times of great stress such as the Temptation.[16] There was nothing wrong with turning a stone into bread and eating when hungry, yet it was a temptation then because it would mean acting on his own divine ability. That would break his mission to live under the same restraints that we do, yet without sinning. He was willing to forego genuine physical needs for a higher purpose. He would eventually sacrifice his physical life completely on the cross.

Jesus came first and foremost in order to reconcile us with God, yet he also gave of himself in caring for people's physical needs. Not surprisingly, crowds gathered for such healing. He was careful in the midst of that to direct people to focus on their spiritual needs first. Forgiving people their sin and calling them to a new way of life often preceded physical healing. He cautioned those healed not to tell anyone of their physical healing, lest that service crowd out the things he had to say. The physical was important. One's body needed shelter and food. Jesus had compassion on those who were ill. But such needs were always recognized as secondary to far more important ones. The greatest sacrifice and rejoicing were reserved for that which transcends the human body.

In the Gospel of John Jesus gives the commission that as the Father has sent him, so he now sends his disciples.[17] His followers are to care actively for people's physical concerns as he did. In fact, such care is evidence that one is indeed his. Jesus describes the final day of judgment as a time when he will

15. John 4:5-34.
16. Matthew 4:1-4; Luke 4:1-4.
17. John 17:18.

separate those who belong to God's kingdom from those who do not. The method of recognizing those who belong is according to how they treated their neighbors. Those who fed the hungry and clothed the poor, cared for those who were sick and visited those in prison, are members of God's kingdom. Those who did not are not. Good works of caring for people do not earn a place in God's family, but rather they are characteristic of people who are in God's family. Such actions do not achieve salvation, but they do reflect it.

The second chapter of Ephesians reads that God's people are "saved by grace through faith, lest anyone should boast." The next sentence is sometimes neglected. It reads: "for we are his workmanship, created in Christ Jesus for good works." These good works are a result of reconciliation with God, not the cause of salvation. Reconciliation with God is a free gift that if fully received will affect how one treats others. If one is being shaped by God as a child of God, one will come to care about what God cares about. That will prominently include people. If one cares about people, one will care about the physical form on which they depend. The examples given in Matthew 25 of expressing love for one's neighbor, while not all-inclusive, are each in reference to physical and social needs, such as feeding the hungry and caring for those who are sick. The physical world is not ultimate, but it is valued as foundational to human existence. Healing, setting right what is not best supporting human life, is a Christ-like calling for all his people.[18]

Suffering

This healing includes the restoration of physical capacity and the relief of suffering. Does such intervention thwart what God ordains? Why would God allow such harms to people in the first place? This is the question of theodicy. It is sometimes stated as when the innocent suffer either God cares but cannot stop it or God can stop it but does not care. Process theology affirms the first option: God always wishes the best for his creatures but simply does not have the power to set all things right. Human beings are in a joint struggle with God to relieve suffering. For John Calvin and Martin Luther, God is quite capable of controlling everything, and does. That the good and holy God causes evils such as pain is mysterious. This is just another part of reality that human

---

18. This theme can also be found in James J. Walter, "Playing God or Properly Exercising Human Responsibility?" *New Theology Review* 10 (Nov. 1997): 39-59, and in Ronald Cole-Turner, *The New Genesis: Theology and the Genetic Revolution* (Louisville: Westminster/John Knox, 1993), pp. 80-97.

beings do not yet understand. Someday they will. For scholars such as Bruce Reichenbach and Richard Swinburne, God could but chooses not to determine everything. As God self-limited himself in the incarnation, so God self-limits himself in other ways as well. Suffering is a direct result of God's sovereignly choosing to give his creatures space to make genuine choices.

Trying to understand physical suffering is not a new issue for the Christian tradition. In parts of the West where the church is well established, suffering is not automatically associated with church membership, yet persecution has been the case through most of the church's history and even today in many parts of the world. Most of church history has been characterized by great suffering. In the Roman Empire it was not even legal to be a Christian until the Edict of Milan in 313, almost three hundred years after the church began. To this day in many parts of the world Christians are severely persecuted for their faith. The accounts in the newspapers as this is being written include slavery in Sudan and execution and burnings in East Timor. This is not a surprise to members of the tradition. The central symbol is the cross, an instrument devised to inflict the greatest possible pain. Since Christ was allowed to suffer, it is not surprising that such might befall his people as well. This is not perceived as in some sense a deserved suffering as in *karma*. Karma was assumed when people brought a man born blind to Jesus. They asked, "Who sinned, the man born blind or his parents?" The puzzle was in their assumption that surely someone had sinned for the man to suffer in this way. If it was his parents, it was not fair for their child to suffer the ill consequences. If it was the child's sin, how could he be guilty at birth? Jesus replied that it was neither.[19] At another point he noted that a recent building accident had befallen people who were no more guilty than others.[20] Bad things happen to people whether good or bad.

Why then does God allow suffering that God could stop? There have been countless opportunities to reflect on that question. They have generated as many responses. The following is a brief list of ten possibilities, none of which is exclusive of the others.

1. A great deal of suffering may be self-inflicted in varying degrees. If one embezzles at work and then is caught, life in jail is a great loss, but it is a self-inflicted one. One cannot blame God for one's incarceration. If one commits adultery and loses one's spouse and children, that is not an affliction from God. That is the result of one's own poor choices.

2. In the two examples above the perpetrator is not the only one hurt.

19. John 9:1-3.
20. Luke 13:4-5.

Many people are hurt by someone else's abuse of free will. This includes God. Why then allow it? Free will exacts a high price. It must be endemic to what is most important to be worth its cost.

3. Other writers attribute some suffering to chance. If individuals want God to leave them alone, God honors that choice and leaves them alone. In that case they cannot blame God for not protecting them in an often dangerous world.

4. Suffering may serve to get our attention. Many people would be happy to sign up for a treatment that would eliminate all sensation of pain from their bodies. They might be surprised to discover that such is the most dangerous sequela of leprosy. Leprosy is a feared disease in that it is associated with disfigurement and the crippling loss of body parts. The damage is rooted in the inability to feel pain. When pain works properly, it lets the body know that something is wrong and needs attention before the damage is worse. It serves as a warning system that one should stop and correct the problem immediately. Suffering can serve this function as well. It may warn that something is wrong and needs correction before the damage becomes worse.

5. While suffering can obscure everything else, if it overwhelms or is allowed to dominate one's life, it can call the sufferer to take stock of what most matters. Clear and right priorities are worth a significant price.

6. Søren Kierkegaard wrote that suffering is a powerful medicine that either kills or heals the patient. One's response is all-important. Suffering can build character in a way that nothing else can. To strengthen a muscle one must exercise it. Suffering is not always only an enemy. It can sometimes be a teacher.[21]

7. Over the centuries Christians have taken some consolation from the belief that God suffers too. While some theologians have argued that God is impassable, incapable of suffering, at least in the person of Jesus Christ and in his people, God suffers. This was central, for example, to Julian of Norwich's meditations on the cross as so many people around her died of the plague.[22] In the cross she could see God's firsthand understanding of human suffering and the lengths to which he went to relieve it.

8. Suffering can be on behalf of others. This has its most clear example in Jesus Christ. When he faced the time of his crucifixion, Jesus prayed for the cup to be taken away from him if there was another way to achieve God's purpose. But he was willing to pay the necessary price to free his people. His later followers often counted it a privilege to share with him in everything, includ-

21. 1 Peter 1:6-7.
22. Julian of Norwich, *Showings*.

ing that price.[23] It appears that almost all the original disciples died for their faith, as did countless others.

9. Suffering on this earth is temporary. Abraham Lincoln used to keep perspective through celebration and trial by repeating to himself that "this too will soon pass away." Somehow things are more bearable if one is confident that their duration is finite. The Christian tradition is confident that there can be horrendous moments, but they are but a fraction of time compared to the eternity promised to God's people.

10. The most prominent response to suffering in the scriptural tradition is to trust in God's character. God knows what he is doing and that is enough. This is the response to suffering found in the book of Job. When Job asks God why he is suffering so, God begins to reminisce with him about creating the great creatures of the sea and laying the foundations of the continents. The point is not that God is more powerful than Job. That is true. The point is that Job does not have the big picture. Job cannot even conceive of all that God has done or is doing. It might be akin to the first time I took my daughters in for their immunizations. They looked at me with disbelief and a feeling of utter betrayal when I allowed them to be stabbed with a needle by the strange woman in a white coat. Yet they had nowhere else to turn for comfort and still toddled back to me. I wished that I could explain that the shot was necessary to protect them from great harm in the future, but they could not yet speak any language, let alone understand the concept of immunization. I would have far preferred to suffer the pain in their place, but that would not have helped them. They simply needed to trust me.

Suffering can serve an important role, yet it is suffering. Following Jesus' example, his people should look for ways to relieve it. If suffering is relieved, is some good lost that would have been achieved through it?[24] Human life is fragile. There will always be sufficient opportunities to suffer to achieve those ends. Healing on all human levels of being was central to the ministry of Jesus Christ and remains characteristic of his people. The physical world is a place to choose reconciliation and healing.

---

23. Philippians 1:29-30.

24. Hans Ruh, "Die ethische Verantwortung des Naturwissenschafters," *Reformatio* 32 (Apr. 1983): 156.

Transformation: A Place to Grow and Develop

To Grow

God is creatively active. God has, is, and will create. With regard to each of his children there is a life-giving creative pattern of first giving physical life. Second, reconciliation brings a new quality of life. "If anyone is in Christ, he is a new creation; the old has passed away and the new has come."[25] Third, living out this new life requires a daily empowerment often called "sanctification." Some traditions speak of a sudden start of sanctification, others of a long process, but whatever the terminology it is broadly agreed that there is a process of learning to live up to one's new status in Christ.[26] One becomes a member of God's family by accepting this new status as a gift, as freely as a child receives. Having begun as a child, one is then expected to grow up.

This takes time as one is set free from the old way of thinking and led to a new one.[27] Intellectual assent is not enough. Nicodemus, a highly educated Pharisee and member of their ruling council, said that he knew Jesus was "a teacher sent from God." Jesus still called him to experience new birth.[28] This new birth is the first step of straightening out old ingrained patterns.[29] Transformation is needed to become someone who can know and enjoy God. The point is not only obedience. It is essential for a child's survival and growth that the child be obedient. But the goal is not for the child rotely to follow precise instructions for every detail of the day. Most parents want their children to become the kind of people who will choose rightly and beautifully, and in some ways uniquely, without a constant drone of commands. The end goal is to thrive in and enjoy a freely cooperative relationship of life in and with God and his people.

To get there requires a marathon, not a sprint.[30] To run a marathon one must train. It is precisely the metaphor of a disciplined athlete training over time that is used in both letters to Timothy and the first letter to the Corinthians.[31] The trainer is God's Holy Spirit.[32] Each day, in each choice, one takes a step further in spiritual growth or decay. Choices become habits. Habits be-

---

25. 2 Corinthians 5:17.
26. Ephesians 4:1.
27. Galatians 5:1; Romans 6:18; 8:9.
28. John 3:3, also James 2:19.
29. Romans 8:1-2.
30. Ephesians 4:15.
31. 1 Timothy 4:7-8; 2 Timothy 3:16; 1 Corinthians 9:24-27.
32. Hebrews 12:5-11.

come virtues or vices. The sum total of one's virtues and vices are one's character, and character is destiny. This pattern is basic to human life. To be successful in a particular moment requires a life of wise and thorough preparation. A famous concert violinist was once praised by a listener, who said, "I would give my life to play like you." He replied, "I have." Following Christ, being transformed to be like him, is at least as demanding as learning to play the piano or to speak a foreign language.

Dallas Willard says, "Some people would genuinely like to pay their bills and be financially responsible, but they are unwilling to lead the total life that would make that possible. . . . We cannot behave 'on the spot' as he [Jesus] did and taught if in the rest of our time we live as everybody else does."[33] To become what we are meant to be, takes sustained commitment and effort. That is the fulfillment of grace, not its contrary. The apostle Paul writes in his letter to the Ephesians that "it is by grace that you have been saved through faith, not by works, lest anyone should boast. For we are God's workmanship, created in Christ Jesus to do good works."[34] This transformation of life is not contrary to being welcomed into God's family by grace; it is the expected result of now being part of God's family by grace.

The transformation involves becoming more and more like God. The second letter of Peter says, "He has given us his very great and precious promises, so that through them you may participate in the divine nature."[35] The Eastern Orthodox tradition writes of this partaking of the divine nature as "theosis."[36] It is often summarized as "God became human so that we may become divine." The language sounds close to the collapse of difference between the Creator and the created. Actually the Eastern Orthodox tradition is in no danger of confusing God and his creation. There is a long tradition of apophatic theology that emphasizes God as qualitatively, totally, beyond even our comprehension. The "divinization" is in fulfilling God's image, being like God, never actually being God. God's children are to learn to love the way that

---

33. Dallas Willard, *The Spirit of the Disciplines: Understanding How God Changes Lives* (San Francisco: HarperSanFrancisco, 1990), pp. 6-7. See also the other two books of Willard's insightful and wise trilogy: *In Search of Guidance: Developing a Conversational Relationship with God*, recently released under the new title *Hearing God* by InterVarsity Press, and *The Divine Conspiracy: Rediscovering Our Hidden Life in God* (San Francisco: HarperSanFrancisco, 1998).

34. Ephesians 2:8-10.

35. 2 Peter 1:4.

36. Robert Prevost has pointed out to me that there is also an allusion to this theme in the western writer Boethius, *The Consolation of Philosophy*, trans. Richard Green (New York: Macmillan, 1962), p. 63.

God does. Jesus Christ is described as the perfect image of God[37] and his disciples are to "be conformed to his likeness."[38] In being like him they are in God's image as well. Living up to his example and call, leaving behind the old way of life for a new one like him, is described as being renewed in God's image.[39]

The transformation does not mean that children of God become less individual. God's people come in all shapes and sizes. Most of them would not be believable if described in a work of fiction. The point is transformation, not replacement. Gregory of Nazianzus writes, "If the spirit takes possession of a fisherman, he makes him catch the whole world in the nets of Christ. Look at James and John, the sons of thunder, thundering the things of the spirit." It is the grace of God that sees Saul aggressively persuading, pursuing, cajoling, to stamp out the church, and then calls Saul to be the apostle Paul using the same temperament and energy to spread its presence. God's love is too much to leave human beings well enough alone. It is God's grace to give life and then transform his people to really live it.

The cost of transformation, discipleship, is great, but not as great as the cost of nondiscipleship. T. S. Eliot writes,

The dove descending breaks the air
With flame of incandescent terror
Of which the tongues declare
The one discharge from sin and error.
The only hope, or else despair
Lies in the choice of pyre or pyre —
To be redeemed from fire by fire.

Who then devised the torment? Love.
Love is the unfamiliar Name
Behind the hands that wove
The intolerable shirt of flame
Which human power cannot remove.
We only live, only suspire
Consumed by either fire or fire.[40]

God's people are God's creation, God's workmanship. What used to be feared requirements of the law, becomes a promise of what is to come. Yet human beings do play a part. There is a sense of cooperation with what God is

37. 2 Corinthians 4:4; Colossians 1:15.
38. Romans 8:29.
39. Colossians 3:9-10.
40. T. S. Eliot, *Little Gidding*, IV.

doing. "Continue to work out your salvation with fear and trembling for it is God who works in you to will and to act according to his good purpose."[41] One enters the family by God's grace and then learns to live as a member of God's family by grace.[42] One learns to reflect God's image, to choose and care wisely as God does.[43] The result is that his people "are being transformed into his likeness with ever increasing glory, which comes from the Lord."[44]

To Develop Ourselves and Our World

The material world is the place where God's people live as this process of creation, redemption, and transformation begins. It is a school for souls. The physical creation is the stage in location and process where human beings have the opportunity to choose freely and learn to live. When Paul speaks of "subduing the flesh," he is not trying to escape the body. Adam and Eve are described as having bodies before the brokenness of their rebellion. God's children will have imperishable bodies in heaven. When "the flesh" is named negatively, it is in reference to embedded patterns of sinful perspective and choice. These are associated with body-flesh because that is where human beings live, where human beings have most of their small sphere of influence. The physical world in general and our bodies in particular are where human beings can make choices. The flesh is to be consciously, consistently, yielded to God. Yielded it can be filled with God's Holy Spirit. The prophet Joel gives a promise from God that in the last days "I will pour out my Spirit on all flesh."[45]

The material world is a place to choose to know and enjoy God and his people. It is then a place to learn commitments, attitudes, and skills. Here one can practice actually carrying them out. Not only once, but repeatedly over time. Here one can become more of who one is meant to be. One can reflect God's image in creatively sustaining, healing, and transforming the physical world. This is the little sphere of influence entrusted to each to learn in. Here one can learn to reflect Christ who perfectly reflects God's image and reveals restored humanity. Here one can be part of the ongoing body of Christ. What God has created is to be sustained by God's stewards, who act the way God does. As God has redeemed one, one is to redeem and heal the world. As God transforms one over time, one is to transform the world for the better.

41. Philippians 2:12-13.
42. Romans 8:1-2.
43. Colossians 3:7-10; Ephesians 4:22-24.
44. 2 Corinthians 3:18.
45. Joel 2:28 and Acts 2:17.

Providence But what is the human role if God has already ordained all that is to be? Some Christian scholars argue that God has already decreed in detail all that will happen, so why should people bother to try to influence the direction of events? There is clear agreement in the classic Christian tradition that God is able to control every detail of everything that happens and that God does make plans that are carried out. Texts such as Psalm 33:11 can be cited that God has consistent purpose and that "the Lord's purpose will prevail."[46] Specific and concrete plans that God made and carried out can be cited from numerous texts.[47] That God makes plans and carries them out is not controversial in the classic Christian tradition. These passages, however, do not address whether God plans every detail of everything that happens, if the route to God's end is always set, or if some plans are conditional.

R. C. Sproul, in his book *Not a Chance,* states that God chooses to control everything precisely. There are a number of texts cited to support this point.[48] For example, Proverbs 16:33 states that "the decision of the lots is from the Lord."[49] Sproul reads this as an affirmation that apparently chance acts such as the cast of lots or the roll of the dice are actually guided to a specific outcome by God's constant involvement. In contrast, the same text has been read by others as referring to the *urim* and *thummim,* sacred lots in the temple, which were cast only with prayer for guidance at important turning points. By that reading, the passage would be referring to God's gracious intervention in a special case *in contrast* with the usual chance results of casting lots. There are also standard questions for Sproul's view, such as if God does ordain everything that happens, then God plans that we harm ourselves and one another. If God does choose to cause all events, these events include people abandoning or abusing their babies, committing rape, and racism. Millard Erickson offers the variation that human beings have free will, but God manipulates circumstances so that we always choose as God ordains.[50] In an otherwise comprehensive and lucid text, it is not immediately clear that this distinction addresses the moral concern. By Erickson's account God is still determining precisely what happens. The logical contradiction also re-

46. Proverbs 19:21; Isaiah 14:24-27; 46:11.

47. Isaiah 22:11; Isaiah 37:26; Luke 22:22; Acts 2:23; 1 Corinthians 2:7-8; Ephesians 1:3-6; and Galatians 1:15.

48. R. C. Sproul, *Not a Chance: The Myth of Chance in Modern Science and Cosmology* (Grand Rapids: Baker Books, 1999). Examples of texts include Psalm 139:16; 1 Samuel 2:1-10; Amos 3:6; and Acts 17:26.

49. Proverbs 16:33.

50. Millard J. Erickson, *Christian Theology,* 2nd ed. (Grand Rapids: Baker, 1998), pp. 412-57.

mains. If sin is acting contrary to God's will, but one always by necessity does God's will, how can any action be sin?

Others argue that while God could choose to control everything, God has revealed that he chooses not to.[51] At great cost to God and creation, God has created a universe where it is possible for a brief time to survive without acknowledging God. This is to make possible a genuine choice to serve and follow and know God, and hence the possibility of relationship and joy for ever that is worth the costly but temporary risk. God is intimately and powerfully involved in the world. "I ordained it and brought it to pass,"[52] but he sovereignly chooses not to control every detail. Human beings have genuine choices and consequences because God sovereignly grants them that freedom. Keith Ward offers a succinct and compelling list of quotations from Isaiah and Jeremiah.[53] He quotes, for example, from Isaiah 30:1-2, "Woe to my obstinate children, who carry out plans that are not mine, heaping sin upon sin," and Isaiah 54:15, "If anyone does attack you, it will not be my doing." From Jeremiah, he cites 14:14, "The Lord said, these prophets are lying. I did not send them and I have not spoken to them," and Jeremiah 7:30-31, "The people of Judah have done evil, declares the Lord . . . they burn their sons and daughters in the fire, something I did not command nor did it ever enter my mind."[54]

This later view holds then that God gives human beings genuine freedom in the temporary earthly sphere. We do make choices that can move the physical world in different possible directions. But even in the view that God chooses to ordain everything that happens, God still chooses to work through human beings. Either way, human beings are making formative decisions and need to make them carefully. We do not have the option of sitting this one out.

Calling In an attempt to act with efficiency and due humility some Christian circles emphasize that there is no point in trying to guide technology or any other society-wide concern. By this view, prophecy has revealed that the world will become more violent and decadent until Christ returns to establish his kingdom by force. This is a popular interpretation of the premillennial tradition. Christ returns before the millennium, a thousand years of God's reign on earth. The conclusion then is that we should focus on

51. John Sanders, *The God Who Risks: A Theology of Providence* (Downers Grove, Ill.: InterVarsity Press, 1998).
52. Isaiah 37:26.
53. Keith Ward, *Religion and Creation* (Oxford: Clarendon, 1996), pp. 13-19.
54. 2 Kings 20:1-6; 2 Chronicles 12:7; Isaiah 3:9; 5:4.

what can be done, not on lost causes. If the material world and human culture are the Titanic, we should focus on getting people into the lifeboats, the salvation of the gospel. There is no point in rearranging deck chairs. The ship is going down. This view is often tempered among premillennialists by the recognition that Jesus clearly and repeatedly called for concern for human physical needs. While eternal salvation will not be achieved by technology, how we use technology does affect people's lives and hence is of importance.

Another concern remains for many that even if God plans our daily transformation for the better, such transformation should not be extended to the physical world. Attempting to heal or improve God's creation would be "playing God." Traditional theists in Islam and Christianity emphasize their responsibility to live up to God's call while vigorously rejecting any attempt to take God's place or role. The very word *Islam* means submission. God is Almighty Creator and Lord. Any manipulation of human beings, as any other part of life, must be at God's direction or at least allowance. To claim the authority or ability to intervene contrary to God would be a claim to equality with God which is the most heinous of sins, in Islam called *shirk,* in the Christian tradition, "pride."

Within the classic Christian tradition, trying to presume God's capability or place is often considered the center of sinful debasement. From the first creation account human beings are created in God's image, in some sense like God, yet not God, always rightfully in submission to God's lordship. According to John Hammes, the first great temptation that caused the fall of human beings and their being cast out of the garden is not unlike the present temptation of human genetic engineering.[55] In each case human beings are tempted to try "to be like God." When they tried to assume God's place for themselves, they fell from what they were created to be. Created in God's image, they were to reflect God in their behavior, not try to take God's place by doing what only God should do.

Such a limitation is anathema to many modern movements that have placed human beings at the center of meaning. In such systems human beings set their own standards as they see fit. That is directly contrary to the Christian tradition expressed by theologians such as Augustine, Aquinas,[56] or Emil Brunner.[57] With different nuances, each one states explicitly a divine-command theory: God's will is the ultimate standard of what is good. Brun-

---

55. John A. Hammes, "Psychological, Philosophical, and Moral Aspects of Biogenetic Engineering," *International Catholic Review: Communio* 5, no. 2 (1978): 181.

56. Thomas Aquinas, *Summa Contra Gentiles,* Q19, A5. See also chapter 17.

57. Emil Brunner, *The Divine Imperative* (Philadelphia: Westminster, 1937), p. 83.

ner, for example, states explicitly that "the Good consists in always doing what God wills at any particular moment."[58] There are many subtleties in how God's will is perceived and carried out in these traditions, yet the central theme remains that human beings are fulfilled and just only in obedience to their Creator.[59]

For changing our physical world or form, the charge is then raised that such alteration is presumptuous. It is playing God by assuming an authority to intervene and an ability to choose wisely that only God possesses. C. Keith Boone emphasizes that this sense of "playing God" is an abhorred attitude against faith, not a rejection of power and creativity.[60] For Boone the charge of "playing God" is a religious objection to an attitude, the attitude of blasphemy, rather than a moral objection to certain actions. While his point is well taken that few traditions reject all power or creativity, his other point that only the attitude is at fault is not as well substantiated. Those who reject "playing God" would probably agree to the abhorrence of blasphemy, as Boone suggests, but would also probably object to the actions that result from "blasphemy" as well. While power and creativity can be put to needed and excellent use, when power and creativity are motivated by pride, harmful actions are likely to follow. Pride would be seen as both a blameworthy attitude and as leading to wrong actions. The presence of pride would be a concern in its own right, as well as a dangerous motivation.

The warning against the prideful taking of God's place in genetic intervention assumes that God has forbidden intervention or reserved it for God alone. Those who argue that genetic intervention is part of the God-given mandate for human beings to share in creation and the redemption of creation, would see the danger not in an attitude of pride but one of sloth. Not fulfilling the responsibility to turn genetic intervention to service would reflect a dangerous and destructive attitude of disobedient apathy.

Allen Verhey has written an insightful and wise article on the use of this phrase "playing God."[61] He describes the many ways it has been intended. The President's Commission, in its 1982 *Splicing Life Report*, reduced the phrase in

---

58. Brunner, *The Divine Imperative*, p. 83.

59. For a current defense of divine command ethics (in this case in the Calvinist tradition), see Richard J. Mouw, *The God Who Commands: A Study in Divine Command Ethics* (Notre Dame: University of Notre Dame Press, 1990).

60. C. Keith Boone, "Bad Axioms in Genetic Engineering," *Hastings Center Report* 18 (Aug./Sept. 1988): 10.

61. Allen Verhey, " 'Playing God' and Invoking a Perspective," in *On Moral Medicine: Theological Perspectives in Medical Ethics*, ed. Stephen E. Lammers and Allen Verhey, 2nd ed. (Grand Rapids: Eerdmans, 1998), pp. 287-96.

effect to "new technology is powerful, we should be careful with it." That is appropriate counsel but grants little direction. Another use emphasizes not interfering with nature, but that would rule out surgery and antibiotics. We regularly and appropriately intervene. Such is basic to human survival. Verhey goes on to describe how Joseph Fletcher enthusiastically proclaimed that we *should* play God. Playing God is acting as God would. For Fletcher, God is a "one rule" utilitarian. No action is of itself good or bad, right or wrong. What matters are the consequences. We should do whatever will produce the best consequences. That is what is most loving. In contrast, Paul Ramsey argues that if we are busy playing our role as human beings, we will not have the time or inclination to try to be God. Yet Ramsey too is open to living out the phrase, if it means imitating God's priorities. This is the sense that Verhey also welcomes. If we are to play God, it should be as God does. Verhey then enumerates several characteristics of God that we should imitate. God is the Creator, one who heals, and one who cares for the vulnerable.

If we are to imitate God, reflecting God's image, that might well include God's action as Creator, Redeemer, and Transformer. It may also be that as God transforms one, one is to take part in his ongoing creation. God has created, is creating, and will create. This material world will someday be completely transformed into a new heaven and a new earth.[62] God creates, redeems, and transforms not only the human beings of his creation, but also all the rest of his creation. One can see God's plan for creation in three primary ways.[63] It may be that God created the world in its highest state of perfection. God is perfect and would only make that which is perfect. Redemption then is to return to the prefall state. Origen and Augustine write of it in this way. A second possibility is to see human sin as a wrenching yet positive and necessary step toward what God intends for us. This is the view of supralapsarianism. God ordained the fall into sin as part of his purpose before the world was even created. The fall was not a tragic mistake. At its deepest levels it was part of God's design. This leaves us with a mystery concerning the logic and morality of sin. A third view would see the creation as good, as God declared it to be, yet not complete. It would have been bettered by Adam and Eve successfully passing their time of probation. They could have learned what it was to experience the difference between good and evil by choosing the good and not the evil. Instead, to their detriment and ours, they chose badly. God has since graciously intervened to bring about redemption rather than simply

---

62. Revelation 21:1.

63. Colin E. Gunton has recently enumerated these possibilities in *The Triune Creator: A Historical and Systematic Study* (Grand Rapids: Eerdmans, 1998), pp. 11-12.

starting over. Redemption restores our prefall opportunity to grow toward God and into what we should be. The good creation is the starting point, not the finale we seek to regain.

God declares the world that God has created "good." Indeed as God's handiwork one would expect it to be perfect. Yet for something to be perfect does not mean that it cannot grow or develop or be improved. The classic tradition describes Jesus Christ as always perfect, wholly and completely God among us, yet he is described as growing in wisdom and stature.[64] It may be that the world is quite good, but not yet finished. It may be incomplete. The God who could have made it instantly chose to take more time than that — by some counts six days and by others six billion years. The Genesis 1 account describes God declaring the creation good at the end of each day of creation before going on to add to it. There might be yet more intended in the process. The Eastern Orthodox tradition goes so far as to speak of *theosis,* divinization, not just of human beings but of the material earth. In the incarnation, God lived among us in the person of Jesus Christ. This has brought human beings and the material world into a union with God for which it was intended.[65] Yet human beings, part of God's creation, may be called to have a part in further development. The world is not now all that it could be. That falling short is sometimes described as God's curse on sin, or the material world could simply be incomplete, or both might be the case. Augustine and Irenaeus discussed these possibilities at length.[66] That the material world could be better is clear. That we are to make it better is as well.

We are first and foremost mere creatures compared to the One who was and is and always shall be. Yet God has sovereignly chosen to make us in such a way that we can reflect God's image. That seems to include creating within our small sphere in a way somehow akin to how God creates. We are designed to be creative creatures. God is not threatened by our pale imitations. In the eleventh chapter of Genesis, the irony is clear when God says to the heavenly court, "Let us go way down there to see if we can make out in the distance below what those people are doing." The people of Babel thought that they were building a tower into the highest heavens.

Human beings are to make their choices as the only earthly beings uniquely in God's image. Our priorities and choices are to reflect God's since

64. Luke 2:52.

65. Thomas Sieger Derr, "The Complexity and Ambiguity of Environmental Stewardship," in Calvin B. DeWitt, *Caring for Creation: Responsible Stewardship of God's Handiwork* (Grand Rapids: Baker, 1998), p. 78.

66. For a standard work on this discussion see John Hick, *Evil and the God of Love* (San Francisco: Harper & Row, 1977).

we are God's people in fellowship with God. This would fit the first command to care for the earth. Human beings have to alter the material world just to survive. They are welcome to do this, not to destroy the world that is not theirs, but to use it as good stewards. Granted, this care for the garden has been taken to an extreme by some. Stewardship does not mean that the earth has to be kept exactly as it is found. Supporting the rose bushes often means war on the aphids. Nature struggles within itself. To nurture one part often means restraining another. The earth itself appears to have changed dramatically many times. Ice-age glaciers leveled ancient forest across much of the northern plains that are now part of Canada and the United States. Ninety-nine percent of the species in the fossil record were extinct before widespread human influence. The shaping work of a gardener is not arrogantly supplanting God or playing God; it is fulfilling a God-given responsibility. Human beings need not feel regret over wiping out smallpox from the face of the earth or making measles an endangered species.

Consider more concrete examples of sustaining, healing, and improving. It would be fitting for human beings to appreciate and care for an intact forest. As stewards understanding the natural cycles we would still be faced with choices. Should we allow a forest fire to take part of it as a process that encourages other parts of nature to thrive in the resulting meadows? If it begins to succumb to a fungus, we could make the choice as stewards to favor and save the trees or the fungus. Nature has many manifestations. We might also choose to transform some of the trees into paper for books like this one that you are reading, a worthy use! The physical world is to be appreciated as God's creation, cared for, healed, and transformed as best fits God-given priorities.

Or think of our own bodily health. It is right as stewards of our teeth to brush them regularly. After all, if you ignore your teeth, they will go away. This is valuing and sustaining them as given. When cavities occur, we should fill them. This is healing, a necessary part of sustenance in a damaging world. Even better would be to place a sealant on them. This is a technology that renders them relatively impervious to cavities. We thereby improve their resistance beyond what they would have on their own. It is best for us to eat proper nutrition and sleep adequately. That is sustaining our God-given bodies. When we have difficulty fighting off a bacterial infection, the healing of targeted antibiotics is welcome. A vaccine that quickly and cheaply transforms part of our bodies' immune system, so that we do not fall sick to a particular disease in the first place, is even better.

Matthew 25 has several themes directly applicable to our question of calling. In it Jesus is instructing his disciples on what to do in the time between his resurrection and his return at the last day of judgment. That time is the period in which we live. The chapter contains three stories. Each one raises issues answered by the next one. The first story is of ten maidens watching into the night for the coming of the bridegroom. The wait was long enough that five of them ran out of oil for their lamps. While they were away buying more, the bridegroom came and they missed the wedding feast. The story reminds us to be always ready for the promised return of Jesus Christ. It could be before the completion of this sentence. On the other hand, it may be a long time. For those who first heard Jesus tell this story, it has already been almost two thousand years. One must have the foresight to prepare for the long haul (bringing enough oil) and the perseverance to endure. So what then are we supposed to do in the meantime?

The story that immediately follows answers that question. It tells of a master who left five talents with one servant, two with another, and one with a third servant, to each according to his ability. When he returned, he was pleased to see that the first servant had doubled his talents to ten and the second had doubled his to four. The master was outraged, however, to discover that the third servant had simply buried the entrusted resource in order to return exactly what he had received. He had failed in his responsibility to multiply the resources the master had given to him. The first story warns that we should always be ready, but we may have to wait some time before the second coming. The second story tells us that the intervening time should be spent wisely in God's service, producing fruit from what he has entrusted to us. We are responsible to multiply and employ our God-given resources. In other words, we are to have our suitcases packed and ready to go, but not merely hang out at the airport. So what would be a godly multiplication and use of resources?

The third story answers that question. It is a description of the last day of judgment. Jesus returns on his throne and his angels separate those who are his people from those who are not. His people are recognized by how they cared for the physical needs of other people. Technology can be of great service in fulfilling that mandate. There was a time when polio crippled countless children. If the disease progressed far enough to weaken or paralyze breathing, all that could be offered was an iron lung. All but the child's head lay inside the massive machine, which pumped air in and out to repeatedly collapse and expand the child's lungs. The technology was primitive, costly, and severely limiting. The better technology that developed from years of concentrated effort was the polio vaccine. A simple oral dose or single shot prevented the child from ever having a case of the disease. The higher technology was inexpensive and freeing. When cars are sliding off a mountainside road, it is admirable to start an ambulance service but even better to install guardrails.

In all three of these sections of Matthew 25, errors of omission (not bringing enough oil, not multiplying talents, not caring for those in need) are treated as seriously as acts of commission. Pride is a serious danger, but so is sloth. While the Lord tarries, we are to do our best here. That includes our best effort to multiply and apply our service of others, including their physical needs.

The question for any technology is, how can we develop this to best love God and our neighbors? Asking that question is not trying to be God; it is following God's orders, fulfilling a God-given mandate to maximize our service while we are here. Such development and intervention is not playing God. It is fulfilling a God-given mandate to serve. Whether our current physical nature is a starting point God intends us to improve upon, broken in the devastation of the fall, or both, it is clear that we could be physically better. We are responsible to do the best we can with what we have. As God's people we are being created, redeemed, and transformed by God. Part of our calling is to participate in that process by sustaining, restoring, and improving what has been temporarily entrusted to us.

Chapter Summary

Since technology exists to shape the material world to our purpose, the foundational question is, what should be our goals? What are we trying to do? At root, why are we here? The Christian tradition recognizes God at the center. The purpose for human beings is to glorify and enjoy God forever. The physical world is our temporary place to be, choose, and grow in that relationship with God and one another. The physical world plays an important but always subsidiary role. Human genetic intervention is an opportunity to make life better, body and soul.[67] Its most direct effects are on the body; its most important on the soul. Affecting the soul is most important, because the soul is why we are here. Long after the body has returned to dust, the soul, our per-

---

67. This point remains whether understanding the soul from the perspective of a dualist or that of a nonreductive materialist. Warren S. Brown, Nancey Murphy, and H. Newton Malony, eds., *Whatever Happened to the Soul? Scientific and Theological Portraits of Human Nature* (Minneapolis: Fortress, 1998), have been influential in describing the human soul as a distinct yet emergent property of the physical body, while J. P. Moreland and Scott B. Rae have recently responded in *Body and Soul: Human Nature and the Crisis in Ethics* (Downers Grove, Ill.: InterVarsity Press, 2000), that the body and soul are two distinct entities intimately related to each other. For our discussion here, both views agree that there is a "soul."

son or self, will continue, and those who belong to Jesus Christ will forever enjoy his gift of eternal life.[68]

Creation, redemption, and transformation is what God is doing in the world at large, the church, our families, and our individual lives. This is happening, most importantly, spiritually. What matters most is not that which is from the dust and that will return to dust. What is central is the kind of persons that we become, at best ones capable of and choosing life forever glorifying and enjoying God. Created in God's image, following the model and direction of Jesus Christ, and motivated by love for God and one another, we are to carry on the pattern of creation, redemption, and transformation in our physical world. That includes the larger environment and our bodies. We are to sustain our physical bodies, which for our focus here prominently includes our genetics. We should restore our bodies when they are damaged, and we should improve them as we can to serve God and our neighbors better.

As human genetics offers us increasingly formative interventions in research, testing, pharmaceuticals, and surgery that directly alters a person's genes, what would their best use to the above ends look like? That is what the following chapters will endeavor to ascertain. The first genetic intervention to be widely practiced is genetic research. That is the point of application where we will begin.

68. John 3:16.

PART I

GENETIC RESEARCH

CHAPTER 4

Searching for Genes and the Individual

Worthy Goals and Significant Dangers

As many children as could abandoned the city each summer. As late as the 1950s, children in the United States and elsewhere moved to stay with friends or family in the country during the summer months. They were fleeing polio. People associated the disease with the heat and proximity of summer in the cities. To develop a case of it could mean life-long crippling or even the inability to breath on one's own. Many families considered it worth the significant sacrifices of cost and long separations to try to avoid it.

Teams of research scientists systematically studied the disease until it was understood well enough for two workable vaccines to be developed. Now children need not fear or flee polio. As discussed in chapter 1, understanding is already a worthy goal for scientific research. The technological application of what we learn can also help us with our physical lot. The research that makes that possible is a worthy vocation both for the one guiding it and for the human participants who cooperate to make it possible. It can be part of a fitting response to the Christian tradition's call to love and serve one's neighbor. As human beings made in the image of God and for Christians to follow Jesus Christ, we are responsible to sustain, restore, and improve ourselves and our world. Developing a vaccine to improve the body's ability to fight polio is a welcome contribution to that mandate.

Finding a vaccine is not the only service that research can provide. Research can also be pursued to learn how to best implement possible treatments. Once a treatment looks promising, unknowns still remain. For example, new pharmaceuticals need to have dose and side effects established.

"With every new technological product, whether it is a word-processing program or telephone-switching software, a steam engine or a nuclear power plant, there is a learning curve. The designers, the manufacturers, and the users must all put time into learning about the product's capabilities and quirks."[1] The assurance that an intervention is achieving its intended end and an awareness of its possible side effects requires careful systematic research.

In medicine, finding what works has traditionally occurred in the midst of trying to find help for particular patients, one at a time. "For most of medical history, the experimental was folded into the therapeutic; patients were experimental subjects only as their doctors worked to heal them."[2] New methods were tried to help particular people with specific problems. But as medicine tackles more complex problems, more systematic experimentation is necessary. "Experimental science means the collection of verifiable, quantifiable data in a controlled way, usually to test an hypothesis derived from prior observations and reasoning. In clinical medicine, this requires that the patient become a subject as well as a patient and the doctor must simultaneously be a scientist as well as a helper. In addition to being an ailing person, the patient becomes a complex experimental system in which multiple variables must be controlled and manipulated if valid information is to be derived."[3] Treating the participant in research as merely an object becomes a constant danger.

One project began as a simple "study in nature." Experiments usually change a variable and then look for its effects. A study in nature simply observes without intervention. In 1930 there was no cure for syphilis. Despite the fact that the disease had been recognized for centuries, no one knew much about it. It was known that it was spread mainly by people having multiple sexual partners, but how and why it progressed in an individual had never been documented. If one could discover how it developed, one might then be able to find a step in the process where it could be interrupted,[4] hence ending the disease or at least its typical course. The usual place to begin is to find an animal that can be deliberately infected with the disease and then ob-

---

1. Robert Pool, *Beyond Engineering: How Society Shapes Technology* (New York: Oxford University Press, 1997), p. 138.

2. Albert R. Jonsen, *The Birth of Bioethics* (New York: Oxford University Press, 1998), p. 125.

3. Edmund D. Pellegrino, "The Necessity, Promise, and Dangers of Human Experimentation," in *On Moral Medicine: Theological Perspectives in Medical Ethics,* ed. Stephen E. Lammers and Allen Verhey, 2nd ed. (Grand Rapids: Eerdmans, 1998), p. 891.

4. Granted, an option was already available that would eliminate almost all sexually transmitted diseases in a single generation — faithfulness to one life partner.

serve it from disease start to finish. Some diseases though, such as syphilis, only occur in human beings. One has to observe it in people to learn its full course. The United States Health Service looked for a rural county in the United States with a high incidence of syphilis. A city would probably have too many people leaving or entering the long-term study area. They found the rural county with the highest known concentration of syphilis. It was Macon County, Alabama, with a syphilis rate of 36 percent.[5]

The administration recruited doctors at the Tuskegee Institute in Macon County to help study the disease. They then selected 399 men with the disease who had not received any treatment for it. As the study began there were no treatments that would stop or ameliorate syphilis. No patient was deliberately given the disease nor was any viable treatment withheld, but physicians did deliberately deceive their patients by not telling the patients that they had syphilis. This was to avoid patients seeking out unproven therapies that would not cure the disease, but might obscure observation of its natural course. Instead, the physicians simply told their patients that they had "bad blood." The lack of full disclosure was not unusual practice in the 1930s. The first goal for most physicians was to heal or at least encourage the patient, not necessarily to inform them. For the study in nature to be successful, the patients had to return to the doctor's office at regular intervals for painful spinal taps. It was difficult for the subjects both to take time from their work and to find transportation. To give an incentive for people to come back regularly, the physicians added to the deception by presenting the spinal taps as *treatment* for their bad blood. Actually the taps were just measuring the disease's development.

During World War II, Oxford University faculty found that Fleming's much earlier discovery of penicillin could be helpful to patients as an antibiotic. The limited available quantities were sent to the front to get soldiers back into action as quickly as possible. As the war came to an end and quantities increased, most civilians with syphilis began to receive the new drug. The people in the Tuskegee study did not. The study had been in place for almost twenty years. To interrupt it now would cut short unique and valuable data. Hundreds of thousands of soldiers had been drafted and given their lives as ordered for the good of their home countries. The patients of Tuskegee were expected unknowingly to do the same.

In 1969, the Alabama State Health Officer specifically asked members of

---

5. Gregory E. Pence tells the story of this study with insight and dispatch, "The Tuskegee Study," in *Classic Cases in Medical Ethics*, 2nd ed. (New York: McGraw-Hill, 1995), pp. 231-41.

the Macon County Medical Society to avoid giving antibiotics to anyone in the study, even if appropriate for some other condition. The concern was that receiving antibiotics would skew the study results. The county-wide physicians agreed not to give antibiotics to any subject for any condition. By this point the study had progressed from deliberate and repeated deception, to not offering a proven therapy, to intentionally preventing proven therapy for the studied condition, to actively working to stop proven therapy for other conditions as well. This unconscionable course has often been explained as solely due to racism, since most of the subjects were African-American. That may have been involved, although that attribution is complicated by the fact that the physicians of the Macon County Medical Society who were carrying out the study were almost all African-American as well. A concerned employee at the Center for Disease Control in Atlanta repeatedly called the study to the attention of leadership, with no resulting change. In 1972 he expressed his frustration to a friend who was a reporter. A few days later *The New York Times* ran the story.[6] The ongoing study results had been regularly published in medical journals with no objections. The wider public read of it with horror. The study was immediately ended, with the subjects receiving penicillin and some monetary reimbursement.

At Tuskegee, the valid interest in understanding was allowed to override the even more important interest in the patients themselves. This is a danger endemic to human research.[7] "Experiment objectifies, assigns its subjects to the status of 'thing' — that is the logic of the undertaking. This does not in itself invalidate all experiment on human beings, but it does require a careful structure of symbolic safeguards — requirements of informed consent etc. — which exist to remind us all that the experimenter's perspective on the human subject is an abstraction, and potentially a dangerous one."[8] Experiments tend to treat participants as objects. To counter that tendency, a structure is needed that requires that all participants, whether healthy or patients already under care, be treated as partners, not just experimental material.[9]

6. *The New York Times*, 26 July 1972, p. 1.

7. A long list of examples of ethically questionable research was provided by Henry Beecher in "Ethics and Clinical Research," *New England Journal of Medicine* 274, no. 24 (1966): 1354-60.

8. Oliver O'Donovan, "Again: Who Is a Person?" in *Abortion and the Sanctity of Human Life*, ed. J. H. Channer (Exeter: Paternoster, 1985), p. 131.

9. Jeremy Sugarman, Anna C. Mastroianni, and Jeffrey Kahn have helpfully gathered into one book the full texts of many guidelines and policies developed for the protection of human subjects in research: *Ethics of Research with Human Subjects: Selected Policies and Resources* (Frederick, Md.: University Publishing Group, 1998).

One response has been the development of Institutional Review Boards (IRB). These committees consist of varying combinations of researchers and community representatives. Before research involving human participants is done, the planned methods must be approved by the IRB where the research will take place. The committee's task is to assure that the rights and welfare of human participants are protected. Traditionally their focus has been on protecting research participants from physical hazards and assuring fairness in the selection of participants.

A second response, often emphasized by institutional review boards, is to require that participants give "informed consent."[10] The process of informed consent is most fully realized when participants are invited to be voluntary partners in research. That emphasis is intended in part to protect them from harm. Recruiting voluntary partners helps to avoid direct harm in that most participants, made aware of the plan of an experiment and free to walk away, will naturally be aware of and protect their own person. Also individual participants may well be aware of aspects of a planned experiment that might be uniquely risk-bearing for them. Seeking voluntary partners not only avoids harm, but it is also part of respect for each participant. The following will further outline the reasoning and concerns of this respect, with focus on aspects that are particularly important to genetics.

The Individual as a Voluntary Partner

Respect

The chapter title "Searching for Genes and the Individual" refers both to seeking genetic information for the sake of individuals and to the need to find a partnership with individuals to discover that genetic information. That partnership can be founded on the inestimable worth of each individual known and loved by God. As described in chapter 3, God sovereignly chooses to work through the consent of each individual. In the Christian tradition, the most important decision a person can make and continue to affirm is to welcome or reject a family relationship with God. Human beings do not initiate this relationship. God does. This is not only affirmed by theologians such as Martin Luther and John Calvin, who consider it central, but also by theolo-

10. Tom L. Beauchamp and James F. Childress have written an insightful, concise, and influential discussion of "informed consent" in *Principles of Biomedical Ethics,* 4th ed. (New York: Oxford University Press, 1994), pp. 142-69.

gians such as Thomas Aquinas, Jacob Arminius, and John Wesley, who are not as often associated with it. Calvinism in particular emphasizes that no one would accept God's gracious invitation unless God first changed the person's will, but across the Christian tradition there is agreement that whether ordained or in some sense free, the process of reconciliation with God includes the human being choosing to welcome God's love and grace. If God deigns that a person's choice be central to the most momentous decision one could ever make, how much more should mere human beings be deferential to a person's choice in lesser matters? With different given reasons, this same call to yield as much as possible to a person's free choice is found in the work of philosophers such as Immanuel Kant and John Stuart Mill. They too conclude that it is fundamental that we should respect the choice of persons to shape their own lives.

The freedom and ability to shape one's own life is often described by the word *autonomy*. Some philosophers have lauded autonomy as an end goal of self-creation. One should make one's own standards to live by. I am using autonomy here in a political sense of protecting space where people can make formative life decisions. That does not mean that to be autonomous they must make decisions in isolation or utter independence. "Autonomy simply means that a person chooses and acts freely and rationally out of her own life plan, however ill defined. That this life plan is *her own* does not imply that she created it *de novo* or that it was not decisively influenced by various factors such as family and friends."[11] An individual can act autonomously in welcoming the counsel of her community or even in choosing to submit to it. The physician who thoughtfully accedes to the code of her profession may autonomously choose to do so. The possibility of connection between individual consent and professional or social community is highlighted by the nature of genetics. While our genes are intimately personal, they are shared in families and across communities. An individual may seek community counsel and a community may be deeply affected by individual choices. These shared consequences have led to calls for community consent that will be discussed in chapter 6.

Whatever the individual participant's context of commitments and community, an informed consent should not be assumed or merely the mechanical signing of a legal document. It should embody respect for the participant by being part of establishing a clear mutual agreement between the researcher and participant on a joint project. Such an agreement will not occur

---

11. James F. Childress, *Who Should Decide? Paternalism in Health Care* (New York: Oxford University Press, 1982), p. 60.

without conscious effort. It cannot be assumed that the researcher and participant understand each other simply because they share a location or general
culture. Even if the researcher is a physician and the participant a patient, the
long and intimate history between doctor and patient that used to assure a
developed understanding has been lost in most cases.

Technology has forced two profound transformations, the consequences of
which are still to be understood. First, the physician is called upon to increasingly be a scientist as well as a helper of the patient, and second, medicine itself promises to become increasingly an instrument of applied biology and sociology. These two trends are altering the traditional character of
the relationship of the physician to the person and to society. They impart
features of a technological transaction to what must remain an intensely
human relationship and they intrude features of a public and social transaction into what must also be a personal and individual confrontation.
These transformations challenge the ethical codes and values as well as the
decision-making mechanisms which have served medicine for so long.[12]

The physician-patient relationship is further distanced by third-party
payers. Medical costs have become so exorbitant that few people can pay
them on their own. With third-party payers come managed-care guiding options and choices by what will be funded and what will not. Also the sheer
complexity of modern medicine tends to place patients under the care of cad-
res of briefly encountered specialists, rather than one physician who knows
them well. In most cases, these multiple pressures have long severed both the
private relationship of one physician with her patient and any assumed understanding that might have stemmed from it.

A further challenge to mutual agreement is the power differential between the researcher and participant. The researcher will probably have
greater education, wealth, and social status than the participant. Dispropor-
tionate knowledge and status encourage deference. The participant may well
be in uncomprehending awe of science and its symbols such as the white lab
coat. Further, the work is likely to occur on the researcher's turf. One learns
from an early age to defer to one's host. When in the researcher's lab or hospital one follows directions. Then as well, if one is going to trust the researcher
with invasive procedures or ingesting unknown substances, one needs to
deeply trust that person. One wants to believe in that person and certainly
does not want in any way to offend. The slightest impatience or frustration on

12. Pellegrino, "The Necessity, Promise, and Dangers of Human Experimentation,"
p. 890.

the part of the researcher can be perceived as a dire warning to back off from any questions or reservations. Finally, time is at a premium for the researcher. One knows that one should not slow someone in a hurry. All of these influences that hinder communication are multiplied if the researcher is also one's physician. It is difficult for a patient to differentiate between therapy and research, especially if the usual treatments have been exhausted. Any intervention can be perceived as a last-chance effort. To communicate freely such a context requires that the researcher consciously create a space where mutual agreement is welcome and sought. The participants would do well to write down questions ahead of any meetings and be prepared to courteously, but possibly tenaciously, ask their questions until they understand what is important to them.

Competence

To achieve the cooperative effort of true volunteers, one must also have participants who understand the relevant information and can choose according to their life plans to take part or not. No one is omni-competent for every task. Most readers of this book can drive a car, but far fewer will be qualified to fly commercial jet aircraft. To be competent to make the decision to take part in research will require at least understanding relevant information and the freedom to walk away without penalty. The ability to do this does not imply the ability to handle decisions competently in other areas. Being able to make decisions in other areas also does not guarantee ability to handle the question of participation at hand. Competence is task specific. It depends on the complexity and nature of the particular decision. It also depends on when it is asked. Some people quite competent after their first cup of coffee are not competent upon first waking. Drug interactions, pain relievers, and what simply appear as good or bad days, can make a difference in what an individual is able to do. It has been argued that increasing the risk of a decision increases the degree of required competency.[13] One has to be more able to decide whether to take part in a potentially life-threatening phase-1 toxicity drug trial than to agree to help test proper dosage of an already proven safe pain reliever. Beauchamp and Childress argue that it is not necessary that the participant actually *be* more able, but rather that greater risk calls for greater *assurance* that the participant is able.[14] For

13. For example, see Dan Brock, "Decision Making Competence and Risk," *Bioethics* 5 (1991): 105-12.

14. Beauchamp and Childress, *Principles of Biomedical Ethics*, pp. 138-41.

Beauchamp and Childress the point is not greater participant ability. It is clearer verification of that ability. For either reading, ascertaining that the potential participant is sufficiently competent to understand participation before taking part is foundational to mutual agreement and protects those who are least able to protect themselves.

Understood Disclosure

To be voluntary partners requires not only freedom from coercion and competent participants, but also the participants' understanding of relevant information.[15] One can manipulate another into almost any action if one controls the information available to that person. If the participant is respected as a free person, to make her own decisions to take part or not requires that the applicable information must be available to her. In the fall of 1999, eighteen year old Jesse Gelsinger died in a gene therapy trial. During the ensuing investigation charges have been made that the investigators gave insufficiently complete disclosure of the risks, hence undermining Gelsinger's informed consent. Concerns have also been raised about conflict of interest in that research program and others when an investigator will reap substantial financial rewards from a resulting new product. The incentive can be substantial to push ahead too aggressively at greater risk to the subjects.

There is genuine risk in most research, but for informed consent the point is not merely presentation of those risks. The goal is that the participant comprehends the information needed to make a free informed decision to participate or not. Of course the subject cannot be expected to understand every detail of a study. What the participant needs to understand is what a "reasonable person" would want to know, plus what is uniquely important to that individual. This is where a process of open discussion is necessary, so that the researcher can welcome questions and provide information uniquely important to the particular participant.

Robert Levine lists sixteen points as basic to adequate disclosure. They are parallel to the requirements used both by the United States Department of Health and Human Services and the Food and Drug Adminis-

---

15. For example, Matthew Miller points out that it is not unusual for participants in phase 1 toxicity trials to not realize that the dosages are expected to be too small to have any positive effect. "Phase I Cancer Trials: A Collusion of Misunderstanding," *Hastings Center Report* 30, no. 4 (2000): 34-42.

tration.[16] To make a free and competent choice, the participant will need to know:

1. that participation is by invitation, not request or demand
2. over-all purpose of the study
3. basis for selection of participants
4. procedures
5. discomforts and risks
6. therapy and compensation available in case of injury
7. benefits to the participant and others
8. alternative treatments that might be advantageous to the participant
9. confidentiality
10. any additional costs to the participant
11. access to answer questions
12. benefits from outside consultation
13. the voluntary nature of participation, with no penalty for not taking part
14. the possibility of incomplete disclosure if that is necessary to the study
15. the number of study participants
16. the availability of new information relevant to participation, as the study progresses.

Genetic Research in Particular

While all of the above are important, specifically genetic research requires more careful note of numbers 2, 5, and 9.

Description of Purpose

Clarity about the purpose of research comes to the fore for genetics in at least three ways.

1. While genes are intimately personal, they are shared with families. Genetic research often requires the involvement of multiple family members. Since many genetic conditions are late onset, hidden as carried, or triggered by the environment, the likelihood is high that family members recruited for

---

16. Robert J. Levine, *Ethics and Regulation of Clinical Research,* 2nd ed. (New Haven: Yale University Press, 1988), pp. 100-119.

inheritance studies may have no idea of their potential risk or its implications. It becomes particularly important to explain the study's purpose.

2. In some widely used consent documents, "gene therapy treatments for patients" would have been more accurately described as "gene transfer research with participants."[17] Terms such as "therapy," "treatment," and "patient" implied greater effectiveness than had yet been shown.

3. If a commercial goal is involved, as is often the case in current genetic research, participants should have the opportunity to know that.[18] They may opt out rather than be subject to a possible conflict of interest between the researcher's concern for them and financial reward.

Discomforts and Risks

Sometimes participation in research involves no measurable risk at all. It might be inconvenient or involve minor discomfort without risk of any substantial harm or other effect. The potential for a small bruise from venipuncture would be a risk, but not a substantial one. Robert Levine has argued that standard phase 1 drug testing for toxicity has a risk only slightly greater than that for secretarial work, and one-ninth that of coal mining.[19] Yet testing is important and takes place because significant unknowns remain. If a new pharmaceutical is safe in animals, it is likely to be safe in human beings, but that is not assured until it is tested in human beings. Drug effects vary in different people and classes of people as well. What might have no adverse effect in adult males might be harmful to women or children. Aspirin is relatively safe for adults but can cause Reye's syndrome in children. These variations require that drug introduction be incremental and carefully monitored to spot any untoward effects. Some effects can be distant, but significant. "It took many years to learn that the administration of diethylstilbestrol (DES) to

17. Nancy M. P. King, "Rewriting the 'Points to Consider': The Ethical Impact of Guidance Document Language," *Human Gene Therapy* 10 (1 Jan. 1999): 133-39. M. Therese Lysaught describes why therapeutic language originally dominated such consent documents in "Commentary: Reconstruing Genetic Research as Research," *Journal of Law, Medicine and Ethics* 26 (1998): 48-54.

18. Stuart E. Lind, "Financial Issues and Incentives Related to Clinical Research and Innovative Therapies," and John La Puma, "Physicians' Conflicts of Interest in Post-Marketing Research: What the Public Should Know, and Why Industry Should Tell Them," both in *The Ethics of Research Involving Human Subjects: Facing the Twenty-first Century,* ed. Harold Y. Vanderpool (Frederick, Md.: University Publishing Group, 1996), pp. 185-202 and 203-19.

19. Levine, *Ethics and Regulation of Clinical Research,* p. 39.

pregnant women to prevent spontaneous abortion could result in the development of a rare form of cancer of the vagina in their daughters."[20] Animal studies lessen risk but cannot eliminate it.[21]

Traditionally the potential for harm from research has been associated with physical hazards. While those can be present in genetic research, currently the most common risks are in newfound information. These will be discussed at length in part II — Genetic Testing. Examples include discovering misattributed paternity or information about family members not in the study but sharing the studied genes. An informed choice requires awareness of such risks.

Confidentiality

"Privacy" is the right to control access to oneself. Edmund Pelligrino writes emphatically that

> every person has the right to choose what parts of his interior life and personality he will expose to others and under what conditions. This right is fundamental to his dignity and integrity as a person; his social effectiveness and emotional health depend upon it. Protection of this right is a mandate in medical and other professional ethical codes. Without it the person cannot confidently enter transactions with lawyer, physician, or psychologist.[22]

To reiterate, privacy is often valued as a human right that is necessary to the function of human dignity, integrity, social life, and emotional health. Further, it is both specifically named for protection by the professions and seems to be necessary to their effective practice.

"Confidentiality" is protecting access to *private* information that has been entrusted to one. This is difficult to achieve in modern interdisciplinary medical practice. During just one hospital stay tens of people from radiology to billing have full access to one's chart.[23] Yet in research, it should be possible

20. Levine, *Ethics and Regulation of Clinical Research*, p. 43.

21. For Peter Singer, testing on animals first can be a sign of "speciesism." Singling out humanity for better treatment is a form of unjustified speciesism, a variation on racism. Only degree of sentience, not species, should carry moral weight. *Animal Liberation* (New York: Hearst Books, 1991).

22. Pellegrino, "The Necessity, Promise, and Dangers of Human Experimentation," p. 893.

23. Onora O'Neill estimates typically fifty-two people have full access to a patient's chart during a single stay in a British hospital. "Genetic Information and Ignorance," An-

to sequester information on a separate data base. Granted, this is difficult in many research university hospital settings.[24] Research participants should be informed of how data will be protected. Will it be gathered with no original identifiers or protected by a separate key code? If encoded, who will have access to that code?

"Anonymous" samples are never connected with a particular source. Most samples used in research are not anonymous, rather "anonymized." That means that the sample was collected with identifying information but has now been stripped of those connections for the researcher using it. A great deal of genetic research is done on such stored samples.[25] The largest source of such identifiable material is in the archives of pathologists who have been storing pathology specimens meticulously for decades. Thomas Murray estimates that as of 1999 there were 282 million identifiable samples of genetic material in storage in the United States.[26] These come from procedures such as biopsies and surgeries and are kept to check lab work and base future treatments. Much of this material has been made available to researchers in anonymized form. No permission from sources has been sought as long as the sources cannot be connected with the tissue.[27] These anonymized materials have provided essential information for epidemiological studies. Yet as genetic testing becomes more specific and information rich, anonymizing becomes more difficult. Having the genetic code of an individual uniquely identifies the individual by its pattern and so has become the standard of identification for the military and increasingly the criminal justice system. As our ability to interpret the code improves, it can also predict physical traits. Since samples are often kept for long periods, future capability to translate increasing detail is relevant. With such rich information, it becomes increasingly difficult to separate the genetic code from the person. It is replete with identifiers by its very nature.

---

nual Meeting of the American Society for Bioethics and the Humanities, Philadelphia, 29 Oct. 1999.

24. Robert Weir et al., "IRB Guidelines for Genetic Research," Ethical, Legal, and Social Issues Committee of the Cooperative Human Linkage Center at the University of Iowa, 1996, pp. 12-13.

25. American College of Medical Genetics, "Statement on Storage and Use of Genetic Materials," *American Journal of Human Genetics* 57 (1995): 1499-1500.

26. Thomas H. Murray, "Genetic Privacy and Discrimination," Ethical Boundaries in Cancer Genetics conference, St. Jude Children's Research Hospital, Memphis, 27 May 1999.

27. "Pathologists Enter Debate on Consent for Genetic Research on Stored Tissue," *Journal of the American Medical Association* 275 (1996): 503-4. Robert Weir has led important work on developing better informed-consent methods for such stored tissue samples.

Such genetic information can be particularly sensitive[28] in that it may reveal information not even expected when consent for a test was given. This may include likely conditions that will not have onset for a considerable time or may never appear since manifestation of a genetic condition often depends on interaction with other genes and the environment. Participants should have a choice at induction to a study whether they want to be informed about the results of any present or future genetic tests. Understanding what they are possibly gaining or losing in this decision can take substantial (expensive) time at induction and later disclosure. They also need to know that confidentiality is not absolute. For example, if certain results are life-threatening to a sibling, the duty to warn may override the responsibility to keep confidences. Without a guarantee of absolute confidentiality some participants will not take part. That is an expected and acceptable result of participants only being in a study as free partners.

Chapter Summary

Human research can be a worthy endeavor. It can yield information instrumental to sustaining, restoring, and improving the world entrusted to us. Yet as research gains much of its power in objectifying its subjects, care must be taken not to abuse the human beings who participate. To assure that participants are treated as people and not things, it is best to do research with voluntary partners. True partners in research will be competent, free to walk away, and understand the full disclosure of information relevant to their participation. For genetic research, it will be particularly important that partners understand the purpose of such research and the information risks it often entails.

---

28. Madison Powers, "Privacy and the Control of Genetic Information," in *The Genetic Frontier: Ethics, Law, and Policy*, ed. M. Frankel and A. Teich (Washington, D.C.: AAAS, 1994), p. 81.

CHAPTER 5

Searching for Genes and Family

Volunteering Someone Else

We are all to one degree or another only the temporarily able. We do not start out as fully competent adults, and we usually do not end as fully competent adults. In between there are often varying levels of awareness and understanding. Ideally research could proceed with only adult voluntary partners as described in chapter 4, but the human condition is that people are not able to speak for themselves for much of their lives. Important decisions some times have to be made on their behalf. If potential research participants cannot grasp the involved choices, who should decide for them? To begin we will focus on considering consent on behalf of adults. Then we will consider the special cases of children, pregnancy, and before.

The role of deciding on behalf of someone else is acting as a surrogate or proxy. The surrogate is usually a competent spouse, parent, or adult son or daughter. Different state legislatures have created varying orders of consultation. If the individual has duly signed a form designating someone to choose for her, that person, whether related or not, speaks with the same authority the patient would have. The form required to make this designation varies with the state but is generally simple and short. The surrogate is expected to decide according to what the potential participant would want if she could decide for herself. That is called "substituted judgment." It is intended to respect the wishes of the possible participant. Substituted judgment can lead to decisions contrary to what the surrogate would choose for herself. A not infrequent example is of a surrogate who is not a Jehovah's Witness refusing a life-saving blood transfusion for a patient who is a Jehovah's Witness. Members of that group have long interpreted a prohibition against eating blood found in Levitical law[1] as a prohibition of blood transfusion.[2] The surrogate may not agree with this unique interpretation, but if deciding as the patient would, would forbid the transfusion.

If one does not know what a patient would want if she could speak for herself, the next appeal is to the patient's "best interest." Best interest asks what the surrogate believes to be in the best interest of the patient. The surrogate is to advocate for the potential participant who cannot speak for herself. This does not mean, however, that the surrogate should assume that the patient is or prefers to be treated as an all-consuming egoist. If anyone believes that the surrogate is not acting sufficiently as the patient would or according to the patient's best interest, he or she can appeal to an institutional review board, institutional ethics committee, or the court system.

For Children

For Paul Ramsey, a child can never participate in research that is not for the child's possible benefit. This is because participation requires volunteering and children are not of sufficient age to volunteer.[3] This offers the highest degree of protection for children from experimental risk but increases general risk. For example, children often respond differently to medication than adults do. They are not physiologically just small versions of adults. Aspirin that would be reasonably safe for adults can cause the devastating Reye's syndrome in children. If a new drug is never carefully and gradually tested with children, great harm can result to many children. The harm can come from improper dosage using the drug, not using the drug due to uncertainties when it might have been quite helpful, or failing to prove that a drug in use is actually ineffective.

Richard McCormick has argued that we can volunteer children if the risk is small and the need great. McCormick reasons from the natural-law tra-

---

1. Leviticus 3:17.

2. If a member of the group accepts a blood transfusion in a moment of weakness to save his or her life, the member can repent and be restored to fellowship. Someone who persists in believing that blood transfusions are acceptable cannot become or remain a member whether he or she has had a transfusion or not. The variety of situations affected by abstention from blood products has become so numerous and complicated that Jehovah's Witnesses have developed an extensive network of well-prepared health-care liaisons to help members and nonmembers cope.

3. Paul Ramsey, *The Patient as Person* (New Haven: Yale University Press, 1970), p. 14.

dition that all human beings recognize and pursue certain goods such as life and knowledge. We know that these goods are part of human flourishing for all human beings and so have a duty of justice to meet them when they do not require substantial risk or discomfort on our part. This obligation extends to children as well. As social beings we can reason that they should volunteer, so they would volunteer if they could. Their presumed consent is "vicarious" yet sufficient to take part.[4]

Thomas Murray suggests that McCormick is right to call for the careful limited participation of children in research that does not directly benefit them. However, Murray does not find McCormick's appeal to vicarious consent convincing. Murray suggests that we do expect our most vulnerable dependents to endure minimal risks for the good of others. He cites the example of giving a crib to a destitute neighbor when your own two-year-old could have benefitted from having it a few months longer.[5] Your two-year-old's loss was small. The benefit to your neighbor's newborn was great. Small risks for the sake of the wider community may be acceptable, even praiseworthy. "Where the risks are minimal and the study competent and potentially significant, parents are permitted morally, but not obliged, to allow their child to participate: not because the child would consent or should consent — consent being an ill fitting metaphor in this case — but because such participation does not violate the parents' duties to their offspring and because it enhances other goods and values prized by the community."[6] Of course a central problem that remains for this view is the definition of "minimal risk." Federally sponsored or funded research in the United States uses the standard that "the risks anticipated in the proposed research are not greater, considering probability and magnitude, than those ordinarily encountered in daily life or during the performance of routine physical or psychological examinations or tests."[7] Kopelman and others have argued that this definition merely shifts the problem from describing "minimal" to the equally vexing "ordinary and routine."[8]

---

4. Richard A. McCormick, "Proxy Consent in the Experimentation Situation," *Perspectives in Biology and Medicine* 18, no. 1 (1974): 9.

5. Thomas H. Murray, *The Worth of a Child* (Berkeley: University of California Press, 1996), pp. 83-87.

6. Murray, *The Worth of a Child*, p. 87.

7. William G. Bartholome, "Ethical Issues in Pediatric Research," in *The Ethics of Research Involving Human Subjects: Facing the Twenty-first Century*, ed. Harold Y. Vanderpool (Frederick, Md.: University Publishing Group, 1996), p. 349.

8. L. M. Kopelman, "When Is the Risk Minimal Enough for Children to Be Research Subjects?" in *Children and Health Care: Moral and Social Issues*, ed. L. M. Kopelman and J. C. Moskop (Boston: Kluwer Academic, 1989), pp. 89-99.

Children are not property to be disposed of in any way that parents wish. Parents are acknowledged as surrogates because of the expectation that they wish their children well and know them well. If these two factors are not present due to factors such as estranged relationships, parents may not be the best surrogates. Substituted judgment is not directly applicable in the case of children in that they have not yet been competent adults, but parents can give permission that considers the child's best interest. When parents do give permission on behalf of their children, they should take care to protect their children. Children are dependent and vulnerable. They need provision and care. Yet parents may also consider the significant needs of others, if meeting them does not harm their children.[9]

A further consideration is that even if legal consent for such participants is not possible until the age of eighteen, depending on maturity, a child can take an increasingly substantial role in reasoning through decisions. This would be most prominent in the case of adolescents, who can be quite competent even if they are not old enough to give legal consent.[10] Yet at even earlier ages proceeding with physical invasion that a child does not want can be threatening and stressful. A child does not have to be able to reason competently to be psychologically harmed by having an invasive procedure forced upon him or her.[11] The child's assent is not definitive, but it is important from the earliest ages of communication.

For Zygote, Embryo, and Fetus

Surrogate decision making can be further complicated by the vexing question of when in human development a fellow human being is present and hence in need of consideration. That question warrants an extended analysis since there is no current consensus and so much of genetic, developmental, and reproductive research takes place in the earliest days of human life. Conclusions in this regard will also be important to the discussion in chapter 8 concerning parental use of genetic testing for selective zygote implantation or abortion.

9. On the role of the family in the Christian tradition see Stephen G. Post, "Marriage and Family," in *Christian Ethics: Problems and Prospects*, ed. Lisa Sowle Cahill and James F. Childress (Cleveland: Pilgrim, 1996), pp. 265-83, and Rodney Clapp, *Families at the Crossroad: Between Traditional and Modern Options* (Downers Grove, Ill.: InterVarsity Press, 1993).

10. Robert F. Weir, "Genetic Research, Adolescents, and Informed Consent," *Theoretical Medicine* 16, no. 4 (1995): 347-73.

11. Bartholome, "Ethical Issues in Pediatric Research," pp. 356-66.

Is it ever appropriate for genetic research to risk or deliberately destroy developing human life? Well-intended research or application can still be unacceptable if it requires abhorrent means. The Enquete Commission report to the German Bundestag has rejected some genetic research as depending on means that it deems morally repugnant.[12] According to the commission, experiments that sacrifice large numbers of human fertilized egg cells should not take place regardless of potential benefits. Unless the research can find other means to proceed, it may not continue. While Germany has banned such research, the United Kingdom has welcomed it during the first fourteen days after conception. When research involves risk for life in the womb or petri dish, is the status of that life different from life after birth? If so, when? The conceptus is alive and human, but so are some of the millions of skin cells one sloughs off each day. When is that human life a person?

This discussion is related to the emotionally charged issue of abortion. There are about one and a quarter million abortions in the United States each year.[13] Forty percent of them were not first abortions.[14] There are many individuals who now have a vested interest in finding it an appropriate or inappropriate decision. Abortion has been the focus of countless extended discussions. In this book we will only have time to introduce the aspects most relevant to genetic research and intervention, along with footnoted bibliography for further investigation.[15] If one has already carefully worked out when to recognize the presence of a fellow human being in human development, one could well skip ahead to chapter 6. The discussion that follows is detailed because the involved arguments are complex and the implications for genetic intervention are great.

Standard terminology is that the new life developing in the womb is called a "zygote" from fertilization until implantation. Implantation begins on about the sixth day after fertilization. Once implanted, the zygote is called an "embryo." The embryo develops until all of the basic organs have formed, by about day twenty-eight post conception. From then on the embryo is called a "fetus," Latin for *child*, and growth is in size and refinement.

---

12. Enquete Commission, "Prospects and Risks of Gene Technology: The Report of the Enquete Commission to the Bundestag of the Federal Republic of Germany," p. 259.

13. The federal government tracks and publishes the number annually from the Center for Disease Control in Atlanta.

14. Ted Peters, *For the Love of Children: Genetic Technology and the Future of the Family* (Louisville: Westminster/John Knox, 1996), p. 101.

15. A fine collection of the most influential arguments can be found in Francis J. Beckwith and Louis P. Pojman, eds., *The Abortion Controversy: T 'enty-five Years after Roe v Wade: A Reader*, 2nd ed. (Belmont, Calif.: Wadsworth, 1998).

Beginning with implantation, the embryo makes hormones that have a powerful influence. These hormones orchestrate an intricate series of changes to support the embryo's growth. Yet life in the womb has no audible voice. How the embryo is treated is determined by the pregnant woman. She decides whether research may put the developing life at risk. She has this power in American society because we highly value autonomy. Any person should be able to control her own body and body products. Granted, we do not treat this as an absolute right. Men have long been drafted to risk and sometimes lose their lives in war to protect people other than themselves. Men and women can be called at any time to present their bodies for jury duty at a required time and place. Control of one's body is highly valued, but it can be tempered by the needs of others. A key question then becomes, when is another present whose needs should also be considered?

Supreme Court Justice Harry Blackmun wrote in the decision that legalized abortion in the United States (*Roe v. Wade* 1973), "If the suggestion of personhood of the unborn is established, the appellant's case, of course, collapses, for the fetus' right to life is then guaranteed specifically by the Fourteenth Amendment."[16] There have been arguments such as that of Judith Jarvis Thomson that even if a person is present, it is acceptable to withdraw support knowing that such action is lethal for the dependent. While her argument has been often repeated, it is not eventually persuasive for a number of reasons.[17]

Thompson argues that pregnancy is like organ donation. One should not be obligated to donate an organ even if someone else's life could be saved by that donation. As with any argument by analogy, the question then becomes if organ donation is the best parallel to pregnancy. In both organ donation and pregnancy, a life is at stake and the life depends on someone sharing organs. On the other hand, there are relevant differences that render the analogy problematic. The two cases are not alike in that pregnancy is temporary. Organ donation involves the permanent loss of an organ. Pregnancy is a natural function of the body, while cutting the body open to remove an organ is not. In pregnancy there is only one person who can sustain the life at risk. Organ donation can have more than one prospective donor and often alter-

16. As cited by Francis J. Beckwith, "From Personhood to Bodily Autonomy: The Shifting Legal Focus in the Abortion Debate," in *Bioethics and the Future of Medicine,* ed. John F. Kilner, Nigel M. de S. Cameron, and David L. Schiedermayer (Grand Rapids: Eerdmans, 1995), p. 187.

17. Judith Jarvis Thomson, "A Defense of Abortion," *Philosophy and Public Affairs* 1, no. 1 (1971): 47-66. See critique by Francis J. Beckwith, *Politically Correct Death: Answering the Arguments for Abortion Rights* (Grand Rapids: Baker, 1993).

natives such as kidney dialysis. In organ donation one might be considering helping someone one does not know personally. In pregnancy the one in need is not a stranger. In the Christian tradition one has an obligation to help whomever one can. Within the ordering of love, there is a substantial obligation to help the vulnerable and a special obligation to one's children.

If someone besides the pregnant woman is involved, that should make a difference in how the situation is understood and weighed. Personal liberty is dear, but protecting life may at times trump liberty interests. It is not controversial that the pregnant woman plays a central role in choices that profoundly affect her body. Her interests and choices matter. What would be most helpful to discuss here is the more contested question of the status of the one developing in the womb. Does that one's survival warrant consideration as well? If the one in the womb is at some point a person, from then on research should only be carried out for that person's therapy or at minimal risk, as with any other child.

Thompson's argument also does not help us in the increasingly frequent case of *in vitro* fertilization, fertilization outside the body. Then we have developing life that is dependent on others for survival, but not on any one particular person. What is its status? Whether a person is present matters, but when does that occur?

Scripture

While one tradition within Christianity may emphasize liturgy and another the teaching authority of the ongoing church, all in the classic Christian tradition take interest in what is taught by Scripture. We will begin there. According to the tradition God is aware of everything in creation and has the power to intervene. This was described in chapter 3. That awareness of course includes the womb. God cares about human beings wherever they are. God's people are to be quick to recognize and care for their neighbors whenever they can. That includes looking for the least, the poor, and the vulnerable ones easy to miss.[18] On the final day of judgment described in Matthew 25, God's people are recognized by how they cared for those in physical and social need. When Jesus is asked how many people have to be included under the command to love one's neighbor, he responds with a story that reverses the question from limitation to extension. One's neighbor is whomever one is able to help. Neighbor love is about inclusion, not qualifications to warrant

18. Deuteronomy 10:18, 1 John 3:17.

concern.[19] Scripture speaks of the rightness of helping, not the right not to get involved. The base-line attitude, the default position, is to care.[20] Yet while these central themes tell Christians to look for those they can help, they do not tell when there is someone else present to help.

A steady tension for neighbor love is that there are often more needs present than resources to meet them. The ordering of love is the ever-present question of how to weigh the competing needs and desires of different people that one is able to serve. With modern telecommunications awareness of needs can be overwhelming. The traditional ordering of love has been by proximity and need. The closer the relationship and greater the need, the more responsibility to intervene. Both proximity and need are at a maximum with one in the womb. The one in the womb is totally dependent for survival and could not be more closely present. Yet could the same be said of a sperm cell or ovum? Each depends for its survival on the one carrying it and could not be closer in physical distance or relation, yet most do not argue for an obligation there. The question remains in regard to development in the womb, when is there a neighbor present to care for?

There are specifically three types of passages that are often quoted as relevant to recognizing when a neighbor is present in the womb. None of them actually establish a time. One set refers to God's knowledge of each person from the beginning, but not when they began.[21] Jeremiah 1:5 is often quoted. In the New International Version it reads, "Before you were in the womb I knew you, before you were born I set you apart; I appointed you as a prophet to the nations." The text is not about human embryology or even about humanity at all. It is about the eternality and surety of God's plans. God has called Jeremiah to this particular vocation and has been planning this even before Jeremiah was in the womb. God's plan and knowledge are complete.[22] There is nothing in this text that designates when Jeremiah became a living human being. If the point of the text was instruction about Jeremiah's existence, it would indicate that he was alive in some realm before be-

---

19. Oliver O'Donovan, "Again: Who Is a Person?" in *Abortion and the Sanctity of Human Life*, ed. J. H. Channer (Exeter: Paternoster, 1985), pp. 135-37.

20. Gene Outka finds this one of the key starting points in considering the status of the fetus in "The Ethics of Love and the Problem of Abortion," James C. Spaulding Memorial Lecture, 27 Sept. 1999, University of Iowa, pp. 17-18.

21. Isaiah 49:1, Galatians 1:15, Ephesians 1:3-4.

22. The text has also been taken as a reference to Augustine's understanding of God, the Creator of time, as not being limited by the time God has created. God could know Jeremiah before his birth, by always being able to see him post birth. For Augustine, every instance of time is always present before God.

ing in the womb. "Before you were in the womb I knew you." That is not the point any more than for Ephesians 1:4, which states that "[God] chose us in Him before the creation of the world." The texts are marveling at God's knowledge and choice, not human existence before time.

A second set of texts refers to God forming human beings in the womb. Examples include Job 31:15 and Psalm 139:13. The latter reads, "You knit me together in my mother's womb." The metaphor of knitting conveys God's intimate involvement in David's life from the beginning. However, again it does not say when that form knit in the womb became a human being. God's presence and care are the subject of the passage, not human physiology. God is intimately involved in the lives of his people from the beginning, but that does not tell us when a person's beginning is. God is involved in the formation of the body that will be David. That does not tell us when the developing body *is* David. Trying to indirectly extrapolate the timing of human presence from this text is again reading in affirmations that are not present.

The third text is an excerpt from the law for the people of Israel. The reading, Exodus 21:22-23, differs markedly from one translation to another. In the New International Version it reads, "If men who are fighting hit a pregnant woman and she gives birth prematurely but there is no serious injury, the offender must be fined whatever the woman's husband demands and the court allows. But if there is serious injury, you are to take life for life." After the phrase "gives birth prematurely" an asterisk refers one to the alternate reading "she has a miscarriage." The NIV text translation is that the fight has caused labor, but the delivered baby is healthy, hence minimal penalty is appropriate for putting the baby at risk.[23] The NIV alternative reading is that causing a miscarriage is a serious offense, but not at the level of taking a human life. If a human life had been lost, the death penalty would have been required. Translation uncertainty is at precisely the point where the passage might have shed some light on the question before us.

Scripture is clear that God is aware and cares about all the stages of human life. While Jesus' disciples were anxious to send away children that wanted to meet Jesus, Jesus welcomed them. Luke uses the same word to describe the children brought to Jesus for blessing, the newborn Jesus, and the not-yet-born John the Baptist.[24] There is concern for all, including the most vulnerable, but Scripture does not specifically address when any particular human life

---

23. John Jefferson Davis succinctly lays out and evaluates three possible readings in *Abortion and the Christian: What Every Believer Should Know* (Phillipsburg, N.J.: Presbyterian and Reformed, 1984), pp. 49-52.

24. Luke 1:41, 44; 2:12, 16; 18:15.

precisely begins. Harold O. J. Brown, who vigorously advocates the presence of each person beginning with conception, is scrupulously honest in his careful exegesis. He states that Scripture does not directly describe the human person as being present at the time of conception.[25] Donal O'Mathuna in the midst of his argument for the presence of a person from conception states it this way: "The problem is that the Bible does not clearly state that the fetus is the image of God from the moment of conception."[26] Pope John Paul II, a fierce opponent of abortion, writes that "the texts of Sacred Scripture never address the question of deliberate abortion and so do not directly and specifically condemn it."[27] The Lutheran theologian Gilbert Meilaender unfailingly advocates nurture for the unborn and summarizes, "We cannot, I think, claim that the Bible itself establishes the point at which an individual life begins, although it surely directs our attention to the value of fetal life."[28]

The Christian tradition calls for care of the seemingly least human being, but when is a human being present to care for? Scriptural principles call for care for the vulnerable and extending neighbor love as widely as possible, but Scripture does not specifically mark a particular point of development in the womb as when a person becomes present. How else might we recognize the presence of a fellow human being? To begin to answer that question we will trace individual human development on back from infancy to its beginning. At each step we will describe reasons given for recognizing the presence of a human being at that stage. As those steps are delineated we will also note some of the strengths and weaknesses of each as a line of demarcation.

Developmental Steps

Infancy

The philosopher Michael Tooley writes, "A newborn baby does not possess the concept of a continuing self, any more than a newborn kitten possess as such a concept. If so, infanticide during a time interval shortly after birth

---

25. Harold O. J. Brown as quoted by John Jefferson Davis, *Evangelical Ethics: Issues Facing the Church Today* (Phillipsburg, N.J.: Presbyterian and Reformed, 1985), p. 148.

26. Donal P. O'Mathuna, "Abortion and the 'Image of God,'" in *Bioethics and the Future of Medicine*, p. 205.

27. Pope John Paul II, *Evangelium Vitae* (The Gospel of Life) (New York: Random House), p. 108.

28. Gilbert Meilaender, *Bioethics: A Primer for the Christians* (Carlisle, U.K.: Paternoster, 1996), p. 29.

must be morally acceptable."[29] For Tooley, being a person requires certain functional characteristics. Most prominently, to be a person requires a continuing concept of self. Since a baby does not yet have this, the baby can be killed. At birth a human baby has little if any self-awareness, and hence it is disposable if the parents do not wish to keep it. Babies should be born on approval. If one is handicapped or otherwise unsuitable, it is not too late to kill it before a person is present. This maximizes parental freedom and is not a loss to the baby. The baby is not self-aware enough to know its loss; hence it can have no loss.

Most people find this a horrible prospect. A human baby is to be nurtured as a fellow human being already present. If a child is irreversibly dying, one need not aggressively extend the dying process. But a stable and alive baby, even with significant disability, is still a human being and hence should receive all helpful care.

First Breath at Birth

Like Michael Tooley, Mary Anne Warren argues that to be a person one must have certain characteristics. Warren's list includes the following five attributes: consciousness, reasoning, self-motivated activity, capacity to communicate, and self-awareness. "All we need to claim, to demonstrate that a fetus is not a person, is that any being which satisfies none of these is certainly not a person."[30] For Warren those criteria are not met until at least birth.

Does the first breath coincide with or actually bring about these characteristics? Most babies are born after about nine months of gestation, yet were able to breath air before if given the opportunity. Would a fetus in its thirty-fifth week in the womb be a nonperson, while a less-developed twenty-eight-week old, born prematurely, would be a person? The distinction would be one of location. But location would be an odd criteria for recognizing humanity. Of course location is deeply important both to the fetus and to the one carrying the fetus. Birth is an arrival into a larger world, but also a separation from the mother who has nurtured the child as long as the child has existed. Location matters, but not in recognizing humanity. The reader is just as much a person wherever she is located. One does not slip in and out of existence depending on one's surroundings or level of dependence. The mother's womb

---

29. Michael Tooley, "Abortion and Infanticide," *Philosophy and Public Affairs* 2, no. 1 (1972): 63.

30. Mary Anne Warren, "On the Moral and Legal Status of Abortion," *The Monist* 57, no. 1 (Jan. 1973): 43-61.

provides nourishment and a protected environment before birth. Nourishment and protection are just as needed after birth. Robert Weir emphasizes this continuity between the late fetus and the newborn. The status of one is the status of the other. He agrees with Warren that neither is a person but argues that the potential soon to be a person found in both the newborn and the late-stage fetus warrants general protection from harm.[31]

Six Months Post Conception:[32]
Viability and Present Self-Awareness

It is certainly true that surroundings can make a difference. If they do not support life, life ceases. But not having adequate external support to survive is a description of deadly circumstances, not of whether a person is present or not. "Viability" then is an odd criterion for recognizing the presence of a person. Applied specifically to the fetus, it means that with effective support the fetus could live outside the womb. Such a standard is more a measure of available support than of the fetus. For example, viability can depend on the economic status of the mother. By the viability standard, one fetus could be a person and the other not, even though they were of equal age and health, and located a few feet from each other. One could be a person and the other not, because one is in the womb of a wealthy woman and the other in the womb of an economically poor woman. The wealthy woman could afford the best neonatal intensive care unit support for her child, while the poor woman might not be able to obtain as effective care at a crowded charity-based unit. Although the fetuses are of equal age and both equally desired, at that moment the wealthy woman's child would be viable, while the poor woman's child would not be. Hence one would be a person and the other not. Or location could be determinative for viability. A woman drives to Logan Airport in Boston. On the way she passes the neonatal intensive care unit at Massachusetts General Hospital. The fetus in her womb is viable and hence a person because there is sufficient care available in a timely enough fashion for the fetus to survive if born at that moment. The woman boards an airplane and it takes flight. Flying over "Mass General" the child has reverted back to a nonviable fetus, hence not a person, because there would not be adequate care available quickly enough for it to survive if

31. Robert Weir, *Selective Nontreatment of Handicapped Newborns* (New York: Oxford University Press, 1984), pp. 190-94.
32. By "conception" I am referring to the traditional definition of when one sperm and egg begin to combine. Increasingly some in the scientific and medical communities have begun to define conception as when implantation begins, about six days after the egg and sperm first join together.

born at that time. Two hours later, landing at O'Hare International Airport in Chicago, the fetus has again become a child, since the baby is now close enough to a neonatal intensive care unit to be viable. Do the economic resources or location of the mother determine whether a fetus is a person? The viability standard proceeds as if it does.

Supreme Court Justice Sandra Day O'Connor has also noted a further problem with the viability standard. As neonatal intensive-care units become more effective, the age of viability decreases. The legal recognition of viability is already considerably later than the age of actual viability in modern neonatal intensive-care units. If the reasoning of *Roe v. Wade* is correct that viability is when the state should become interested, that criterion indicates an earlier date than the law currently attached to it. This is the self-contradiction of current United Kingdom law as well.[33]

Bonnie Steinbock advocates consideration of the individual in the womb beginning at six months, but not based on viability. She argues for the presence of self-awareness, not just the eventual potential for it, as the time when the fetus should be taken into account. It is at about six months after conception that the fetus has interests in enjoying life and avoiding suffering. Hence the fetus would have a claim on our moral attention. The fetus does not have to be rational. That develops long after birth. Sentience occurs about the time that neocortical brain function arises. The fetus in some limited sense is aware.[34] She continues that being aware is basic to having interests. "If an abortion is performed before the fetus becomes sentient (probably toward the end of the second trimester), the fetus is killed, but not harmed, paradoxical as this may sound. For to be harmed is to have one's interests set back or thwarted. Without thoughts or feeling or awareness of any kind, the embryo has no interests. Without interests, it cannot be harmed."[35]

The interest view raises at least two questions. Does the view confuse having an interest with knowing that you have an interest? In other words, can one be harmed without knowing it?[36] Second, the sleeping, the unconscious, and those tired of life may not have current interest in life, but may in the future. Do they warrant care on that basis? If those who do not have conscious interests currently, but one day will, are admitted to have interests, the argument for protecting those with interests might lead to an earlier stage of development than sentience.

33. David Leal, personal communication, 4 Dec. 1999.
34. Bonnie Steinbock, *Life Before Birth: The Moral and Legal Status of Embryos and Fetuses* (New York: Oxford University Press, 1992), pp. 86-87.
35. Steinbock, *Life Before Birth*, p. 6.
36. Steinbock responds to this concern beginning at page 16 of *Life Before Birth*.

Ninety Days Post Conception: Quickening

Some have suggested that quickening is the point where one should recognize that another is present. This is a psychologically important moment in that the woman can feel that something else is moving of its own volition. Independent movement emphasizes its otherness. While such is an important moment psychologically, we do not say that one person's existence depends on the awareness of another. If one is shipwrecked alone on a deserted island, one has not ceased to exist. Whether someone is in the womb does not depend on whether someone else is aware she is there. Quickening reveals more of the pregnant woman's sensitivity than of the fetus's being.

Sixty Days Post Conception: Formation

Sixty days after conception there is still the need for tremendous development. The skull will not knit into solid form until after the tight passage through the birth canal. The lungs will not expand to receive air until the last week or so before birth. Neuronal connections will be created mostly after birth, as will the immune system, which is not completely in place for about six months after birth. Yet with all the growth to come, there will be no more new organs. What is continuing is refinement and increase in size.

Through the early and medieval church, the long-standing consensus among theologians was that God gave a soul at the point when a body had formed in the womb. Or in the perspective of "traducianism," a soul inherited from one's parents develops with the body and is at last fully present when there is a formed body. Both soul creation and traducianism reasoned that one needed a body to have a soul. Abortion before ensoulment was serious. *The Didache,* one of the earliest church manuscripts not in the New Testament, teaches that there are two ways, one of life and one of death. It lists abortion as one of the ways of death. For Tertullian as well, abortion was not an option in the Christian community. It was ending developing life, but it was not ending the life of a human being. This distinction was termed the difference between "unformed and formed." It was used by many early church teachers such as Tertullian (160-240), Lactantius (240–ca. 320), Jerome (347-419), Augustine (354-430) in the *Enchiridion,* Cyril of Alexandria (ca. 375–444), and Theodoret (393-457).[37] Estimates as to exactly when the body was formed varied but centered on about two months post conception. Thomas Aquinas (1224-74), "the

---

37. John Connery, *Abortion: The Development of the Roman Catholic Perspective* (Chicago: Loyola University Press, 1977), pp. 40, 50-52, 56.

angelic doctor" of the Roman Catholic tradition, set it at precisely forty days from conception for boys and ninety days for girls,[38] using what he perceived to be the best science available in his day. These are the numbers given by Aristotle.[39] Aristotle derived these figures from observations that if what was expelled moved on its own, the developing life was considered formed and if formed ensouled. If it did not move independently, it was not yet animated, hence not yet having a soul. Aristotle thought he had observed male and female movement after miscarriage beginning at different points of development.

There were three other sources for the distinction between formed and unformed besides the work of Aristotle. One was in the Septuagint. The Septuagint was the most widely used Greek translation of the Hebrew Bible. Its translation of the Exodus 20:22-23 passage that we discussed earlier makes this distinction. There is a monetary penalty for ending *unformed* life, but if *formed* life is killed, the death penalty is required, life for life.[40] Second, in the Hebrew Bible, human beings are often called *nephesh,* an animated body. Can one be an animated body without a body? Granted, one still has a body after a leg amputation or the removal of a cancerous kidney, but having a substantial body of some sort still remains basic to being a human being in this world. Third, in Job 10:10-11, Job prays, "Did you not pour me out like milk and curdle me like cheese, then clothe me with skin and flesh and knit me together with bones and sinews?" This has been read as a description of life beginning with an unformed state and then later developing to a formed one. By this distinction between formed and unformed, having a body and not yet having a body, miscarriage or abortion before formation is a true loss of what was becoming a body. Miscarriage or abortion after formation is the tragic loss of a present body and person.

Forty Days Post Conception: Brain Activity

Islam teaches as a matter of revelation that a person is present beginning at forty days after conception.[41] Some writers have arrived at this date by an appeal to first brain activity.[42] This appears parallel with the Harvard brain-

---

38. Connery, *Abortion,* p. 110.

39. Aristotle, *On the History of Animals,* bk. 7, ch. 3.

40. Connery, *Abortion,* pp. 50-64.

41. Jacques Cohen and Robert Lee Holtz, "Toward Policies Regarding Assisted Reproductive Technologies," in *Emerging Issues in Biomedical Policy,* ed. Robert H. Blank and Andrea L. Bonnicksen (New York: Columbia University Press, 1990), p. 228.

42. Baruch Brody, *Abortion and the Sanctity of Life* (Cambridge, Mass.: MIT Press, 1975).

death criteria. That widely accepted standard includes that if one has no electrical brain activity in two separate readings at least twenty-four hours apart, one has died. No one has regained consciousness after such a verified reading. The argument then runs that if the fetus has no brain activity, it is not alive. Nervous system activity begins at about forty-two days and a complete brain structure is in place by eighty-four days.

One must be careful in the application of this criterion in that it hinges on the observation that if brain activity is absent for twenty-four hours, it never comes back. The lapsed time requirement is to show that the loss is irreversible. In the case of the fetus the lack of brain activity does not indicate that it will not occur. On the contrary, it is to be expected shortly. If symmetry is intended with the absence of brain activity that will not return, the criteria leads to an earlier time in development when an entity is present that is likely to eventually have brain activity. This is the case as well with using the criterion of heartbeat. One is not declared dead because one's heart has stopped beating. If that were the case, we would have tens of thousands of citizens back from the dead. It is not unusual for someone to lose their heartbeat, even intentionally in many heart surgeries, and have it brought back again. Death is associated with the *irreversible* loss of heart activity. To apply this criterion symmetrically to the beginning of life would call for recognizing life when there is an entity that will have a heartbeat eventually, not just when it actually starts at about twenty-eight days.

Twenty-eight Days: Formation, Sentience, and Heartbeat

We now know that formation of the basic body systems is largely present at twenty-eight days post conception. If the historic reasoning described earlier is correct that a person is ensouled, having a soul, as soon as a body is present, that reasoning may lead to this point. As of twenty-eight days, all organ systems have begun. From then on growth can be described as increase in size and refinement. Now that there is a human animated body, a nephesh, a human being is present.

Helga Kuhse and Peter Singer together propose that twenty-eight days is significant. "To be very very cautious in erring on the safe side," twenty-eight days would be the earliest point of sensing pain, hence the embryo should be protected from suffering from that point.[43] Note that this argu-

---

43. Helga Kuhse and Peter Singer, "Individuals, Humans and Persons: The Issue of Moral Status," in *Embryo Experimentation: Ethical, Legal, and Social Issues* (Cambridge: Cambridge University Press, 1993), p. 74.

ment obligates one to avoid inflicting pain. It does not advocate that the embryo survive. For Kuhse and Singer the embryo has no higher status than any other being capable of feeling the same degree of pain. To single out the human being would be irrational "speciesism." Stringent controls should be in place to avoid the suffering of any being, including the human, to the degree that the being can sense suffering.[44] In other words, Kuhse and Singer's reasoning is calling for some form of anesthesia as necessary after twenty-eight days, not the protection or continuance of developing life.

Fourteen Days Post Conception: The Primitive Streak and the End of Twinning

The first weeks after conception are quite formative for developing human life. For that reason a great deal of genetic research and proposed interventions are proposed for those first days. It is important then for our task of evaluating human genetic intervention to keep refining our analysis even within this narrow time frame. Three physiological changes occur at about fourteen days after conception. One is that the cells within the embryo begin to specialize so that the embryo has a top and bottom, front and back. In particular, the cells begin to form the primitive streak that is the beginning of the spinal cord and nervous system. A second change is that most of the present cells begin to segregate into fetus, placenta, and other supportive tissue. Third, with this differentiation, the possibility of twinning recedes. Up to this point, the number of individuals present was not yet decided.

Referring to these distinctions, the first two weeks of development are sometimes called the pre-embryo stage. British legislation marks this change from pre-embryo to embryo as the beginning of a protected individual. The fourteen-day mark is also the choice of the Embryo Research Panel convened in 1994 by the National Institutes of Health in the United States. Its assigned task was to advise the federal government on the funding of embryo research. The panel concluded that embryos should be accorded "profound respect, but this does not encompass the legal and moral rights attributed to persons."[45] What the panel meant by "profound respect" is not immediately apparent. It did not see any violation of profound respect in creating an embryo

---

44. Peter Singer, "The Moral Status of the Embryo," reprinted in *Classic Works in Medical Ethics: Core Philosophical Readings,* ed. Gregory E. Pence (Boston: McGraw-Hill, 1998), p. 89.

45. National Institutes of Health Human Embryo Research Panel, *Final Report of the Human Embryo Research Panel* (Washington, D.C.: National Institutes of Health, 27 Sept. 1994), pp. 49-50.

for experimentation and discarding the developing life when it was no longer needed. Since it concluded that the moral status of life in the womb gradually increases with the degree of development, one might conjecture that some of the panel members meant the kind of profound respect one should show a corpse. The embryo and the corpse are both due respect because they were or could become associated with a person. Once a body no longer is a person, it can be dismembered and destroyed for pathology studies or education. The embryo is also associated with a human person, one that might exist in the future; hence it deserves special care, yet not treatment as a person.

Since, according to the panel, moral status is gradually acquired as the individual develops, it is a relatively arbitrary policy matter to draw the line at fourteen days, although as stated above, there are distinct physical events about that time. It is odd that the panel claimed to form their conclusions without reference to religion, theology, or philosophy.[46] As Daniel Callahan pointed out, "The report has in fact adopted a particular philosophical viewpoint on the purpose of the law, on the imperative of scientific research and on the moral status of the fetus. If the Panel did not notice that, just about everyone else will."[47]

Professor Patricia A. King served on the Embryo Research Panel. She wrote a partial dissent to the panel's conclusion that experimentation on embryos was acceptable up to fourteen days after conception. Her concern was that in some cases "human life is being created solely for human use."[48] For King, one could destroy in research the spare embryos created for implantation after attempts at *in vitro* fertilization. They would be available because they were not implanted due to genetic defect or number. On the other hand, creating embryos specifically to use and destroy would not show adequate respect. Does it matter if the source of the embryo is excess from fertility treatment or creation for experimentation? *In vitro* fertilization requires harvesting ova. To do so requires a cycle of ovarian hyper-stimulation by daily injection of hormones, repeated blood samples, ultrasounds, and then oocyte retrieval. It requires less cost, time, and discomfort if many ova are harvested at one time. The success rate of frozen eggs is not as good as that of frozen embryos, so the usual practice is to fertilize a large number of available eggs,

---

46. *Final Report of the Human Embryo Research Panel,* pp. 50-51.

47. Daniel Callahan, "The Puzzle of Profound Respect," *Hastings Center Report* 25 (Jan./Feb. 1995): 40.

48. As quoted by The Ramsey Colloquium, "The Inhuman Use of Human Beings: A Statement on Embryo Research," in *Do the Right Thing: A Philosophical Dialogue on the Moral and Social Issues of Our Time,* ed. Francis J. Beckwith (Sudbury, Mass.: Jones and Bartlett, 1996), p. 294.

freeze them now as new embryos, and then implant as needed. If a person is present in each embryo, the embryos should not be discarded because of origin point, any more than adults could be discarded because of origin.

If each embryo is not yet a person, one could argue that as tissue they can be used in whatever way seems helpful. But symbolic respect would still be important. Because a body is so closely associated with a particular human being, it should not be treated merely as a thing. Respect or the lack thereof will tend to spill over to embodied persons. Using and destroying embryos would be more respectful if its aim was to save lives than just to improve cosmetics. Even if embryos are not yet persons, it could still be argued that they should not be available for any use. Corpses are no longer persons, yet still should be used respectfully only as a last resort, because of their association with persons.

The British and American panels included the possibility of twinning as an important landmark. That twinning was still possible for fourteen days post conception implied that an individual was not uniquely settled, hence of less need of support as a unique individual. It can be countered that twinning would only increase the number of individuals involved, not question whether there is one already present.

Six Days Post Conception: Implantation Begins

Others have placed the recognition of someone's presence at implantation. Implantation is when a fertilized egg or zygote becomes firmly attached to the uterine wall. The process begins on day five or six and has been completed by about day nine. Now there is a direct connection with the mother and the zygote is called an embryo. The embryo develops the placenta and remains until birth. Once implanted the chance of birth is much higher than before implantation. "For every 100 eggs subject to normal internal fertilization, 85 will be fertilized if intercourse is frequent, 69 are implanted, 42 are alive a week later, 37 at the sixth week of gestation, and 31 at birth."[49] Rae and Cox cite miscarriage at 20 to 50 percent of that which is conceived even before implantation.[50] It is difficult to confirm precise percentages complicated by factors such as maternal age, but there is a consensus that about two-thirds of conceptions naturally do not survive to birth.[51] It is also clear that after implantation is success-

49. R. J. Berry, *God and the Biologist: Faith at the Frontiers of Science* (Leicester, U.K.: Apollos, 1996), p. 61.

50. Scott B. Rae and Paul M. Cox, *Bioethics: A Christian Approach in a Pluralistic Age* (Grand Rapids: Eerdmans, 1999), p. 166.

51. Alastair Campbell, Max Charlesworth, Grant Gillett, and Gareth Jones cite 42 percent survival from conception to full implantation and many studies concur on a

fully completed, the chances of survival to birth increase dramatically. It would be odd if most of the people who have ever lived did not last for more than one week in the womb. This is not a conclusive argument. One could note that for much of human history the mortality rate of children has been high. More died than survived childhood, yet we do not consider children any less persons. It is also not an argument from location, but rather probability of survival.[52] Gilbert Meilaender does not specify implantation but writes that "at least for the present, it is better to grant that an individual human being comes into existence somewhat later than fertilization."[53]

Conception: Fertilization and Syngamy

A Unique Genetic Individual Fertilization begins when one sperm enters the ovum. Within about twenty-four hours of sperm entry, the genes of the sperm and ovum fuse. At this point the zygote is microscopic, yet in that cell or two there is a combination of genetic instructions that has never been before and will never be again. If allowed to continue in a womb the chance is substantial, above 30 percent, that the zygote will develop to live birth. This is certainly a landmark as the genetic starting point of a unique individual. Paul Ramsey puts it this way:

> Micro-genetics seems to have demonstrated what religion never could; and biological science to have resolved an ancient theological dispute. The human individual comes into existence first as a minute informational speck . . . the unique, never-to-be-repeated individual human being (the soul) was drawn forth from his parents at the time of conception.[54]

For Ramsey the zygote is human, not daisy or tiger, because human beings are the source. Since the zygote is growing, the zygote is alive. Combining

---

roughly 30 percent natural survival rate from conception all the way to birth. *Medical Ethics,* 2nd ed. (Oxford: Oxford University Press, 1997), p. 82.

52. Does location matter whether in dish or womb? Is the embryo worth more or less if dispersed as a source of pluripotential stem cells? Such cells can then be grown into needed human tissues, but that ends the development of the embryo into a particular person. The question can be avoided in that pluripotent cells can now be derived from other sources, eliminating the appeal to the destruction of embryos as a necessary last resort. J. A. Thomson et al., "Embryonic Stem Cell Lines Derive from Human Blastocysts," *Science* 282 (1998): 1145-47.

53. Meilaender, *Bioethics,* p. 31.

54. Paul Ramsey, "The Sanctity of Life: In the First of It," *Dublin Review* 241, no. 511 (Spring 1967): 3-4.

these two observations, the one-celled zygote must be a human type of life. Because its particular combination of genetic instructions is unique, it is an individual. Because it is an individual human life, it is a person-soul and should be treated as such.

The Roman Catholic tradition as well has emphasized conception as an important transition for some years now. Pope Paul VI wrote in the encyclical *Humanae Vitae* that "from its very inception it [human life] reveals the creating hand of God."[55] This statement does not teach yet that there is a human being present from conception. However, a more recent encyclical from John Paul II states explicitly that "abortion is the deliberate and direct killing, by whatever means it is carried out, of a human being in the initial phase of his or her existence, extending from conception to birth."[56] He further writes, "Even if the presence of a spiritual soul cannot be ascertained by empirical data, the results themselves of scientific research on the human embryo provide a valuable indication for discerning by the use of reason a personal presence at the moment of the first appearance of a human life: how could a human individual not be a human person?"[57] The encyclical elaborates that when a unique set of human genetic instructions is present, a person is present. That begins at conception. This reasoning from genetic uniqueness is the most compelling single argument for protecting human life from conception. The following arguments from the presence of the image of God and the soul from conception, as well as the case from "potentiality," are less persuasive.

Reasoning from the Presence of the Image of God and the Soul A second set of arguments for recognizing a person at conception reasons from the presence of the image of God and the soul. In one version, from conception human beings are made in God's image. Bearing the image of God is the essence of being human. It does not require that the human being do anything or be able to do anything. It is not a functional characteristic. The image of God is that God chooses to have a unique relationship with us. The person that God cares about, who hence bears his image, is present from conception.[58]

This line of reasoning admirably emphasizes God as the gracious initiator in relationship. It also maximizes protection for the vulnerable in that our

55. Pope Paul VI, *Humanae Vitae*, English translation: *The Pope Speaks* 13 (Fall 1969): 329-46.

56. Pope John Paul II, *Evangelium Vitae*, Encyclical Letter, 16 Aug. 1993, as reprinted in *Ethical Issues in Modern Medicine*, ed. John D. Arras and Bonnie Steinbock, 5th ed. (London: Mayfield, 1999), p. 330.

57. *Evangelium Vitae*, p. 331.

58. Rae and Cox, *Bioethics*, pp. 130-39.

status comes from God's valuation, not from something that we can do. We cannot lose our status if someday we lose function or if we will never have function. Those of us who are able-bodied are only temporarily so. Due to development, illness, accident, and aging, it is part of the human condition to vary in one's level of function. Yet in *The Republic,* Plato's utopia, Plato founded human dignity on ability to function, so that if one could not take part in the community, one's life was not worth living.[59] Functional definitions of humanity render one at risk to disability in the view of anyone with greater capacity in the chosen area of measure. Might the tone-deaf not be human, in a musician's world? In contrast, if the image of God means that God cares, the image of God is present for those with no consciousness at all, such as anencephalics, children born without most of the brain. They will never have awareness, yet they still bear God's image because God cares about them.

Reasoning from the Genesis description of human beings as created in God's image does have two difficulties. First, the Genesis text describes God creating Adam and Eve as adults. Whether they would already have this image in the womb is not addressed. Second, the argument hinges on a particular definition of the image of God that makes no reference to being like God in capability, task, or action. The image of God is a description of God's love for his creature, not capabilities or actions by the creature. The image of God could be in a daisy, if God so values it. And God does value the daisy, yet the image of God is described as something unique to human beings. That would seem to imply something about human beings that sets them apart. If it is the ability to have a conscious relationship with God, which is something God initiates by his grace, then not having sufficient awareness to have a relationship would mean not bearing his image. If God's image means reflecting God in some way, learning to love as God loves or to seek truth as God is truth, then capability and choice become crucial to reflecting God's image. Immediately upon creation in God's image, human beings are given the task of multiplying and ruling the earth. The image of God might partially reside in how this task is carried out in a godly way. The image of God would then consist of a capacity to know God, a call to act as God does in caring for God's creation, and a relationship with God that empowers one to fulfill that commission. Definitions of the image of God were described in more detail in chapter 3. If the image of God includes one or two or three of the above attributes, it would be available to all human beings but not present in all human beings. Every human being would be valuable and capable of bearing God's image, but not all would actually be doing so. Texts stating that followers of Jesus

---

59. Plato, *The Republic* (New York: Penguin, 1955), p. 97.

Christ are being changed from one degree of glory to another, closer and closer to the image of Christ,[60] who is the perfect image of God,[61] seem to assume that the image is involved in character and action, not simply an automatic status from being human.[62] The image of God would not then be present from conception.

A variation on this argument is an appeal to the presence of a God-given soul or "animation" beginning at conception. In 1977 John Connery could write that the Roman Catholic Church "has never taught immediate animation. Even the Fathers of Vatican II resisted efforts to elicit such a teaching statement in connection with its condemnation of abortion. And the most recent declaration (1974) on abortion from the Sacred Congregation for the Doctrine of the Faith made a similar bypass of the question."[63] The idea that a soul is present from conception, ensoulment unconnected to the presence of a physical body, first became widespread as an implication of a papal pronouncement of 1854 that Mary was immaculately conceived.[64] The implication that some drew was that Mary must have been present from the moment of this miracle. Some Protestants have offered a parallel argument that since God was incarnate from conception, there must be a full human person present from conception.[65]

There are four difficulties for this argument. (1) It is not immediately clear why the incarnation or the immaculate conception of Mary calls for the presence of a human person from conception. Within the Roman Catholic tradition one could affirm the immaculate conception without assuming that the timing of Mary's presence was at conception. For the Protestant version of the argument from the virgin birth of Jesus Christ, one could affirm that God was uniquely present in bringing about the biological beginning of Jesus of Nazareth without assuming that the human person Jesus of Nazareth was present from his biological beginning.

(2) Both cases also depend on the assignment of unique souls, Jesus and Mary, at the beginning of physical development. The persons are then in some sense attached to the conceptus and wait for the body to develop sufficiently to be indwelt. This depends on the soul being a separate metaphysical reality rather than an emergent property of the God-designed human form.

---

60. 2 Corinthians 3:18.

61. Colossians 1:15-20.

62. See chapter 3.

63. Connery, *Abortion*, p. 308.

64. G. R. Dunstan, "The Moral Status of the Human Embryo: A Tradition Recalled," *Journal of Medical Ethics* 10 (1984): 38-44, as cited by Berry, *God and the Biologist*, p. 69.

65. Rae and Cox, *Bioethics*, p. 138.

That assumption is widely contested and has recently received a great deal of careful discussion.[66]

(3) Zygotes are not stable in their first days. Combining and splitting is not uncommon in the first two weeks after conception. If a soul is assigned at fertilization, then are there two souls in one zygote that later forms twins? Or one soul for the first zygote and another created later when the zygote splits? What of two zygotes, with two assigned souls, that later merge for the birth of one normal child? Or do some zygotes not have a soul and hence do not double up when two zygotes merge? The biological fact of zygote instability seems to rule out the possibility of a simple one-to-one assignment of one soul at conception to each zygote.

(4) Immediate soul assignment also assumes that two-thirds of the human souls God has specially created never live on earth, since two-thirds of conceptions do not continue to birth. The standard response to this is that many children have died in birth or at a young age, but that does not make children any less full human beings. Granting that important point, the question would still remain why God would design a biological system where most of the human souls he creates never live on earth. One possible response is that the high rate of zygotes failing to implant is not God's design but a result of sin embodied in nature. That would in turn raise the question of to what degree the natural world as we see it now reflects God's intent or the corruption of the fall (or God's intent through the fall for supralapsarians). Speculations on the timing and nature of soul creation or the image of God do not offer the desired clear base to establish personhood at conception.

Potential A third set of reasons often given for recognizing that a person is present from conception works from ideas of potential. One version emphasizes the continuity from conception to adulthood. Since development is continuous, the conceptus should have the same status as the adult. However, continuity does not imply identity. An acorn has a continuous connection leading to an oak tree. They are both oak, but we do not call an acorn an oak tree. A foundation course of cement block will be in continuity with the resulting building, yet we call it the foundation, not a building. Continuity does not necessarily embody identity. Continuity between the conceptus and the

66. Patrick Lee, *Abortion and Unborn Human Life* (Washington, D.C.: Catholic University of America Press, 1996), and J. P. Moreland and Scott Rae, *Body and Soul: Human Nature and the Crises in Ethics* (Downers Grove, Ill.: InterVarsity Press, 2000), propound the dualist view. Warren S. Brown, Nancey Murphy, and H. Newton Malony, eds., *Whatever Happened to the Soul? Scientific and Theological Portraits of Human Nature* (Minneapolis: Fortress, 1998), advocate an emergent view of the soul.

adult human being does not imply that they are of the same standing. A continuum of development does not imply equal presence or value. "The fact that an entity can undergo changes that will make it significantly different does not constitute a reason for treating it as if it had already undergone those changes. We are all potentially dead, but no one supposes that this fact constitutes a reason for treating us as if we were already dead."[67]

One could vary this argument from equal presence or value to one of indistinguishable connection. If it is acceptable to kill a human being at one point in life, it is acceptable at any point, because human development is on a continuum. If we can discard a zygote, why not an embryo? If we can deliberately end the life of a thirty-five-week-old fetus, why not a twenty-eight-week-old premature birth? The concern is not without precedent. It was not that long ago that the country that gave us Bach and Brahms began with ending the lives of the severely handicapped and then kept expanding the category of "lives not worth living" until millions of people were deliberately executed. The Nazi experience is discussed in more detail in chapter 15. This concern that one choice that might seem innocuous can lead to another until society is involved in horrific choices is called the slippery slope. It is not confined to the European continent. Chapter 4 has a description of the Tuskegee syphilis experiments that began with simple observation, progressed to deception, then failure to offer treatment, and finally active work to keep patients from an effective treatment.

The slippery slope is found in two distinct forms that often overlap. Resolving one does not automatically resolve the other. The conceptual type of slippery slope is when there is no logical difference between the first step and a later one. If the first is acceptable, there is no logical reason not to do the later. The social version of the slippery slope is an observation of society, that if the first step is taken society will continue into the horrific even if there are good logical reasons to stop. The slippery-slope concern calls for care to set limits not just at the immediately acceptable, but to take into account the momentum of decisions. Where will the decision lead in the future?

When abortion was first legalized with *Roe v. Wade*, there was no discussion of what is often called "partial-birth abortion." It is the current practice in the United States that one may do such abortions without legal restraint up to nine months and beyond. The only limitation is that the skull of the one to be aborted is still in the birth canal. The fetus is withdrawn until the body is outside and only the head is still inside. At this point a hole is cut

---

67. John Harris, *Clones, Genes, and Immortality: Ethics and the Genetic Revolution* (Oxford: Oxford University Press, 1998), p. 50.

in the skull and the brain is suctioned out. Then the empty skull is pulled the rest of the way out of the birth canal. If the procedure was done after the head left the birth canal, it could be prosecuted as murder. By doing it before the head is withdrawn, it is a legal abortion. Situations where this procedure would be medically necessary to protect the life of the mother are hard to imagine if present at all. The main advantage is to be sure that the one delivered is dead. Both houses of the United States Congress have repeatedly passed bills outlawing the procedure by wide majorities, but not the sufficient two-thirds in the Senate needed to override presidential veto. The law hinges on location as the determinative distinction between being and not being a person. Advocates of keeping the procedure legal argue that it is important to protect the practice, lest there be a slippery slope of new limitations on doing abortions. Opponents consider partial-birth abortion an example of how the slippery slope progresses, with after-birth infanticide probably coming next.

The strongest argument from potential is probably that of Don Marquis. He reasons that what makes it wrong to murder an adult is that the victim is deprived "of all the experiences, activities, projects, and enjoyments that would otherwise have constituted one's future."[68] "What makes killing any adult human being prima facie [obligatory unless over ridden by a higher moral rule] seriously wrong is the loss of his or her future."[69] The loss of future to a fetus is as great as the loss of future to an adult. "For any killing where the victim did have a valuable future like ours, having that future by itself is sufficient to create the strong presumption that the killing is seriously wrong."[70] Marquis extends this argument to embryos when he says, "Of course, embryos can be victims: when their lives are deliberately terminated, they are deprived of their futures of value, their prospects. This makes them victims, for it directly wrongs them."[71]

Marquis makes a point of saying that the argument does not apply to contraception.[72] To deprive an entity of future prospects, an entity must be present. Before conception, there is not yet a particular entity with a future. He admits that there is a particular ovum and sperm that while currently separate will meet and continue on into the future, but argues that their numbers are so vast and the circumstances of which survive so varied, that they cannot be recognized in advance. It is wrong to kill an adult. It is not wrong to kill an

---

68. Don Marquis, "Why Abortion Is Immoral," in Pence, ed., *Classic Works in Medical Ethics: Core Philosophical Readings*, p. 189.
69. Marquis, "Why Abortion Is Immoral," p. 189.
70. Marquis, "Why Abortion Is Immoral," p. 193.
71. Marquis, "Why Abortion Is Immoral," p. 197.
72. Marquis, "Why Abortion Is Immoral," p. 198.

individual human cell. But following his reasoning it would seem that if the cell has valuable future prospects, it is wrong to deprive the cell of those prospects.

The microscopic dot of one fertilized cell is not presently self-aware, but such developing life is potentially in unbroken continuity with a later person. But if it is potentiality that matters, not presence, then separate egg and sperm have potential. Why only recognize the potential starting at conception? If conception, then why not the sperm and ovum before they connect? Sperm and ovum have potential to become an embryo. Now granted, a sperm and ovum have the potential *to become an individual,* while the embryo has the potential *of an individual.* But "whatever has the potential to become an embryo has whatever potential the embryo has."[73] The odds in intercourse do change dramatically for any one sperm, from a one in a million chance at being part of a healthy baby to one in three once it unites with an ovum. This dramatic change in odds is not the case, however, in some types of *in vitro* fertilization where one sperm is selected for injection. Once conceived there is less contingency than in a given ovum and sperm, but the zygote is still contingent. Most will not survive to birth. If the difference is the likelihood of reaching adulthood, that increases over time. The Dominican theologian Albert the Great (1206-1280) affirmed that it was worse to be sterilized than to have an abortion. An abortion ends one life. Sterilization keeps many potential children from being.[74] *Humanae Vitae* makes several arguments against contraception but never that it stops the development of potential life. In fact, the papal encyclical directly calls upon Roman Catholic scientists to improve the rhythm method, which is a conscious effort to avoid potential lives.[75]

Now the word "potential" is used by some in a specific technical sense to refer to the unfolding of something that already exists. One example would be a fertilized egg with a newly formed genetic code that will direct much of what the embryo becomes. The substantial merits of the genetic argument for personhood have already been raised in the previous pages. The more common meaning of "potential" is "not yet." We usually do not consider potential as the same as actual. A seed may have the potential to become a flower but is not actually a flower yet. If it was a flower we would call it a flower, not a potential flower. A student has the potential to be a college graduate but is not one until successfully completing the required course of study. With cloning technology every nucleated cell in your body (you have about three billion)

73. Harris, *Clones, Genes, and Immortality,* p. 50.
74. Connery, *Abortion,* p. 107.
75. Pope Paul VI, *Humanae Vitae,* pp. 329-46.

has the potential to be nurtured into a pregnancy, birth, and adult human being. If we do not clone each cell with such potential, have we murdered all those people? If one removes one cell in the first few days from a conceptus, that cell can be nurtured into an identical twin. This occurs naturally on occasion. If instead of reimplanting that one separated cell, one dissolves the cell to read the DNA code, has one murdered an identical twin? Blurring the distinction between potential and actual leads to countless quandaries. George Annas appeals to common moral intuition with the following story. If a fire broke out in a fertility lab and there was only time to save a two-month-old baby there in a bassinet or a rack with seven embryos, most would save the baby without hesitation. Yet carrying out the test-tube rack instead could have saved seven people, if indeed each embryo was a person.[76] Thankfully that is not the usual choice that we face. But if Annas is correct about what we would do, what is guiding our choice? Could it be that we make a clear distinction between potential and present?

Finally on potential, some thoughtful scholars such as Rae and Cox are concerned that "if essential personhood is determined by function, it follows that essential personhood is a degreed property. After all, some will realize more of their capabilities to reason, feel pain, self-reflect, etc., than others. Moreover, it is undeniable that the first several years of normal life outside the womb include an increasing expression of human capacities."[77] On this basis one's personhood could increase and decrease and along with it one's corresponding rights. There are philosophers who have come to precisely that conclusion. However, it need not follow. Personhood could be seen not as a degreed property, ebbing and flowing with degrees of mental capacity. It could be understood as a threshold property. What would matter would not be the degree of ability but whether there was any present ability at all. Rae and Cox affirm a threshold. A person is not present until conception. When a full complement of human genetic material is present, the essence or substance of a person is there.[78] That is their threshold. If there is no conception yet, there is no person yet. Above we have examined reasons given for seeing that threshold at conception or other points.

76. As quoted by Bonnie Steinbock in *Life Before Birth*, p. 215.
77. Rae and Cox, *Bioethics*, pp. 168-69.
78. Also propounding this view are Lee, *Abortion and Unborn Human Life*, and Moreland and Rae, *Body and Soul: Human Nature and the Crises in Ethics*.

Burden of Proof

There is certainly a woman central to pregnancy, and both a man and woman as sources of gametes. They should be loved as neighbors. An essential part of neighbor love is in its extension, not exclusion.[79] When is someone else also present to be cared for? Careful thinkers on both sides admit that they do not know for sure. To some degree then the discussion shifts to burden of proof.

On the one side, since at some point a person's life is at stake, the burden of proof is on those who would withdraw support from a pregnancy because they believe there is no person present at that time. John Jefferson Davis cites an illustration from Harold O. J. Brown. If a hunter shoots at movement behind a bush, without first being sure that the movement was caused by an animal and not a human being, the action is reprehensibly irresponsible.[80] The teaching authority of the Roman Catholic Church, the Sacred Congregation, uses the same reasoning in stating that there is immediate ensoulment at fertilization because no one can prove the contrary. It would be culpable to take a chance on assuming that the soul is not present.[81]

However, there are those who argue that the burden of proof is on those who would withhold help to universally recognized people in order to protect possible people. John Harris writes,

> When we bear in mind that . . . most of the secrets of the development of life are obtained in early embryos, and that we are extremely likely to be able to use what we learn from such embryos to save many lives and ameliorate many conditions which make life miserable, we would not only be crazy but wicked to cut ourselves off from these benefits unless there are the most compelling moral reasons to do so.[82]

For Harris, the choice is between saving lives that might be present but are probably not, and saving lives that we know are present. Does the burden of proof lead to favor protection of developing life or provision of life and health for people that we know already exist? We should not kill people to benefit others, but we should also not let people die to protect only human tissue such as sperm and ovum, even though such gametes do have great potential.

---

79. O'Donovan, "Again: Who Is a Person?" pp. 135-37.

80. Davis, *Evangelical Ethics: Issues Facing the Church Today*, p. 148.

81. Carol A. Tauer, "The Tradition of Probabilism and the Moral Status of the Early Embryo," *Theological Studies* 45 (1984): 3-33.

82. Harris, *Clones, Genes, and Immortality*, p. 65.

Laws usually require set lines so that the weight of the state falls consistently where clearly expected. Sixteen-year-olds know that they can receive a driver's license in the United States even though some were responsible enough to drive a car a year or two before and some will never be mature enough to properly drive a car. The law sets one clear line so that there is less conflict over whether a line has been transgressed and so that all are treated equally. This issue of personhood does not lend itself to such a thin line. Public policy has to draw one, but in this case ethics is not as simple as we might wish.

At this point it might be helpful to review what resources we have from Christian Scripture, experience, tradition, and reason. *Scripture* calls for inclusive neighbor love, especially for the vulnerable. Christians are to be characterized by hospitality, but Scripture texts simply do not provide a statement of precisely when in development a neighbor is present to love. The actual *experience of pregnancy* has brought many women to recognize the presence of another at least beginning with the independent movement of quickening at about ninety days. Our *social experience* of slippery-slope progression has made it clear that wherever one draws a line it will be contested and pushed. That calls for an early and clear line of protection for human life in the womb. The Christian *tradition* has long discouraged abortion but for most of its history has not considered it as serious as killing a person until after a body has formed. Often that has been understood as being in place by forty days after conception. *Reason,* by simple observation, has given us new information that body systems are all present at twenty-eight days, including a beating heart. That is where the logic of the early church's appeal to formation would now lead. Symmetry with our definition of death as the permanent loss of consciousness would also lead to this point or earlier since if allowed to continue the fetus will come to consciousness. The likelihood of birth dramatically rises with implantation beginning at six days and the presence of unique genetic instructions at fertilization call for protection for life from the earliest stages. There are important grounds then for not deliberately discarding or otherwise jeopardizing developing human life.

Those who are most persuaded by the genetic evidence or by the need to be cautious in the face of uncertainty over the status of the zygote will oppose throwing away or otherwise threatening human life beginning at conception. Even those who find the line more convincing at a later stage such as implantation should still be loath to end life as it begins. For example, if one holds the implantation view that the person or soul is not yet present in the first days after conception, after a rape one could take a "morning-after pill" to shed the uterine lining. If any conception has taken place, the developing life would not then implant. According to the view that a person begins at implantation, one would not be destroying a human being because a person would not yet be present. However, even by that perspective, in light of all else considered one should not do so lightly. It would be disparaging to the worth of new human life to use "morning-after pills" merely for regular birth control when more respectful alternatives are available. Thinking through all of the above considerations in this chapter moves one decisively toward protection of life in the womb.

Chapter Summary

Human research can be a worthy endeavor. It can yield information that brings about understanding, healing, and newfound ability. Yet as research gains much of its power in objectifying its subjects, care must be taken not to abuse the human beings who participate. To assure that participants are treated as people and not things, it is best to do research with voluntary partners. If important research requires human participants that are not able to speak for themselves, surrogates should decide on their behalf with great care. Children who are not old enough to legally consent may assent at minimal risk and with parental concurrence to do what only children can do, such as test the effects of a drug on children's physiology. Research on human life in the womb operates under similar constraints carefully reasoned in the chapter. The earliest transitions in pregnancy have received the most detailed discussion because they are decisively relevant to the ethics of many kinds of human genetic intervention. Whether a particular early intervention is a help or harm will often depend substantially on whether a person is present when the intervention takes place.

CHAPTER 6

Searching for Genes and Community

Genes combine uniquely in individuals, yet are held in common by families, ancestors, and descendants. One could think of the letters DNA (which stand for deoxyribonucleic acid) as reminding one of "Descendants 'N Ancestors." While our genetic heritage is quintessentially intimate to each of us, it is shared with a whole history of ancestors in the past and possibly descendants to come, potentially extending out in ever widening circles. Our genes are held in common by families and then dispersed among untold numbers of descendants. This genetic link with others reminds us of our interdependence with fellow human beings. Our very genes are part of a community of human beings and the wider community of life. Whether we seek or dread the responsibility, our choices have effects beyond ourselves. Our actions help to decide what genes will be present in future people, as the actions of others have deeply shaped what genes we inherited. As genetic individuals we are inextricably part of a community of family, humanity, and life.

These communities and the rest of the environment are as influential to the individual's form as genetic heritage. Our physical form and health is the result of an intricate interaction between genes and their surroundings within and outside the body. We shape our environment and the environment shapes us. An important part of that environment is found in the communities that we belong to by intentional affiliation, acceptance, or no choice of our own. Our communities range from churches to the federal government, and our life in them is complex. In this chapter we will highlight three points of interaction particularly important to genetic research. They are group consent, gene patents, and social investment.

Group Consent

The freedom and ability to shape one's own life is often described by the word *autonomy*, as discussed in chapter 4. Autonomy does not necessarily mean isolation or utter independence. "Autonomy simply means that a person chooses and acts freely and rationally out of her own life plan, however ill defined. That this life plan is *her own* does not imply that she created it *de novo* or that it was not decisively influenced by various factors such as family and friends."[1] An individual can act autonomously in welcoming the counsel of her community or even in choosing to submit to it. The physician who thoughtfully accedes to the code of her profession may autonomously choose to do so. This connection between individual consent and professional or social community is highlighted by the nature of genetics. While our genes are intimately personal, they are shared in families and across communities. An individual may seek community counsel and a community may be deeply affected by individual choices. If a community may be affected by genetic research results, what role should it have in consent to research on individuals of that group?

Population geneticists study "demes." Demes are groups of individuals more genetically similar to each other than to any other individuals. Demes are of interest for anthropological studies of how populations have moved and interacted geographically. For example, Stanford University researchers have found a rare variation in DNA sequence that only occurs among the Inuit and people indigenous to the Americas.[2] This is evidence for modeling historical migration patterns. Population studies can also be useful for finding groups at risk for a particular allele. To reach such goals "The Human Genome Diversity Project" was founded. The project would coordinate gathering genetic samples from populations around the world, particularly ones that have been more isolated and hence are more distinctive. The project was not well received by many groups, who distrusted the effort as probably leading to prejudice, commercial exploitation, or neocolonial control.[3]

A number of indigenous-people representatives in North America have insisted that their leadership can give or refuse research participation for the people of their nations. For example, Judy Golbert of Salish/Kootnay College

1. James F. Childress, *Who Should Decide? Paternalism in Health Care* (New York: Oxford University Press, 1982), p. 60.

2. Eric T. Juengst, "Groups as Gatekeepers to Genomic Research: Conceptually Confusing, Morally Hazardous, and Practically Useless," *Kennedy Institute of Ethics Journal* 8, no. 2 (1998): 186.

3. P. Kahn, "Genetic Diversity Project Tries Again," *Science* 266 (4 Nov. 1994): 720-21.

claims to speak for several indigenous people groups when she says, "We see genetics as a battle, the same battle begun five hundred years ago with Columbus. We are closing doors to research of all kinds. We remember contaminated blankets. We do not want to be known. That would make us more vulnerable. We were exploited before. That is the likely end of this too."[4] The concern is that genetic research will be used against communities, whether by creating or reinforcing stereotypes, or other forms of exploitation. Since the risk falls on the group, not just those who would individually take part, the group should have some say.

However, Eric Juengst points out three major problems with group consent.[5] The first is that social groups that could grant permission are not identical with the demes under study. It is rare for genetic connections to coincide completely with political or social groups. Second, when social groups are consulted as if they are demes, that reinforces the stereotype that they are genetically set apart. Such a characterization encourages and enables prejudice and discrimination. It gives impetus to the too common idea that "they," whoever "they" may be, "really are different from us." Third, groups nest within groups. The permission of one might not address the concerns of another. Could Hmong leadership in Laos decide whether Hmong in Minneapolis could take part in a study? Juengst proposes that the best available course is to include risk to social group identity in the informed consent process for individuals.[6] The individuals can then decide whether to take part.

Informing individuals of possible group impact from their participation should be included in informed consent. Since genetic studies of groups would require substantial numbers of participants, such informed consent would be in some sense a minor referendum on group participation, yet that may not sufficiently address group impact. It is reasonable to consult and take into account the needs and concerns of all affected by research as much as possible. That would include consulting with a group's leadership when research is designed to describe their group. Characterizing a group is fraught with difficulties since heritage and allegiance is so complex among human beings, but if that is the attempt, more than the immediate participants have a stake.

One example of a positive interaction between researchers and group risk can be seen in the investigation of Familial Alzheimer Disease (FAD).

4. Judy Golbert, "Genetic Research on Indigenous People: Community Consent and Refusal," Annual Meeting of the American Society for Bioethics and the Humanities, Philadelphia, 28 Oct. 1999.

5. Juengst, "Groups as Gatekeepers to Genomic Research," p. 196.

6. Juengst, "Groups as Gatekeepers to Genomic Research," p. 197.

FAD is characterized by early dementia and is pervasive in a few families among an ethnic group of Germans who migrated to the United States from the Volga region of Russia. The American Historical Society of Germans from Russia (AHSGR) became concerned that researchers were describing research in such a way that people might think the entire Volga German population was afflicted with early dementia. In response, the ASHGR met regularly with the principal investigators, and the investigators made a point of clarifying in their publications that early onset AD was characteristic of only a few specific family lines, not all Volga Germans.[7] Since effects of gene research may be felt by groups, it is at least wise for investigators to avoid group harms and to seek counsel from group leadership when it is available to that end. This would supplement participant consent, not replace it.

Gene Patents

The complexity of modern science usually requires a wide network of individuals and labs working together to find and confirm new information. That makes singling out one person or organization for a patent problematic. It can be difficult to clearly locate who "discovered" a given gene. The pattern in the past has been to freely share new discoveries. No one attempted to patent crucial developments such as the Southern Blot or the recognition of trisomy 21 as the cause of Down syndrome. That has recently changed. Biotechnology companies and universities have begun to claim proprietary rights to new developments and found them quite lucrative. The flurry of patents and licensing agreements has enriched some and incensed others.

On May 18, 1995, a press conference was called to publish a brief statement signed by almost 180 religious leaders. The breadth of support was impressive. Signers included Kenneth Carder (the chair of the United Methodist Genetic Science Task Force), Abdurahman Alamoudi (the executive director of the American Muslim Council), Rabbi David Saperstein (director of the Religious Action Center of Reformed Judaism), Richard Land (executive director of the Christian Life Commission of the Southern Baptist Convention), and ninety-one bishops of the Roman Catholic Church. The statement was only three sentences long.

---

7. Thomas D. Bird, "The Chromosome 1 Type of Familial Alzheimer Disease," in *Genetic Testing for Alzheimer Disease: Ethical and Clinical Issues*, ed. Stephen G. Post and Peter J. Whitehouse (Baltimore: Johns Hopkins University Press, 1998), pp. 20-21.

We, the undersigned religious leaders, oppose the patenting of human and animal life forms. We are disturbed by the U.S. Patent Office's recent decision to patent human body parts and several genetically engineered animals. We believe that humans and animals are creations of God, not humans, and as such should not be patented as human inventions.

The rally was admirable for gathering such an august group and for its accompanying press release that recognized God as "the center of all that is." However, the resulting statement seems to misunderstand what has actually been patented. The US Patent Office grants to a patent holder the exclusive right for twenty years from filing date to prevent others from making, using, or selling a process, manufactured machine, or composition of matter that has been made possible by the patent holder. What is patented must be novel, nonobvious, and useful. These criteria would not seem to apply to human body parts, so what is the above proclamation referring to as "human body parts" patented in "a recent decision"? Ted Peters has traced the reference to patent number 5,061,620. The patent covers a nearly pure collection of pluripotent hematopoietic stem cells and a method for obtaining the concentration.[8] Note the patent has not been granted for your cells and mine. It is for a unique process developed by a particular group of researchers. The technique is to concentrate a particular type of cell outside the body. In such a form the cells might be helpful to the treatment of leukemia or AIDS. The patent is not on the existence of the cells. It is on a new method of gathering them for therapeutic use. Peters goes on to cite the parallel of the patent on the vitamin-pill version of B12. B12 occurs naturally in the body but in minute amounts. A patent was granted for a purification process that makes it available as an inexpensive and pure dietary supplement. The patent is on a particular method of making it available and useful, not the substance itself.

Patenting such novel developments gives an incentive to investigators and those who make their work possible financially to pursue useful inventions. Without such patent protection many new inventions would not receive the investment needed to create them. Much of the current rapid progress in therapy development can be traced to the tremendous investments of private capital. As large as the National Institute of Health (NIH) and Department of Energy (DOE) federal genetic programs are at $200,000,000 a year, private companies have been investing each year ten times that figure.[9]

The patent for "a genetically engineered life form" might refer to patent

8. Ted Peters, *Playing God: Genetic Determinism and Human Freedom* (New York: Routledge, 1997), p. 119.

9. Peters, *Playing God,* p. 132.

number 4,736,866 on a cell line of oncogenes that can make a mouse susceptible to breast cancer. The patent is not on mice or even a breed of mice. It is on a method of consistently inducing cancer in an animal. This gives researchers a way to test various treatments without beginning first with human cancer patients. In citing the patent there may be welcome concern for animal comfort and life. Animal experimentation is a significant issue. Suffering should not be deliberately caused or allowed where it can be avoided. On the other hand, patent ownership is not central to that concern. If it were, the ownership of companion animals would be tightly associated with abuse.

Where there is cause for considerable concern is in the patenting of genetic sequences. This first became an issue when the NIH filed patents on human gene fragments in June 1991. It was controversial for at least three reasons. First, while the sequence was known, the function of the code was not. This does not meet the requirement that the patented process or object be useful. Second, many scientists were concerned that such patenting would impede the free flow of information and cooperation central to the intricate and interdependent work of the Human Genome Project. The third commonly voiced concern was that such patents commodify the involved animals and human beings. Patenting a genetic sequence might imply that the source of the genes is merely an object. Mark Hanson nicely parses out this concern in a special supplement to the *Hastings Center Report*.[10] Respect and market forces do not need to be mutually exclusive. One can pay a pianist to perform a concerto and still respect the pianist's artistry and skill. Yet commodification is an important concern. People should never be treated as merely things, nor should we make choices that would likely lead to that end. Patenting of processes and applications are welcome to encourage investment and further development of tools of genuine service.[11] That is different from allowing people to be treated as mere objects.

Under new leadership in 1996 NIH filed a disclaimer forfeiting all previous patent claims. There are still, however, thousands of patent claims made by other individuals and labs on gene sequences. As this book goes to press, such patents are being contested in the court system. Ted Peters suggests that the validity of such patents depends on the way three remaining ambiguities are resolved.[12] First, do patents refer to a process of gene isolation or to the physical product of the process or both? The United Methodist Church, for

10. Mark J. Hanson, "Religious Voices in Biotechnology: The Case of Gene Patenting," *Hastings Center Report* Special Supplement (Nov.-Dec. 1997).

11. Donald M. Bruce, "Patenting Human Genes — a Christian View," *Bulletin of Medical Ethics* (Jan. 1997): 18-20.

12. Peters, *Playing God,* p. 129.

example, approves of process patents in the case of genetic sequencing, but not product patents.[13] The second ambiguity is whether genetic fragments are discovered or are the product of human invention. Some patent attorneys have argued that this is a key distinction between patentable and not patentable. Third, is the genetic code a pattern or a particular chemical structure? This distinction is crucial to whether a patent on a man-made copy of a genetic sequence can patent the original sequence appearing in nature as if it were man-made. If copies can carry such authority, human genetic sequence information could be controlled by relatively few people. That would probably be a tremendous loss to further research and application. Audrey Chapman raises a fourth concern as to whether such patents put developing nations at a significant disadvantage.[14] The information holds the prospect of great value for all but as patent protected would be yet another resource controlled by the more developed nations, exacerbating economic disparity. On the other hand, proponents of patenting warn that if the economic incentive of patenting is not available, the information and resulting products will not be developed at all or much more slowly.

The traditional economic reward of patents for the innovation of new products does encourage efforts that can be eventually helpful to all. Trying to patent the sequence of human DNA, however, seems a presumptuous and damaging claim.

Social Investment

While a person's genome has pervasive ramifications for one's body, it does not control all the factors that affect human health. Environmental insult can interact with genetic susceptibility.[15] Personal choices are formative. Social factors may play a prominent role as well.[16] The government of the United

---

13. "New Developments in Genetic Science," in *The Book of Resolutions of the United Methodist Church* (Nashville: United Methodist Publishing, 1992), pp. 332-33.

14. Audrey R. Chapman, *Unprecedented Choices: Religious Ethics at the Frontiers of Genetic Science* (Minneapolis: Fortress, 1999), pp. 137, 144-45.

15. Stuart L. Shalat, Jun-Yan Hong, and Michael Gallo, "The Environmental Genome Project," *Epidemiology and Society: A Forum on Epidemiology and Global Health* 9, no. 2 (1998): 211-12.

16. Examples of articles emphasizing this aspect include N. Kreiger et al., "Racism, Sexism, and Social Class: Implications for Studies of Health, Disease, and Well Being," *American Journal of Preventative Medicine* 9, no. 6 (1993): 82-122; and J. W. Frank, "The Determinants of Health: A New Synthesis," *Current Issues in Public Health* 1 (1995): 233-40.

States has and is continuing to invest a large percentage of medical research funds in the Human Genome Project. The hope is that this will eventually bring therapeutic benefits well worth the investment. Some critics are concerned that it might be drawing funds away from more effective research and direct social services. For example, concern has been raised that the hypothesis of a "thrifty" gene in some minorities drew attention away from social factors that were causing high rates of obesity. Public health services could have more quickly and directly helped through education and employment efforts than through genetic studies.[17]

Paul Edelson has expressed the concern that the quick adoption of phenylketonuria (PKU) testing supplanted more effective interventions for the vast majority of the mentally challenged. The state-mandated testing helped to avoid mental retardation for those who carried the genetic anomaly. Prevention is better than attempts at repair after the disease has had devastating effects. But the vast majority of cases of mental retardation are not PKU related. Edelson is concerned that social policy should not be distracted from education and services for those who are already with us and continue to arrive.[18]

Philip Ferguson writes,

> The point is not so much whether . . . a blind person cannot enjoy a Rembrandt . . . but whether social arrangements can be imagined that allow blind people to have intense aesthetic experiences. . . . People in wheelchairs may not be able to climb mountains, but how hard is it to create a society where the barriers are removed to their experiences of physical exhilaration? . . . Someone with Down syndrome may not be able to experience the exquisite joy of reading bioethics papers and debating ethical theory, but . . . that person can experience the joy of thinking hard about something and reflecting on what he or she really believes. . . . The challenge is to create the society that will allow as many different paths as possible to the qualities of life that make us all part of the human community.[19]

---

17. R. McDermott, "Ethics, Epidemiology and the Thrifty Gene: Biological Determinism as a Health Hazard," *Social Science and Medicine* 47, no. 9 (1998): 1189-95.

18. As cited by Diane B. Paul, "The History of Newborn Phenylketonuria Screening in the U.S," in *Promoting Safe and Effective Genetic Testing in the United States: Final Report of the Task Force on Genetic Testing,* ed. Neil A. Holtzman and Michael S. Watson (Baltimore: Johns Hopkins University Press, 1998), p. 141.

19. Philip Ferguson as quoted by Erik Parens and Adrienne Asch, "The Disability Rights Critique of Prenatal Genetic Testing," *Hastings Center Report* Special Supplement 29, no. 5 (1999): S13-S14.

In some ways a genetic impairment of a function is not a disability unless the structure of society requires that function to participate. Sometimes the most effective service may be in removing the structural demand.

Genetic research has great promise to enable better understanding of one facet of who we are. It cannot bring utopia, but it is already enabling us to make life better for many people. That is a worthy endeavor, yet one that needs to be balanced with research into other promising ways of service.

Chapter Summary

Genes combine uniquely in individuals, yet are held in common with families, ancestors, descendants, and larger communities. The role of community in genetic research raises questions in regard to group consent, gene patents, and social investment.

1. In genetic research, care should be taken not only for individuals, but also for the impact and input of groups when they are implicated.

2. Private patenting should be allowed for man-made processes. That will encourage their development and service, but patenting the raw data of human genetic sequence is problematic. Such would probably hinder more than help human genetic discovery and service.

3. Devoting societal resources to the basic science of human genetics is already bearing much fruit, but it cannot account for or meet every societal need. Funding research in human genetics should not eclipse other evident social needs and services. The best proportion of social investment in human genetics compared to other pressing needs is another book.

PART II

GENETIC TESTING

CHAPTER 7

Testing Genes and the Individual

The Proliferation of Genetic Tests

It has long been recognized that information about your genes can be useful for maintaining health. That is why when you begin with a new physician there is usually a long set of questions about your parents and siblings. The physician's interest in the health of your relatives is a crude genetic test. By examining their health history he or she can develop a general picture of the gene pool that you share with them. While your genes could not be more intimately personal, they came from just two people. The medical history of the people who gave you a genetic start and the health of others from the same two sources is revealing of your probable future health.

However, family history does not reveal which particular genes you inherited from each parent. Each parent carries two copies of most genes. The copies may be quite different from each other as distinct variations of the same basic gene. They might be expressed in combination, or one might be hidden by the action of the other. That means that you might have inherited gene variations from your parents that they did not express. Each of your dark-haired parents may have also carried the genetic code for blond hair. If it is the hidden genes involved in blond hair that you inherited from both parents, you could be the blond child of two dark-haired parents. As described in chapter 1, phenotype does not fully reveal genotype. Family health history reveals a context for understanding your genetics but remains imprecise as a way of estimating exactly which genes you inherited and how those genes might uniquely interact.

Direct genetic testing of the individual is an opportunity to see precisely

149

which genes were inherited. That information can be used sometimes to diagnose and better treat disease or to predict with some probability likely future health. Both of these uses will be described in this chapter. We will then look at Christian resources for using such information. The implications of genetic testing for family members such as siblings and children will be the focus of chapter 8. Chapter 9 will consider community-wide effects of genetic testing, such as challenges for the medical insurance industry.

The variety and availability of genetic tests are developing rapidly. For some time now the number of specific tests has been roughly doubling each year. As of 1998 in the United States alone already "over 500 commercial, university, and health department laboratories provided tests for inherited and chromosomal disorders, and genetic predispositions."[1] The University of Washington and the National Institutes of Health provide a constantly updated list at www.genetests.org.

Companies that develop genetic tests are under financial pressure to encourage wide use of the tests as early as possible. It costs approximately 30-40 million dollars to develop and bring to market one genetic test.[2] Such sums motivate corporations to make tests available quickly. The amount of capital involved also means that numerous constituencies in these corporations have concurred on the safety and efficacy of the test or drug. Companies will rarely want to risk this level of investment on a drug or test that will end up being rejected for any reason, let alone predictable ethical ones.[3] Yet the degree of risk they are willing to take varies.

In January 1994 Genica Pharmaceuticals Corporation gave physicians throughout the United States a mail-in packet for a genetic test (for APOE) requiring 2.5 milliliters of whole blood and a payment of $195. A cover letter indicated that while Medicare would not reimburse the test, since it was classified as investigational, the provider would be "free to seek payment directly from the patient."[4] Some argue that this test has proven effective for confirm-

---

1. Task Force on Genetic Testing of the NIH-DOE Working Group on Ethical, Legal, and Social Implications of Human Genome Research, *Promoting Safe and Effective Genetic Testing in the United States: Final Report of the Task Force on Genetic Testing*, ed. Neil A. Holtzman and Michael S. Watson (Baltimore: Johns Hopkins University Press, 1998), p. 7.

2. John Varian, "Genetics in the Marketplace: A Biotech Perspective," in *Genetic Testing and Screening: Critical Engagements at the Intersection of Faith and Science*, ed. Roger A. Willer (Minneapolis: Kirk House, 1998), pp. 64-65.

3. There are charges that the mass marketing of Fen-phen is a tragic exception.

4. Stephen G. Post, "Education for a Too-Hopeful Public," in *Genetic Testing for Alzheimer Disease: Ethical and Clinical Issues*, ed. Stephen G. Post and Peter J. Whitehouse (Baltimore: Johns Hopkins University Press, 1998), p. 226.

ing the diagnosis of Alzheimer's disease in a patient experiencing dementia, but there is consensus that it is still not a predictive test for most people. The test proposal did not qualify what the test could actually achieve.[5] The Genetics and IVF Institute of Fairfax, Virginia, began offering a genetic test for one type of breast cancer risk when there had been a general consensus among laboratories that the test was not assured enough yet for public use. In England, cystic-fibrosis carrier screening was offered in a direct marketing campaign to the public in December 1994 by a commercial laboratory called University Diagnostics.[6] There is marked disagreement over whether "do no harm" should be the leading principle in when to present a new test or whether people have a "right to know" and can sort out for themselves if a test is sufficiently accurate.[7]

Physicians have numerous reasons to offer genetic tests. Possible patient-care benefits, liability limitation, financial rewards, and patient requests all encourage use. When physicians offer a test, it will sound to most patients as if it has been recommended, especially if they are asked to sign a refusal form if they do not take the test. Requiring such statements has become common practice in the United States for pregnant women over thirty who do not accept amniocentesis. As genetic tests become widely used, they are established as the standard of care because they are widely used. A government task force has recommended changing that quick dissemination pattern to one where the accuracy of the test is proven by standard trials before it is made available in clinical practice.[8] This recommendation is not currently in force. Users of genetic tests need to exercise discernment as to their accuracy and meaning.

The precision and ease of many of the tests is increasing daily. For example, the first test for Huntington disease (HD) required genetic samples from relatives. While samples from affected relatives can still be helpful in borderline cases, the current test for HD requires a DNA sample only from the person seeking the test. When specific tests are referred to in this book, the information will be accurate for the time of writing, but probably already dated by the time the book is in print. References to specific tests that follow are to help describe consistent patterns and concerns that continue. The latest

5. Post, "Education for a Too-Hopeful Public," p. 226.

6. Peter S. Harper, "Direct Marketing of Cystic Fibrosis Carrier Screening: Commercial Push or Population Need? *Journal of Medical Genetics* 32 (1995): 249-50.

7. Kimberly A. Quaid, "Implications of Genetic Susceptibility Testing with Apolipoprotein E," in *Genetic Testing for Alzheimer Disease*, pp. 118-19.

8. Neil A. Holtzman, Patricia D. Murphy, Michael S. Watson, and Patricia A. Barr, "Predictive Genetic Testing: From Basic Research to Clinical Practice," *Science* 278 (24 Oct. 1997): 603.

information for readers on the capability of particular tests will have to come from professional sources that are updated daily such as www.geneclinics.org.

To Treat Disease

A *diagnostic* test is used to discern the cause of a present set of symptoms. Testing to see why a particular newborn appears jaundiced would be diagnostic. In contrast, a *screening* test is applied for an entire population without presenting symptoms. A screening test for phenylketonuria (PKU) is part of standard care for every newborn in all fifty of the United States.[9] It is a rare disorder found in only one in eleven to fifteen thousand births, but its effects are devastating. It used to be that about 90 percent of those affected ended up with IQs of less than 50. The benefit of the screening test is that when children are found to have the genetic anomaly, they can be placed on a diet that avoids phenylalanine. This keeps the phenylalanine from building up to toxic levels in their bodies. The required food is often described as expensive and bland, but staying on the diet makes a striking difference. On such a diet the individual with PKU may still experience some learning disabilities but can be expected to have close to a normal IQ. Screening does not cure the condition. What the test provides is a warning that the newborn's diet needs special attention. Physicians can be alerted by genetic testing to a number of conditions in time to reduce disease damage. This can be accomplished by triggering heightened vigilance and by better targeting treatment.

Vigilance

The first genetic screening test to be widely used to prevent harm from a genetic condition was the just-mentioned test for PKU. Recently a test for the more common condition of hereditary hemochromatosis has become available. Hemochromatosis causes excessive iron storage and can be lethal. The disease occurs in about four out of one thousand Caucasians. Treatment is simply draining off blood. If that is done regularly at an appropriate interval, the disease does no damage. The underlying condition is not cured, but vigilant treatment called for by the genetic test results can protect the patient from harm.

9. Diane B. Paul, "The History of Newborn Phenylketonuria Screening in the U.S.," in *Promoting Safe and Effective Genetic Testing in the United States*, p. 137.

Genetic test results can also affect how people time the start and frequency of disease screening. The following are three examples. First, breast cancer can be effectively treated about 95 percent of the time if found early and less than 20 percent of the time if not discovered until later stages. The American Cancer Society recommends that a woman have her first mammogram at forty and annual examinations starting at fifty. In the families that carry a genetic predisposition to breast cancer, half of the breast cancer cases occur by the age of forty-six.[10] The generally recommended screening pattern begins too late for such families. Being part of a family with a history of breast cancer and carrying particular gene variations associated with such is a strong indication for beginning mammograms at an earlier age and more frequent pace. One could make the further case that if a patient is considering bilateral mastectomy (surgical removal of both breasts) and oophorectomy (surgical removal of the ovaries) as prophylactic measures because breast cancer has been common in the family, a surgeon should know that the patient indeed has a positive breast-cancer allele test before doing the extensive surgery. Without the gene, the potential surgical patient still has a risk for breast cancer, but not the high familial one that may be motivating such a radical effort at prevention.[11]

A second example is that of familial adenomatous polyposis (FAP). It is characterized by the presence of hundreds of polyps in the colon and rectum. Each polyp can be the starting point for cancer. FAP is inherited as an autosomal dominant with greater than 90 percent penetrance. Penetrance is the rate at which having a specific genetic variation leads to a particular disease. So there is almost a 50 percent chance of manifesting the disease if one parent has it. If one carries that allele, regular endoscopy is worth the expense and discomfort to catch and remove polyps before they become cancerous. However, if the genetic test shows that the individual did not inherit the FAP variation of the gene, there is no need for frequent, uncomfortable, and expensive endoscopy.[12]

A third instance is that of retinoblastoma, a cancer of the retina. If un-

10. Varian, "Genetics in the Marketplace," p. 67.

11. Another set of questions and concerns is raised if the patient knows that she has tested negative for the heightened family risk of cancer but wishes to proceed with the radical procedure anyway, having seen what cancer did to her family members and not wanting to accept even the general population risk.

12. S. M. Powell et al., "Molecular Diagnosis of Familial Adenomatous Polyposis," *New England Journal of Medicine* 329 (1992): 234-39; G. Petersen, "Knowledge of the Adenomatous Polyposis Coli Gene and Its Clinical Application," *Annals of Medicine* 26 (1994): 205-8.

treated, it leads to at least blindness and frequently death. When a genetic test alerts caregivers to the presence of the allele for retinoblastoma, monthly scans of the retina can be done to detect the minute start of expected cancers. They can then be removed by laser. Again the underlying condition is not cured. Lifelong vigilance is required, but one can see and live. Without the test life would probably be cut dark and short.

Targeting Treatment

Genetic knowledge can also affect daily behavior. The alpha-1 antrypsin "z" allele increases the risk of emphysema in those who smoke. "In Sweden one in ten teenagers told that they have the zz genotype smokes, compared to one fifth of those in the control group."[13] Genetic diagnostic testing can sometimes help physicians to select from the start the treatment that will be most likely to help the individual. This treatment targeting may be used in at least three ways.

First, genetic tests can be used to help confirm a diagnosis. For example, the test for variations of a gene dubbed APOE is useful to confirm Alzheimer's disease (AD) as the source of dementia but inadequate to predict AD in someone asymptomatic.[14] One can carry the APOE4 allele associated with AD, live to an advanced age, and never develop dementia. This appears to be because the disease is the result of the interaction between several genes and the environment.[15] While certain variations at APOE are associated with the disease, having an allele associated with the disease does not confirm that you will have the disease or when.[16] Such a test is only predictive for the 1 to 2 percent of cases that are in families with an autosomal dominant type that usually has onset in the forties. Now if someone has clear signs of dementia already present and carries a double dose of the associated APOE4 allele, the possibility that the dementia is from AD in-

13. Steve Jones, *Genetics in Medicine* (New York: Milbank Memorial Fund, 2000), p. 15.

14. Allen D. Roses, "Genetic Testing for Alzheimer Disease," *Archives of Neurology* 54 (Oct. 1997): 1226-29; Stephen G. Post et al., "The Clinical Introduction of Genetic Testing for Alzheimer Disease: An Ethical Perspective," *Journal of the American Medical Association* 277, no. 10 (1997): 832-36.

15. Quaid, "Implications of Genetic Susceptibility Testing with Apolipoprotein E," p. 124.

16. Eric S. Lander, "Scientific Commentary," *The Journal of Law, Medicine and Ethics* 26, no. 3 (1998): 187.

creases.[17] In the United States three to four million people are alive today who have been diagnosed with AD. Autopsies have confirmed that about 10 to 20 percent of cases diagnosed as AD have actually been caused by other pathologies.[18] A more accurate diagnosis might help avoid ineffective treatment and aid future planning. Such a genetic test is not by itself adequate to predict whether or when dementia will occur, but it may be useful to improve the accuracy of diagnosis for someone who has signs of dementia. Whether this particular test serves that purpose effectively is currently a point of much discussion.

Second, genetic tests can help discern the distinct physiology of particular persons. While human beings have the same genes, variation within genes can make a difference in how an individual's body responds to a particular drug. For example, 10 percent of Caucasians and 2 percent of Asians do not respond to codeine due to a genetic mutation that disables a particular enzyme. Prescribing codeine to these people will be ineffective. Some blood pressure medications work well for 30-35 percent of the population while having virtually no effect on the blood pressure of other patients.[19] The people who are not affected by one drug may be greatly helped by another. Prescription is often a trial-and-error search for the right fit. Reading an individual's genetic code could be helpful to select directly or design the drug that would be most efficacious for that particular person. It can also warn against the use of some interventions. Women who carry the blood-clotting factor variant V Leiden are at much greater risk than others of blood clots when taking oral contraceptives.[20] A genetic test that recognized the presence of the allele would warn against using that contraceptive. Already in standard use is a genetic test to protect patients scheduled to use suxamethonium, a common muscle relaxant in surgery. The drug is safe for most people, but lethal for a few "who have a rare version of a gene involved in nerve transmission. Now all are tested before the drug is used."[21]

Third, genetic testing of attacking organisms might also lead to antibiotics effective on the first try. Rather than use broad spectrum antibiotics

17. Allen D. Roses argues to 95-99 percent. "A New Paradigm for Clinical Evaluations of Dementia: Alzheimer Disease and Apolipoprotein E Genotypes," in *Genetic Testing for Alzheimer Disease*, p. 51.

18. Roses, "A New Paradigm for Clinical Evaluations of Dementia," p. 37.

19. Varian, "Genetics in the Marketplace," p. 71.

20. I. Martinelli et al., "High Risk of Cerebral-Vein Thrombosis in Carriers of a Prothrombin-Gene Mutation and in Users of Oral Contraceptives," *New England Journal of Medicine* 338 (1998): 1793-97.

21. Jones, *Genetics in Medicine*, p. 11.

or a series of more focused agents, one could analyze the bacteria or para-sites, and prescribe the most effective drug to attack that particular organism the first time. Genetic testing might also eventually provide a way to genetically recognize and target cancerous cells specifically. That would be vastly superior to our current techniques of broad-scale poison, slash, or burn. Diagnostic genetic testing is increasingly useful for vigilance and targeted treatment.

To Plan for the Future

Currently most genetic tests do not lead to treatments but do help predict future health. Such information can have positive uses and significant dangers. That mixed picture is compounded by the test results coming in probabilities. Probabilities can be difficult to explain and comprehend.

The Problem of Probabilities

Predictive Results Come in Probabilities

Genetic test results are characterized by probabilities. This is due first to the challenge of accurately reading the genetic sequence. Tests vary both in their sensitivity and specificity. *Sensitivity* refers to catching all the actual cases of a gene variation, an allele, that causes the disease. For a highly sensitive test if one has the allele, it will be noticed by the test, but there may be many false positives. In contrast, *specificity* is the test characteristic of marking only actual cases. A test that is specific will have few false alarms. If the test says that one has the allele, one probably does, but the particular test may not be sensitive and hence may miss many actual cases. For example, the alpha feto-protein (AFP) screen, which is part of the "triple test" often given to pregnant women, is very sensitive. It will pick up almost all cases of neural-tube defects. However, it is not very specific. Fifty out of every thousand women will have test positives, but only one of those fifty positives will have a confirmed diagnosis. The ideal is one inexpensive test that gives an alarm for every actual case and only an actual case. Short of that, if the concern is serious, one would want high sensitivity to be sure that there is an alert if the allele is present and then a follow-up test with high specificity to rule out the false alarms.

The second reason that genetic test results come in probabilities is the inherent nature of the complex and interactive genetic system. Even a perfect reading of sequence does not so much reveal what will happen as that one possibility is more likely than another. A test that is quite right that you bear a particular allele associated with a disease does not mean that you will get the disease. An allele can be present without being manifested. The degree to which it is expressed is its penetrance. Huntington disease (HD) was long assumed to be fully penetrant. If you had the autologous dominant allele, you would eventually suffer HD. Now it is clear that HD mutation has to do with the number of sequence repeats at a particular location. Thirty-five or less repeats will not cause the disease. Forty or more repeats will. It is present but not fully penetrant in the thirty-six to thirty-nine repeat range.[22]

The age of onset and severity are other variations. Cases of trisomy 21 (Down syndrome) are clearly diagnosed, but with no sense of how severely it will be expressed, which can range from mild to devastating. A confirmed diagnosis of Down syndrome does not reveal if the child will have average intelligence or minimal ability. While HD, FAP, and autosomal dominant polycystic kidney disease (ADPKD) are all autosomal dominant, they also differ in penetrance, prognosis, and pathology. For example, FAP has earlier onset than HD and can be treated. "Genetic disorders encompass a wide range from mild to severe to totally dysfunctional to lethal, with clinical manifestations ranging from prenatal to late-life onset and from gradual to fulminant in course."[23]

The genetic picture is yet further complicated by the interaction of multiple genes with each other and the environment. Variations of BRCA1, BRCA2, and p53 can all cause breast cancer, but together account for only 5 percent of breast cancer cases. If a woman has a cancer-causing allele of BRCA1 she has a 50 to 85 percent lifetime chance of breast cancer. Without the allele she still has the general population lifetime risk of 11 percent.[24] Genetic tests reveal propensity. They show an increase or decrease in the likelihood that one might suffer a disease. If one twin of genetically identical twins has diabetes mellitus, the other twin has a one in three chance of eventually developing the disease. That is much higher than the risk of the general popu-

22. David C. Rubinsztein et al., "Phenotypic Characterization of Individuals with 30-40 CAG Repeats in the Huntington Disease Gene Reveals HD Cases with 36 Repeats and Apparently Normal Elderly Individuals with 36-39 Repeats," *American Journal of Human Genetics* 59 (1996): 16-22.

23. Charles R. MacKay, "Discussion Points to Consider in Research Related to the Human Genome," *Human Gene Therapy* 4 (1993): 480.

24. Gail Geller, "Genetic Screening for Cancer Susceptibility," Ethical Boundaries in Cancer Genetics conference, St. Jude Children's Research Hospital, Memphis, 27 May 1999.

lation, but two out of three identical twins never develop the diabetes mellitus that affects their twin with identical genes.[25]

Genes are deeply influential but do not act unilaterally. Hence accurate genetic tests usually cannot tell a person conclusively what disease will come, when it will come, or how severely. Eric Juengst creatively describes the predictive power of genetics as being like a weather forecast.[26] This is not as disparaging as it would sound in summer Oxfordshire. Juengst is assuming substantial accuracy that remains carefully qualified. Genetic testing reveals part of what the person faces, but not how she or he will precisely experience it.

Most People Find Probabilities Difficult to Understand

Most people seem to have difficulty comprehending probabilities. This is evident in the popularity of state-sponsored lotteries. The problem is not limited to patients. Alice Wexler writes of seeing a notation on the medical record kept by her gynecologist that "the patient has a high probability of developing Huntington's disease." Actually her chance of the disease was decreasing as she passed the age of usual onset without symptoms and at most her risk had been 50 percent, since she had one affected parent.[27] That is only one instance, but it exemplifies a lack of preparation that concerns many who train and license physicians. "Most practitioners in primary care have not had a single hour of instruction in genetics as part of their formal training."[28] Medical schools are gearing up the genetics portion of medical education, but in an already crowded curriculum it is difficult to give genetics and genetic counseling the attention it increasingly warrants.

The physician Neil Holtzman did a study of physicians and medical students at the prestigious Johns Hopkins School of Medicine. He asked, "If a test to detect a disease that occurs in one of 1,000 people is falsely positive in 5 percent of unaffected people, what is the chance that a person found to have a positive result actually has the disease?" Eighty-four percent of the first-year students gave a correct answer of 2 percent.[29] Seventy-three percent of the

---

25. Daniel Drell, "FAQs," *Human Genome News* 9, nos. 1-2 (1998): 4.

26. Eric T. Juengst, "Gene Therapy in Cancer Treatment and Prevention," Ethical Boundaries in Cancer Genetics conference, St. Jude Children's Research Hospital, Memphis, 28 May 1999.

27. Alice Wexler, *Mapping Fate: A Memoir of Family, Risk, and Genetic Research* (New York: Random House, 1995), p. 231.

28. Francis S. Collins, "Shattuck Lecture — Medical and Societal Consequences of the Human Genome Project," *New England Journal of Medicine* 341, no. 1 (1999): 35.

29. In this example, if 5 percent of those tested receive a false positive, that would be 50

third-year residents were also right, but only 49 percent of the residents, interns, and practicing physicians in continuing education ascertained the answer.[30] Holtzman attributes this downward trend to the observation of cognitive psychology that experience distorts the perception of probabilities. The more experienced physicians are used to tests confirming a set of symptoms that have already brought the patient into the clinic or hospital. Their expectation is for a much higher rate of accurate positives and that tends to affect their calculation. The prominent *New England Journal of Medicine* recently published a study that found that physicians who had ordered a genetic test for colon cancer misinterpreted the results 31.6 percent of the time.[31] Before the difficult process of building patient understanding has even begun, almost one-third of the cases were already misunderstood. The seriousness of this result is compounded by the observation that this is a group of physicians who sought out the test. It would be reasonable to expect that physicians who knew of the genetic test and ordered it would have a better chance of interpreting it correctly than practitioners who did not order it. That implies that there might be an even larger percentage of physicians not ready to use or explain these tests accurately.

Turning to the person tested, Nancy Wexler writes about the perception of probabilities for the person at risk of HD.

> The ambiguous condition of 50% risk is extremely difficult to maintain in one's mind, if not impossible. In practice, a 50-50 risk translates to a 100% certainty that one will or will not develop the disease, but the certainty changes from one to the other from moment to moment, day to day, month to month. It can be helpful to discuss this phenomenon with counselees so that they know that fluctuations in their convictions are a normal part of the coping process.[32]

We described above that the familial genetic variation for breast cancer can increase a woman's risk from the 11 percent risk of the general popula-

---

people out of the 1,000 tested. If only one in 1,000 have the disease, then there are probably 49 false positives surrounding the one positive that will be confirmed as an actual case. One confirmed case out of fifty initial positives is a rate of two per one hundred, that is, 2 percent.

30. Neil A. Holtzman, *Proceed with Caution: Predicting Genetic Risk in the Recombinant DNA Era* (Baltimore: Johns Hopkins University Press, 1989), p. 161.

31. F. Giardiello, J. Brensinger, G. Petersen, M. Luce, L. Hylind, J. Bacon, S. Booker, R. Parker, S. Hamilton, "The Use and Interpretation of Commercial APC Gene Testing for Familial Adenomatous Polyposis," *The New England Journal of Medicine* 336 (1997): 823-27.

32. Nancy Wexler in *Genetic Counseling: Psychological Dimensions*, ed. Seymour Kessler (New York: Academic, 1979), p. 218.

tion to 50 to 85 percent. Without one of those alleles her odds are markedly better than with one of them, but she may still get breast cancer. It would be a false sense of security to think that one was less susceptible than most because one had escaped the high-propensity allele in the family. This would be true as well for other genetic conditions that have both high-risk family lines and general population risk. If one does not have the Alzheimer's disease (AD) allele that runs in one's family with a 50 percent risk of early onset, one still has the general population's risk for other forms of AD.

Cultural misconceptions can also make it difficult to understand probabilities. Rayna Rapp reports that one ethnic group in New York City clinics consistently entered with the conviction that if a daughter was pregnant, the mother could not be.[33] Because comprehension involves integrating new information with what one already knows, mistaken expectations can befuddle attempts to understand accurately. If the knowledge base one is building on has misconceptions, it is even more difficult to conceptualize and evaluate statistical information.

Probabilities Counseling

Takes Time The complexity of genetic testing probabilities and the importance of understanding its implications usually require significant counseling time both for informed consent before testing and later for assimilating the test results. It is not sufficient to simply present the information, even though often in medical care the patient does not have to understand the intervention to benefit from it. This may be the case as well for some aspects of genetic testing that affect how the physician treats the patient, such as a genetic test to discern which antibiotic will be most effective for a particular infection. The patient just needs the prescription, not an explanation of how the physician discerned which antibiotic would be the most appropriate. However, the usual case for genetic testing is that it is precisely the tested individual's understanding that is the goal. There is no purpose in the exercise unless the information is understood by the one tested. If the test is used to help a patient choose between treatments or to plan ahead, it becomes imperative that the patient understand what it means. Otherwise there has been no point in doing it.

The probabilities of genetic test results are not easily absorbed for most people. One of the most effective ways to express probabilities may be by parallels to familiar experiences. Rather than only give straight numbers incom-

---

33. Rayna Rapp, *Testing Women, Testing the Fetus: The Social Impact of Amniocentesis in America* (New York: Routledge, 1999), p. 83.

prehensible to many people, the counselor might add, for example, that a particular risk is about the same as that of having a car accident driving from the counsel site to the state capital. If care is taken that the comparison is accurate, such a familiar example may communicate far more to most people than straight numbers.[34] Seeking genuine understanding takes time. On average, counseling a patient for chromosome or molecular genetic testing takes about one hour.[35] As of 1996, Thomas Murray estimated that there were only about a thousand genetics professionals, M.D. geneticists, and genetic counselors in the United States.[36] Time-consuming provision and explanation of the tests falls increasingly on physicians in general practice. They have little preparation to do it and are not generally reimbursed well for conveying information. Most reimbursement schemes tend to reward procedures. One solution is to expand the number of genetic counselors. They are trained at the masters level to provide time-consuming education and emotional support. This needed service can be intense and draining.

> If the individual knows he or she is at risk [for HD], the years in anticipation of this change can be years of dread, of silent apprehension, of noisy emotional disarray, or of intense productivity. As is frequently the case with a late onset genetic disorder for which there is neither a screening test nor treatment, individuals at risk are disease wise and doctor shy. They know the odds from an early age and often regard the genetic counselor with the same contempt the battlefield soldier shows for the journalist — ticking off in round numbers the daily toll of lives lost, reducing the human struggle to arid statistics. Nor is the psychological counselor seen as a source of refuge. Mental health workers are considered to be only for the mentally deranged while those at risk perceive themselves to be suffering from a physical threat.[37]

Besides education and psychological support, an important part of the work for whoever gives counsel is to anticipate and diffuse moral conflicts be-

---

34. Granted, noting these parallels might make the consultands a bit more apprehensive on their next car trip.

35. L. C. Surh et al., "Delivery of Molecular Genetic Services Within a Health Care System: Time Analysis of the Clinical Workload," *American Journal of Human Genetics* 56 (1995): 760-68; B. A. Bernhardt and R. E. Pyeritz, "The Economics of Clinical Genetic Services," *American Journal of Human Genetics* 44 (1989): 288-93.

36. Thomas H. Murray, introduction to *The Human Genome Project and the Future of Health Care*, ed. Thomas H. Murray, Mark A. Rothstein, and Robert F. Murray Jr. (Bloomington: Indiana University Press, 1996), p. ix.

37. Nancy Wexler in *Genetic Counseling*, p. 200.

fore they happen. We will look at misattributed paternity in chapter 8 as an example of unexpected information genetic testing can reveal. Another part of counsel is to clarify that confidentiality is not an absolute when others are at significant risk. These possibilities should be raised before testing in order to have the informed consent discussed in chapter 4. Also people should have a choice not only in whether they are tested, but in what information they receive. Even good results should usually not be forced on the person tested. If the counselor presses good results but not bad ones, people will know the results by whether the counselor insists that the consultand hear them. Giving good results over the phone and bad results by appointment quickly becomes evident to the consultands through avenues such as the waiting-room grapevine or relatives.

Striving for Neutrality Genetic counseling to date is characterized by an emphasis on being nondirective. This has been particularly important to the development of genetic counseling because of the recent history of forced sterilization in some countries and murderous regimes in others that claimed genetic health as justification. If genetic counseling is completely nondirective, without force in any direction, it cannot be coercive. A common response to requests for actual decision advice is for the counselor to say, "That is up to you and your physician." There have been groups that declared that some lives were not worth living. The genetic counselor just tells the parents what their life or the life of the fetus is likely to be like. Some regimes have killed citizens ostensibly for the genetic health of the wider population. Counselors leave life and death decisions up to patients and parents. Governments have sterilized people against their will.[38] Genetic counselors are invited to give information to people who request it. For current counseling practice the dominant model is to emphatically reject the traditionally more directive physician/patient relationship.

> [T]he traditional model of the physician-patient interaction is, generally, inadequate for the genetic counseling situation. In genetic counseling, the usual goals are to promote autonomy and to give priority to individual rather than societal needs. To achieve such goals, the genetic counselor will need to reconcile competing values as well as conflicting personal and professional beliefs and philosophies. For some counselors, particularly physicians, these conflicts may be particularly keen. Trained as helping professionals, they are often eager to take command so as to be able to carry out

---

38. Avoiding such societal abuses will be addressed in chapter 15.

their helping functions. However, what constitutes "help" in, let us say, reducing a fracture is not the same as helping a person reach a decision about a matter in which it is unclear.[39]

Recently there have been calls to temper the nondirective model as unrealistic and not as helpful as it could be. Patient "response may depend on how the information is presented. Thus, in one British survey, 10 percent of women responded to a letter offering a cystic fibrosis carrier status test by mail, a quarter to a test made during a medical appointment, and almost three times as many when offered an immediate test, face to face."[40] Elisabeth Beck-Gernsheim quotes Dietrich Ritschl as saying that "the concept of an ethically neutral counselor in medicine is a fossil from the positivist age of naive ideals concerning science."[41] Even the facts selected to be taught convey values. The fact that there was a genetic test and an appointment implies that attention is warranted. That a procedure is offered means having to choose whether to use it and that it is within the realm of reasonable choices carried out by the medical system. Genetic services do not offer options they consider immoral. Also the way information is presented can be quite influential. If both parents are heterozygous for a recessive gene, the counselor can say each of their children has a one in four chance of the disease occurring, or the counselor can say each child has a three to one chance in favor of not having the disease. Stating that "there is a one in ten chance of affliction" is perceived differently from "nine times out of ten there is no problem." A conscientious counselor will need to state information a number of different ways to achieve understanding while avoiding undue influence. Even care with word choice is not enough in that voice inflection and body position communicate as well.

Claiming to be completely nondirective while pursuing such personal and formative information camouflages unavoidable counselor perspective. That makes it more difficult for consultants to screen influential convictions contrary to their own. Attempting to be completely nondirective restrains the

---

39. Kessler, ed., *Genetic Counseling: Psychological Dimensions*, p. 5.

40. Jones, *Genetics in Medicine*, p. 14.

41. Dietrich Ritschl, "Die Unscharfe ethischer Kriterien," in *Medizinische Genetik in der Bundesrepublik Deutschland*, ed. Traute M. Schroeder-Kurth (Frankfurt, 1989), p. 136, as quoted by Elisabeth Beck-Gernsheim, *The Social Implications of Bioengineering*, trans. Laimdota Mazzarins (New Jersey: Humanities Press, 1995), p. 94. Beck-Gernsheim also cites on the problems of nondirective counseling Wolfgang van den Daele, "Techische Dynamic und gesellschaftliche Moral," *Soziale Welt* 2, no. 3 (1986): 155ff., and "Das zahe Leben des praventiven Zwangs," in *Der cordierte Leib*, ed. Alexander Schuller and Nikolaus Heim (Zurich, 1989), pp. 221ff.

offer of professional advice that the one tested may want and keeps counselors from calling for the consideration of others besides the patient immediately before them. It also fails to acknowledge that most counseling centers do take ethical stands. For example, many avoid revealing information that might lead to selective abortion for gender, and many will not test children for late onset disorders that have no current treatment. The nondirective model does well in encouraging a comprehensive presentation of the test information, available options, and how they are most often evaluated. This should be pursued with full respect and fairness. However, counselors should also be honest about their own conclusions and practice policy and their reasons for them. This allows clients to gain from the counselor's considered perspective and to filter that perspective if it is founded on convictions contrary to their own. Rather than claiming utter neutrality, a more attainable goal may be to take care to respect those who have come for counsel and give them full available information. Particular communities may find it helpful to sponsor counseling that openly acknowledges their community's worldview, commitments, and resources.

Value and Risk of Predictive Testing

Planning Ahead

There are substantial gaps between recognizing an allele by genetic test and understanding how it works or repairing it. If one already has symptoms of a disorder, a genetic test may confirm diagnosis. Diagnosis does not of itself offer cure or amelioration of the disease, but it can still be useful information. Knowing some sense of the likely disease course can also give a better sense of appropriate and timely medical treatment. Efforts can be more focused. In the case of AD, for example, diagnosis gives a likely time frame for how soon decisions need to be made. This encourages forethought as to what would be the best plan for care, coordination of support, and present use of resources. One can also learn from the experience of others with that disease. One can then begin to plan with those expectations in mind. Surveys of AD advocacy groups consistently reveal that most patients and their families want accurate diagnosis so that the patients can play a major role in determining their own affairs and planning with their families before the dementia worsens.[42]

Other kinds of medical tests can offer information about the likely fu-

---

42. Roses, "A New Paradigm for Clinical Evaluations of Dementia," p. 44.

ture, but genetic testing dramatically increases the number of situations where one can know of a likely future disease long before any actual symptoms. That makes possible planning before the need for it would have been otherwise realized. Genetic tests offer less fate and more responsibility. Planning in the light of diagnostic and predictive genetic tests can affect decisions of whether or how to have children, provision for the future, and completing unfinished business that may be out of reach later. One might move near a major research site or extend oneself less financially. It would be prudent to be sure one has the best available insurance — although the latter is the adverse selection that insurance companies fear. We will look at that more closely in chapter 9.

In such planning it is important to remember the distinction between being a carrier and having a disease. Because genes come in pairs, a person often carries a defective gene that is paired with and covered by a working gene. Individuals might pass the potentially disease-causing allele on to their biological descendants but might very well not have the disease at any point in their life.

Unexpected Information

As in the research discussed in chapter 4 or any other medical intervention, informed consent is important out of respect for the person. Such informed consent will require understanding of the implications of the test before it takes place. One often crucial aspect of genetic testing is unexpected information. A test for one condition may reveal another. The human body is intricately interactive. Phenotype often stems from the interaction of multiple genes, and one gene can have multiple effects throughout the body. There can be many implications from a single test, since one gene can be involved in several diseases. Both heart disease and Alzheimer's are associated with APOE4. One's APOE alleles affect the transport of cholesterol, hence risk for heart disease.[43] When the cardiologist orders the APOE test for heart risk, the result can reveal a possibly higher risk for AD. It is true, as discussed earlier in the chapter, that APOE alleles associated with AD are much more helpful for confirming the source of present dementia than in predicting dementia presymptomatically.

For informed consent the patient must understand that the test at hand is revealing information not only about the presenting condition but also about whatever else it addresses. Eric Juengst emphasizes that this means spe-

---

43. A little more precisely, the apolipoprotein E genotype partially reveals risk for hyperlipidemia and its sequela atherosclerosis and myocardial infarction.

cialties outside of genetics, in the case of APOE cardiologists and neurologists, will need to think about the implications of tests and their patients' understanding of those implications beyond the immediate focused investigation.[44]

This situation is compounded by the development of "gene chips" that do comprehensive genetic screening. In the interest of efficiency, alleles associated with many different conditions will probably be examined on one gene chip at the same cost as testing for one allele. Part of orientation and informed consent should be to ask the consultands if they want all available information, including what was not directly sought. The point is to honor the consultand's freedom to have or not have information. Information can have many implications.

Psychological Impact

There can be great relief when one is tested and found not to be carrying a genetic variation that has afflicted family members. Such news can free one for long-term planning, bearing children, from expensive or invasive monitoring tests, or from building one's life around a disease that will never occur. Even good news, however, can have its stresses. This is particularly true if the recipient hoped good news would have more impact than it did, had made major decisions that are now recognized as mistaken and irreversible,[45] or feels guilt at surviving what has afflicted family members. One study showed that 10 percent of those found to be free of HD risk had difficulty adjusting to their new status.[46] They had worried or made sacrifices for nothing. They were free, while people they cared about were not. Being free from risk did not meet their expectations of how much better life would be.

When the news is bad, self-deception can be the method of coping. Alice Wexler tells how her mother, despite her own laboratory work in genetics and extensive contrary literature, clung to the erroneous assertion in one neurology text that only males were at risk of inheriting Huntington disease.[47] People often hear only what they want to hear. For those who do accept the high risk confirmed in their genes, normal setbacks and mistakes can become questions

44. Eric T. Juengst, "The Ethical Implications of Alzheimer Disease Risk Testing for Other Clinical Uses of APOE Genotyping," in *Genetic Testing for Alzheimer Disease*, pp. 187-89.

45. Harry Karlinsky, "Genetic Testing and Counseling for Early-Onset Autosomal Dominant Alzheimer Disease," in *Genetic Testing for Alzheimer Disease*, p. 113.

46. Sandi Wiggins et al., "The Psychological Consequences of Predictive Testing for Huntington's Disease," *New England Journal of Medicine* 327 (1992): 1402-5.

47. Alice Wexler, *Mapping Fate*, pp. 223-39.

of whether one is manifesting the disease. Constant self-monitoring can be disheartening and exhausting. Different diseases have different dominant concerns. "Every disease calls forth particular images and fears in its victims and potential victims. Cancer evokes the threat of pain and suffering, unpredictability is the hallmark of multiple sclerosis. For the person at risk for HD, the relevant metaphor is the time bomb."[48] When receiving genetic test results people need factual information but also emotional coping support and skills. Follow-up is crucial in that the full emotional impact of bad news from a genetic test, as for any bad news, is usually well after the news is first heard.[49] The most difficult aspect for some people is guilt at the possibility of having passed on a risk factor to their children and how it will affect them.

Yet even with bad news, it can be reassuring to at least know what one is up against. Alerted to a likely problem, people often find it helpful to seek experience and counsel from people at various stages ahead in the disease progression. Those ahead can pass on ideas and equipment for resolving issues that will no longer be a surprise. For example, increasing depression and enervation can occur in HD long before a clear case develops. Knowing of this possibility ahead of time by genetic testing can allay some of the attendant fears of the unknown when it occurs. The unsteady weaving of someone's walk can be recognized as HD instead of attributed to drunkenness. Understanding what is happening can be the first step in constructively coping both for the person with the condition and for caregivers.

Choosing to Test

Nancy Wexler devoted years of her life to developing a test for HD with the full expectation that she would take it as soon as it was available. When it was, she did not.[50] In Britain only about a tenth of those with a parent with HD choose to take the presymptomatic test.[51] Breast cancer is a much more common condition. In one group of women, 94 percent of those asked said they would be interested in having a BRCA1 test done for breast cancer susceptibility. After counseling, 7 percent actually took the test.[52] The study did not

---

48. Nancy Wexler in *Genetic Counseling: Psychological Dimensions*, p. 201.

49. Karlinsky, "Genetic Testing and Counseling for Early-Onset Autosomal Dominant Alzheimer Disease," pp. 112-13.

50. Alice Wexler, *Mapping Fate*, pp. 223-39.

51. Steve Jones, *Genetics in Medicine*, p. 12. See also Kimberly A. Quaid and Michael Morris, "Reluctance to Undergo Predictive Testing: The Case of Huntington Disease," *American Journal of Medical Genetics* 45 (1993): 41-45.

52. Geller, "Genetic Screening for Cancer Susceptibility."

parse out why the dramatic falloff. It could have involved cost, logistics of multiple trips for pre- and post-test counseling, risk to medical insurance, or other factors. Currently testing tends to be expensive, and once the knowledge is gained, it cannot be undone for the recipient or the medical record. The implications of the latter will be central to the discussion of insurance and employment in chapter 9. Genetic testing can be extremely useful but carries significant risks at this time. In particular, "until reforms are implemented, individuals should seriously consider the implications of a positive genetic screening test, even if only a susceptibility conferring test, on their ability to access both health and life insurance before agreeing to be tested in any setting."[53]

Resources from the Christian Tradition

A text from the Wisdom literature inherited by the Christian tradition states that wise people look ahead.[54] Genetic tests can provide another tool to better care for the God-created and animated bodies that we are. Some might argue that the greatest trust in God may be shown in not pursuing such information. God will provide when the time comes. On the other hand, fully recognizing God's gracious provision, it may be our responsibility to do what we can. This was a theme described in some detail in chapter 3. On that basis one should pursue genetic testing whenever it would help one's assigned task of better sustaining, restoring, and improving the physical form that we each have and our care for others. PKU detection and change in diet to protect the brain from harm would be an example of sustaining. Detecting the genetic variation for retinoblastoma and lasering cancers of the retina as they appear is an example of restoring the body. Attempts to improve physical life by the use of genetic tests will be raised in chapter 8. Of course, how test results might help needs to be balanced with any serious risk such as loss of insurance in the current system.

Christians are called to love their neighbors. Most of us have many people who depend on us. Genetic testing can help us provide for them by informing action to avoid the effects of genetic disease, ameliorate such effects, or account and plan for them. The surgeon who regularly cuts herself, hence

53. Nancy E. Kass, "The Implications of Genetic Testing for Health and Life Insurance," in *Genetic Secrets: Protecting Privacy and Confidentiality in the Genetics Era,* ed. Mark A. Rothstein (New Haven: Yale University Press, 1997), p. 314.

54. Proverbs 14:8.

exposing patients to her blood, should be sure that she is clear of HIV. The airline pilot at risk of HD must think of the people who depend on her judgment and coordination. To move to a less desirable job can be a significant loss, but there is still an obligation to think about the passengers who entrust one with their lives. The more common practice of driving a car is a menace to others if one is afflicted with some conditions. Usually genetic test results do not call for withdrawal but rather heightened vigilance. The disease may never be manifested at all.

When genetic tests reveal the probability of future difficulties, the challenge of why the good and powerful God would allow such is raised. It is not a new question. The suffering and death of Jesus Christ on the cross while he was fully innocent and within God's will is at the heart of the Christian tradition. While the Christian tradition does see some suffering as self-inflicted, there is no karma. Bad things do happen to good people. Also, contrary to the popular health-and-wealth gospel, suffering is not always the result of a lack of faith any more than it was for Jesus Christ, who suffered terribly. Ten classic responses as to why God allows suffering were described in chapter 3.

The responsible use of genetic testing does not require maximum use immediately. Denial of risk is not helpful long term, but postponing a test temporarily until one is ready to hear results may be quite appropriate. If one is already deeply depressed or unstable, it is not a good time to hear formative news, good or bad. It might be better to deal with the most fundamental issues that are troubling one first. Jesus began his ministry proclaiming that the kingdom of heaven was at hand,[55] not, "Behold, I am God among you." He told his listeners more of God and himself as they were able to hear it.

There need not be despair when genetic tests indicate probable difficulty in the future. This life is temporary. The classic tradition promises that God will well use everything in the lives of those who love God and live according to God's purpose.[56] That is not a promise that life will be easy or comfortable, but rather that it will be always useable to good ends. Also, one will never be alone. Christians are to pray to be equal to the task. The promise is that what is needed will be provided. The church is to play an important role of guidance and support through both good and difficult times.

55. Matthew 4:7.
56. Romans 8:28.

Chapter Summary

The number of available genetic tests is increasing rapidly. They can be used for confirmation of present diagnosis or prediction of future health. These results can be quite useful for better medical care and life planning. The information comes in probabilities that are often difficult to comprehend. Great care is needed on the part of those tested and their caregivers that the one receiving the test has given an informed consent to do the test and that the results are accurately understood afterward. Effects on life planning and self-perception can be marked whether the news is good or bad. Follow-up counseling can be crucial. Christians have confidence in the future because they trust its maker and have support in church community. Some of the rich resources of the Christian tradition and community for dealing with genetic testing will be described in more detail in the last pages of part II. Christians might do well to avail themselves of genetic testing when it can help in carrying out their mandate to sustain, restore, and improve their physical bodies and the world entrusted to them, out of love for God and neighbor. However, there will be times when valuable genetic testing might not be advisable, such as if it jeopardizes medical insurance. That risk will be described in the beginning of chapter 9.

CHAPTER 8

Testing Genes and Family

Family News

Sharing Genes

When one investigates an individual's genes, one is learning a great deal about that person's parents, siblings, and other relatives. Genetic testing of an individual tests a family. While genetic information is personal, it is shared with kin. Almost every human being is genetically unique,[1] yet the individual's genes have come from just two other people. Except for a few random mutations an individual has no gene variations (alleles) that were not present in one of two parents. This common heritage is also manifested in siblings, who are unique combinations from the same two-person source. Each parent also received genes from just two people. While our genetic heritage is quintessentially intimate to each of us, we are inextricably part of a family. In this chapter we will discuss ways that genetic testing challenges families and a family's procreation.

Current genetic testing often seeks direct family involvement. BRCA1, a gene that can have alleles associated with a heightened risk of breast cancer, has more than 50 variations. If family members who have the disease can be tested first, that reveals which variation to look for in the family. The first test of a family member currently costs about $2400. For each person thereafter, now knowing which specific allele to search for, the cost is about $200.[2] When

---

1. Through twinning there might be one or two other persons sharing the same genetic combination.
2. Gail Geller, "Genetic Screening for Cancer Susceptibility," Ethical Boundaries in

oligo-nucleotide probe arrays, "gene chips," become widely available, family will be less needed for diagnosis because one can then test all fifty-plus alleles quickly and cheaply. Even with such testing of multiple alleles, there will still be cases where direct measurement of relatives will be desired. Huntington disease (HD) was first predicted by a linkage test that was about 90 percent accurate and required blood from several elderly family members. Based on the actual gene discovered in 1993, HD can now be spotted by the number of repeats in a particular region of the gene. That can be measured with a sample from just one person. However, while HD is assured from CAG repeats of forty or above and there are no recorded cases with thirty-five or less repeats, thirty-six to thirty-nine repeats does not confirm whether one will manifest the disease or not.[3] In such borderline cases, testing family members who manifest the disease gives a better idea of how many repeats are necessary to have the disease in that particular family.

Genetic heritage in common also means that families share risk beyond the physical. For example, what is discovered about one family member can affect the insurance of all in the family. Genetic counselors should help those considering a genetic test to realize the implications for others.[4] The person tested is rarely the only person affected by test results.

Duty to Tell or Hear?

As it becomes possible for individuals to be genetically tested on their own, what obligation do they have to those who may share risk of disease and its implications? The results obtained by one person's initiative can be revealing of others. This can lead to conflict if those affected weigh the benefits and dangers of the information differently from the one considering the test. One individual may consider paramount the insights of the test results for planning, while another might consider the threat to maintaining insurance more important.

Cancer Genetics conference, St. Jude Children's Research Hospital, Memphis, 27 May 1999.

3. D. C. Rubinsztein et al., "Phenotypic Characterization of Individuals with 30-40 CAG Repeats in the Huntington Disease Gene Reveals HD Cases with 36 Repeats and Apparently Normal Elderly Individuals with 36-39 Repeats," *American Journal of Human Genetics* 59 (1996): 16-22.

4. David H. Smith, Kimberly A. Quaid, Roger B. Dworkin, Gregory P. Gramelspacher, Judith A. Granbois, and Gail H. Vance, *Early Warning: Cases and Ethical Guidance for Presymptomatic Testing in Genetic Diseases* (Bloomington: Indiana University Press, 1998), p. 28.

Familial Alzheimer Disease (FAD) is inherited as an autosomal dominant. Each child has a 50 percent chance of early onset dementia. Once one member of a family is so diagnosed, the presence of the allele is significant for other family members. If the consultand is found to have FAD, one of the parents must as well. Who prevails when a parent does not want to know FAD status but an adult child does? If the person tested has an identical twin, the test results will be true for the identical twin as well.[5] Families are often already stressed by the presence of the genetic disease or denial of it. In this charged situation some principles are needed to help family members and counselors sort out who can access what information when.

Out of respect for their own life choices, we usually do not force unwanted information on people. If they would rather not know when offered that option, one would need substantial reason to override their wishes and tell them anyway. Laboratories often produce reports with the entire pedigree visible to aid genetic counselors in interpreting the results to the tested individuals. Of course this means that if consultands see the chart, a great deal is learned about all the other members of the family, not just their own cases.[6] Confidentiality should not be so easily lost. Steps can be taken to focus information. Yet while confidentiality should be pursued, people wanting to know helpful information for themselves should not have that withheld. One would best start by seeking consensus among those most affected. Sometimes a little time is all that is needed for a consensus to develop. If consensus is unattainable, one person not wanting to know should not keep family members from obtaining information about themselves that they deem helpful.

Christians would not override another's wishes lightly both out of respect for the other and out of concern for the other's well-being. Having a right to do something is not the same as its being the right thing to do. One of the principles that the apostle Paul advocated in his first letter to the Corinthians was that one needs to consider not only if an action is acceptable of itself, but also how it will affect other people.[7] The particular case in Corinth was that Paul said meat sacrificed in pagan temples was just meat. It was acceptable to buy it in the marketplace. However, if a Christian sister or brother thought it was wrong to eat meat that had previously been sacrificed to idols and ate some anyway, for him it would be rebellion against God, a sin. The sin

5. Kimberly A. Quaid, "Predictive Testing for HD: Maximizing Patient Autonomy," *Journal of Clinical Ethics* 2 (1992): 239.

6. Task Force on Genetic Testing, *Promoting Safe and Effective Genetic Testing in the United States: Final Report of the Task Force on Genetic Testing*, ed. Neil A. Holtzman and Michael S. Watson (Baltimore: Johns Hopkins University Press, 1998), p. 14.

7. 1 Corinthians 8, 10.

would not be in the action itself but in the person's intent when he chose to do something that he believed God did not want him to do. Paul wrote that while it was not forbidden to eat such meat, one should not eat it if doing so might influence a brother to do what the brother believed to be wrong. A more familiar example might be that even if one sees no ethical problem in drinking some wine with a meal, one should still not offer wine to an alcoholic or drink it in front of him, if that might encourage him to do the same to his harm.

The opinion or preference of others is not a trump card. Such a standard would be impossible to meet since preferences of varied people so often conflict. However, there would need to be significant reasons to choose an action that would harm another in some way. One is to love one's neighbor as oneself. Certainly that includes one's family members. When a Christian encounters conflict within one's family over whether to test or not, every effort should be made to find a way to meet the concerns of all those affected by the decision. One should take seriously the needs of others. Having done that, if conflict remains, one proceeds as seems best.

Children as a Special Case

Almost every newborn is screened for phenylketonuria and congenital hypothyroidism.[8] These are treatable conditions. If knowledge of genetic mutation in a child makes a difference in immediate treatment, such as heightened vigilance, testing is usually not controversial. Even untreatable conditions can still benefit at times from confirming diagnosis. The test for HD could be appropriate for a younger patient to confirm an untreatable diagnosis of HD if that conclusively rules out harmful treatments for otherwise possible disorders. But what if the test is for a condition with no current treatment and late onset such as Alzheimer's disease (AD)? Is it in the best interest of a child with no symptoms to be screened in that case? Notice that an important step has already been taken here in framing the question as one of what is best for the child, the one actually being tested.

The NIH-DOE Genetic Task Force concluded that "genetic testing of children for adult onset diseases should not be undertaken unless direct med-

8. E. H. Hiller, G. Landenburger, and M. R. Natowicz, "Public Participation in Medical Policy Making and the Status of Consumer Autonomy: The Example of Newborn Screening Programs in the United States," *American Journal of Public Health* 87, no. 8 (1997): 1280-88.

ical benefit will accrue to the child and this benefit would be lost by waiting until the child has reached adulthood."[9] A joint statement of the American Society of Human Genetics and American College of Medical Genetics concurs but qualifies the recommendation as a general one with expected exceptions from unique patient and family characteristics and circumstances.[10] The concerns that are usually raised to prohibit testing include psychological ones such as that testing positive for a likely future disease may be misunderstood by children as a current sickness or warrant for low self-worth. Another is that families will reject or overindulge children with such results.

Cynthia Cohen recognizes these dangers but argues that there are no studies available to confirm them as usual outcomes.[11] She notes in particular that the concern that parents might either reject or overindulge such children points out the differences in parental response. For Cohen, there is too much variation in children and their families to make psychological assumptions. The most compelling reason that she sees for withholding test results until adulthood is the argument for adult autonomy. Dena Davis has specifically made the case that obtaining test results for a child's future disease status is to "preclude the child's right and opportunity to make that decision for himself in adulthood."[12] Cohen acknowledges the value of adult autonomy but argues that the child may be sufficiently mature and the family supportive that the best time to begin working with the information would be before legal majority. Children who are too young to give legal consent can sometimes be quite competent. For Cohen, if the child is not legally or personally competent to choose, parents are the best ones to decide if it is the right time to seek such information.[13]

As parents consider, it is relatively easy to justify a test if the results are immediately helpful. That might include understanding, planning, or prevention. If there is no direct benefit to the child, especially for late onset con-

9. Task Force on Genetic Testing, *Promoting Safe and Effective Genetic Testing in the United States,* p. xiv.

10. American Society of Human Genetics/American College of Medical Genetics Report, "Points to Consider: Ethical, Legal, and Psychosocial Implications of Genetic Testing in Children and Adolescents," *American Journal of Human Genetics* 57 (1995): 1233.

11. Cynthia B. Cohen, "Wrestling with the Future: Should We Test Children for Adult-Onset Genetic Conditions?" *Kennedy Institute of Ethics Journal* 8, no. 2 (1998): 111-30.

12. Dena Davis, "Genetic Dilemmas and the Child's Right to an Open Future," *Rutgers Law Journal* 28 (1997): 580-81.

13. Cohen, "Wrestling with the Future," pp. 126-28. See also Lainie Friedman Ross, *Children, Families, and Health Care Decision Making* (New York: Oxford University Press, 1999).

ditions, testing is more problematic. Most adults choose not to be tested for HD.[14] That is evidence that we should not generally take away that option from a child. The Huntington Disease Society of America states explicitly that "minors should not be tested unless there is a medically compelling reason for doing so. . . . Parental anxiety concerning a child's risk does not constitute a medically compelling reason."[15] Combining autonomy concerns with current risks to insurance if an expensive disease is predicted in a child's medical record, it is usually best to forego testing for distant and untreatable conditions unless some important circumstance intervenes. Recognizing exceptions should be up to parents who are seeking what is best for the child.

Decisive Procreation

Rachel pleaded with Jacob, "Give me children or I will die."[16] The desire for children is often powerful. If it were not, probably far fewer people would take on the drastic loss of freedom and increase in responsibilities that is entailed in having them. Thomas Murray describes some of the paradoxes of family life.

> We celebrate individualism, yet we find meaning in family relationships; we cherish freedom, yet we have children whose needs constrain us profoundly; we want the liberty to get up and go whenever it suits us, yet our flourishing depends on lifelong commitments and enduring, steadfast relationships; we exalt choice and control, yet families are built largely on acceptance of people as they are, with all their imperfections; we participate in a vigorous commercial culture, yet we cherish and protect a sphere in which interactions are regulated by values alien to the world of commerce and markets.[17]

Why do we have children? To share our lives? Mindless biological drive? To provide labor or take care of us in old age? Are children a form of immortality, a legacy? Are they a practical escape from current family, proof of virility or

14. Steve Jones, *Genetics in Medicine* (New York: Milbank Memorial Fund, 2000), p. 12. Kimberly A. Quaid and Michael Morris, "Reluctance to Undergo Predictive Testing: The Case of Huntington Disease," *American Journal of Medical Genetics* 45 (1993): 41-45.

15. *Guidelines for Genetic Testing for Huntington's Disease* (New York: Huntington's Disease Society of America, 1994), p. 10.

16. Genesis 30:1.

17. Thomas H. Murray, *The Worth of a Child* (Berkeley: University of California Press, 1996), pp. 5-6.

adulthood, someone to love?[18] Duty? Calling? Surprise? No doubt each of these has motivated some parents. Motivations can be complex and varied within one parent and from one person or society to another. Considering the effort involved, societies and family lines that do not feel compelled to procreate die out. None of us are here without great effort by people before us. Should that effort include using genetic testing to pursue the best possible genetic start for our children? We will think about that question in the order of an individual's development, beginning with his or her parents choosing a mate.

Prospective Mates

Avoiding Genetic Disease

Should genetic testing have a role in our choice of the husband or wife with whom we would like to have children? Nobel laureate Linus Pauling advocates in *The UCLA Law Review* that

> there should be tattooed on the forehead of every young person, a symbol showing possession of the sickle cell gene or whatever other similar gene, such as a gene for phenylketonuria. . . . If this were done, two young people carrying the same seriously defective gene in single doses would recognize this situation at first sight, and would refrain from falling in love with one another. It is my opinion that legislation along this line, compulsory testing for defective genes before marriage, and some form of semi-public display of this possession, should be adopted.[19]

Even if we came to be accustomed to forehead tattoos as a universal requirement, they would be too front and center on our foreheads for the role of genes in our lives. Such a plan would encourage stereotyping and stigma. In contrast, Ted Peters describes the completely discreet *Chevre Dor Yeshorim* (generation of the righteous) program of the Committee for the Prevention of Jewish Genetic Disease:

> Among the concerns of especially Ashkenazi Jews are Tay-Sachs disease and cystic fibrosis, both autosomal recessive traits that appear in one of every

18. Seymour Kessler, ed., *Genetic Counseling: Psychological Dimensions* (New York: Academic, 1979), pp. 30-32.

19. Linus C. Pauling, "Reflections on the New Biology," *UCLA Law Review* 15 (1968): 268-72.

twenty-five persons of this population. To avoid giving birth to children with two copies of either of these genes is the aim of this program. Young men and women of marital age take blood tests to see if they carry the problematic alleles. They are assigned numbers. The test results are filed according to numbers — so the persons remain anonymous — with the Committee. When a proposal for marriage is made, the couple telephones in their numbers. If this particular couple would put a future child at risk, they are told.[20]

When a couple is informed that their genes together bear a high risk of genetic disease, the couple might decide not to marry. Laurie Zoloth-Dorfman writes that in 1993 alone, eight thousand people were tested in the *Chevre Dor Yeshorim* program and sixty-seven marriages avoided.[21] By 1997 the program had expanded to four recessive conditions: Tay-Sachs, cystic fibrosis, Gaucher's disease type 1, and Canavan's disease.[22] As an example, Tay-Sachs is a progressive neurodegenerative disorder that is usually fatal by the second or third year after birth. It is characterized by developmental retardation, paralysis, dementia, and blindness.

Whatever their ethnic heritage or community, a couple with information that any child born to them would have a one in four likelihood of a brief painful life might proceed with marriage anyway. In that case there would still be at least eight possible ways they could respond.

1. They could choose to accept the one in four risk for each child they conceive, leaving the results to God or chance. The one in four devastating result might never occur, or it might happen to every child they conceive. The statistical average of one in four children at risk actually being afflicted is consistent over large numbers. Within the small numbers of one family each child would have the one in four risk, but all the family's children, none of them, or some other proportion of them could end up with the disease. If the parents allow such risk for each child, they would not be availing themselves of everything possible to protect and provide for their children.

2. They could decide not to have children. Having children is not gener-

20. Ted Peters, "Love and Dignity: Against Children Becoming Commodities," in *Genetic Testing and Screening: Critical Engagement at the Intersection of Faith and Science,* ed. Roger A. Willer (Minneapolis: Kirk House, 1998), p. 127.

21. Laurie Zoloth-Dorfman, "Mapping the Normal Human Self: The Jew and the Mark of Otherness," in *Genetics: Issues of Social Justice,* ed. Ted Peters (Cleveland: Pilgrim, 1998), p. 192.

22. Ari Mosenkis, "Genetic Screening for Breast Cancer Susceptibility: A Torah Perspective," *Journal of Halacha and Contemporary Society* 34 (Fall 1997): 14.

ally an obligation for marriage in the Christian tradition. Even the Roman Catholic faith, which might be associated with such a view, accepts both abstinence and the rhythm method to avoid pregnancy. Not having children could free a couple to live and serve in ways otherwise not possible.

3. They could adopt children. This, however, has become increasingly difficult in the United States. With almost one and a quarter million abortions in the United States last year, there are relatively few healthy infants available.[23] Lee Silver writes that "in 1984, two million infertile American couples competed for 58,000 new-born American children."[24] Also some states have moved toward open adoption, which can mean substantially adopting the child's family as well as the child. That can be an intimidating prospect. There are some older children with physical or social handicaps available and in great need. Many of these children are kept from homes by artificial barriers to adoption such as the rules in some states that forbid people of one shade of skin color from adopting children with another.

4. A relatively new possibility is that the couple could implant zygotes from another couple. For a number of years now *in vitro* fertilization (IVF) clinics have brought together more zygotes than they have implanted. IVF refers to conception outside the body. The *in vitro* literally means "in glass." Obtaining ova for IVF requires extensive hormonal intervention and a surgical procedure. While there has been recent success freezing and reviving unfertilized ova, typically zygotes (fertilized ova) have tolerated freezing for later implantation far better, so clinics have fertilized large numbers of ova with the intention of then implanting them only as needed. The result has been thousands of zygotes frozen indefinitely and in many cases eventually thrown away. If one is convinced that a zygote is a person, adopting a zygote not otherwise able to develop is akin to adopting a child in need of a home.

Tristram Engelhardt is attempting to explicate the Eastern Orthodox tradition when he rejects this course. For Engelhardt "the bond of husband and wife" is the "sole appropriate locus for human reproduction. . . . Though the acceptance of zygotes for surrogate gestation could be interpreted as the saving of early human life, one must avoid immoral actions, even if those will save the life of others. . . . 'Rescue' surrogacy is a further step in routinizing forms of reproduction that fall short of the mark,"[25] as does contraception for

---

23. The federal government tracks and publishes the number of abortions each year from the Center for Disease Control in Atlanta.

24. Lee M. Silver, *Remaking Eden: How Genetic Engineering and Cloning Will Transform the American Family* (New York: Avon Books, 1998), p. 160.

25. Tristram Engelhardt Jr., *The Foundations of Christian Bioethics* (Amsterdam: Swets & Zeitlinger, 2000), pp. 257-58.

Engelhardt.[26] Engelhardt expects that a believer rightly immersed in love for God and the Easter Orthodox liturgy will recognize this guidance as of God.[27]

5. The couple could use gametes (sperm or unfertilized ova) from a donor who is not a carrier for the disease. With the continued example of Tay-Sachs, half of the couple's children would probably be carriers from one parent's inheritance, but none would be afflicted with the disease. The couple would be able to raise the child from birth with no question as to who the child's parents are. The child would be genetically a descendent of one parent and the sperm or ova could be chosen to represent the other parent's genetic code for substantial overlap. This is not difficult to achieve in that 99 percent of human DNA is virtually identical among all human beings anyway. Where there are differences, for example in the genetic code for one eye color or another, the genetic method of encoding, say, dark brown eyes would be the same whether from donor or parent. DNA is a patterned strand of four chemicals. Whether the four chemicals are assembled by a machine or one body or another body the resulting piece of DNA can be identical. For a given piece there may be no way to tell by examining it where it was sequenced. Ascorbic acid is vitamin C whether assembled in a lab or a rose hip. The child would appear to be the couple's genetic descendent and would in fact have much in common genetically with each parent. A genetic link with one's children has long been prized. The centrality of that connection is discussed in chapter 13.

There would be some asymmetry in genetic relationship. One parent would be more closely related to the child genetically than the other. Would that produce psychological difficulties? If one takes seriously the biblical description of the couple being one flesh and the apostle Paul's description that the husband's body belongs to the wife and the wife's body to her husband, there would be a close identity of wife and husband with each other that does not emphasize such a distinction.[28] Children often resemble or favor one parent more than the other in temperament or appearance anyway. That need not be a barrier to genuine love and care.

There are about one million people conceived through gamete donation living in the United States today.[29] Studies have found such resulting families at least as healthy psychologically as traditional birth families. "Children conceived by ART [artificial reproductive technology] did not differ from naturally conceived children in emotions, behavior, or quality of

---

26. Engelhardt, *The Foundations of Christian Bioethics*, p. 268.
27. Engelhardt, *The Foundations of Christian Bioethics*, pp. 157-211.
28. 1 Corinthians 7:4; Ephesians 5:28-33.
29. Silver, *Remaking Eden*, p. 181.

family relations."[30] In fact, "adoptive parents and parents conceiving children through ART expressed greater warmth and emotional involvement with children, as well as greater satisfaction with parenting roles, relative to birth parents."[31] The ART families may have a genetic link through one or both parents, adoptive families no genetic link, and most function quite well.

Some traditions object to any technological intervention in the God-given process of procreation. That view is discussed in the next section of this chapter. It is not unusual for others to reject the donation of a gamete as an instance of adultery prohibited in the Ten Commandments.[32] The Ten Commandments do warn people against breaking the marriage covenant by inappropriate relationships or physical intimacy. It is, however, a significant extension to apply the prohibition to the donation of a physical substance. Otherwise, accepting a life-saving blood donation or a heart transplant would be a form of adultery. Since there is no contact with the gamete source, let alone sexual intimacy, donation of a gamete does not seem to fit the violation of the marriage covenant that was known and prohibited when the command was given. It is also not an "irresponsible intrusion" on the part of the donor but rather an invited gift. If one needs an analogy, organ donation would be a closer parallel than adultery. One would rather get along with one's own organs, but if that is not possible, a life-giving donated organ is gratefully received.

Some thoughtful Christian scholars have read the description of marriage in the opening chapters of Genesis as prohibiting the joining of gametes from other than the married couple. This conclusion is not directly taught by the text. The biblical account reads that a man shall leave his parents and a woman her home and the two shall become one flesh.[33] In context the subject is that Adam is alone and that is not good. God forms Eve from Adam's flesh-rib and Adam is pleased to meet one like himself. His first words at the sight of Eve are "I see me." The passage goes on to say that Adam and Eve then became one flesh. The chiastic structure is from one flesh alone to Adam's flesh divided to Adam and Eve one flesh together. The presented problem and so-

---

30. Susan Golombok et al., "Families Created by the New Reproductive Technologies: Quality of Parenting and Social and Emotional Development of the Children," *Child Development* 66 (1995): 285, 295; Frank van Balen, "Child-Rearing Following in Vitro Fertilization," *Journal of Child Psychology and Psychiatry and Allied Disciplines* 37 (1996): 687, 692.

31. Nancy L. Segal, "Behavioral Aspects of Intergenerational Human Cloning: What Twins Tell Us," *Jurimetrics* 38, no. 1 (1997): 61.

32. Exodus 20:14.

33. Genesis 2:18-25.

lution is about companionship. The passage does not directly refer to the genetic heritage of children or to children at all. A husband and wife united to one another for life at every level does provide a beautiful and practical context for procreation, but procreation is not the presented point of the immediate text.

The best case against welcoming a donor gamete is to extend the idea of being "one flesh" to a general principle.[34] Since the covenant of marriage is intended to be a lifelong union between a husband and wife, and in the earlier chapter of Genesis human beings were told to be fruitful and multiply, it is expected that children are to be born through that one-flesh relationship. That is consistent with the worldview and teaching about marriage found throughout Christian Scripture. There is a further step, however, which is the crucial one for this particular argument. It is reasoned by some that the covenant relationship between husband and wife includes that the physical nature of all progeny should come only from the couple's physical nature. Since human beings are a unity of many levels, including the physical, the one-flesh union sets expectations for all other aspects of their life together. Progeny is an important part of that life, so the exclusivity of the one-flesh union between the parents applies to the genetic heritage of the children they procreate as well.

The strictest application of this reasoning excludes the introduction of any genetic material to the children from outside the couple. The one-flesh covenant of marriage includes the entire genome of any children the couple procreates. For example, two parents with cystic fibrosis have no working copy of the gene instructions needed for fully functioning lungs. A prohibition against introducing any genetic material the parents do not possess would forbid giving their child the genetic instructions needed for fully functioning lungs and a more normal life span. Or as another example, a genetic vaccine that grants a greater resistance to cancer than found in either parent would have to be rejected.

An alternative appeal to "one flesh" as a general principle welcomes genes from outside the one-flesh relationship if they are introduced as part of redeeming the gametes, but not to create or replace them. All of the child's genetic material should come from the parents except that which is defective or absent in one or both of the parents. By this view the importance of the genetic connection to parents is emphasized while also accepting the legitimacy of correcting damage or deficiencies attributable to the fall. The argument for

34. Gary P. Stewart et al., *Sexuality and Reproductive Technology* (Grand Rapids: Kregel, 1998), p. 34.

this version depends on at least three contested convictions. First, genetic defects or deficiencies are due to the corruption or curse of the fall. That is a point debated in kind since Irenaeus and discussed in the following pages of this chapter. Second, applying this version would depend on the definition of "damage" or "deficiency." That turns out to be a rather difficult task which is addressed in chapter 10. Third, the reasoning focuses on where the DNA originates, not what the sequence of DNA encodes. DNA can be introduced from outside the married couple in order to heal-redeem their gametes. The key is that the two gametes at the time of conception must originate from the husband and wife. There could be two identical twins, by definition having the same genes. If one identical twin became infertile by accident or environmental insult, the sibling could not donate replacement gametes. There would be no physical difference in the gametes from either identical twin. Coming from the same genetic pool, the gametes would have the same genetic possibilities. Yet by the proposed principle the identical twin's donation would be unacceptable because the gamete did not originate from the husband or wife. If this principle is indeed a God-given requirement for procreation, that gametes must originate from the covenanted couple, then gamete donation is prohibited for those seeking to obey God.

6. Gamete selection is another possibility the couple faces. Polar body biopsy is available to reveal the genetic code carried by an ovum without destroying it. Ova can be selected for exposure to sperm and conception. In that manner half of the genetic code of a zygote could be carefully chosen before conception. The parents could assure that the mother did not contribute a Tay-Sachs gene. The child would have a 50 percent chance of being a carrier from the father but would not have the disease. Selecting gametes requires removing a number from the bodies where they grew, sorting for desired characteristics, and then combining two still outside the body. This of course continues to raise the question of intervening in the natural course of procreation. That will be addressed in the next section.

7. The couple could create zygotes *in vitro* with their sperm and ova and select the ones without the disease to implant for pregnancy. All genetic heritage would be from the parents alone. Some zygotes would probably be deliberately discarded. That is acceptable or not depending on the status of zygotes (chapter 5) and again the rightness of intervention in the procreation process, which will be discussed later in this chapter.

8. They could conceive through sexual intimacy, test any resulting pregnancy, and abort any fetus that would soon die anyway. For the tradition that founded the *Chevre Dor Yeshorim* this is acceptable if the termination is within forty days of fertilization. Forty days is a landmark in that tradition

because they believe a person is not present until that point. As described in chapter 5, when one recognizes that a person is present, ending the progression of pregnancy becomes problematic.

Some day gene therapy may add the possibility of treating a developing zygote or embryo that has a disease allele so that the disease will not occur. That would make it possible for the couple to choose each other and disease-free biological children without choosing against any particular conception. In the meantime, the possibility of the above interventions means more responsibility in choosing to pursue one or not. The ethics of gamete selection, zygote selection, and tentative pregnancy, will be discussed further in this chapter since they each depend on genetic testing.

Misattributed Paternity

Misattributed paternity is not an uncommon complication when a couple considers genetic testing and having children as a couple. Parents often seek genetic counseling because an earlier child born in the family had a recessive genetic disability or disease such as Tay-Sachs or cystic fibrosis. They are considering whether to bring their genetic contributions together again since these particular diseases occur when both parents contributed a disease allele to the child. About 10 percent of the time genetic analysis reveals that there is no risk of the disorder occurring a second time because the father was not the source of the sperm for the afflicted child and does not carry the harmful allele.[35]

Since the couple has come in as a couple for genetic testing, this is important information. They are not at risk of another child having the recessive condition. It is also important medical information for the father. If he ever marries someone else, he is not at risk of passing on the affliction. What is complicated is that by presenting as the source of the sperm for the afflicted child, the father apparently does not know that he was not the male genetic source for the child. This information might very well reveal a previously unknown betrayal of trust.

Lainie Ross reports surveys in the late 1980s that in such a case 96 to 98.5 percent of the surveyed physicians and genetic counselors would not reveal the misattributed paternity to the father.[36] They would assure the couple

---

35. Sally MacIntyre and Anne Sooman, "Non-paternity and Prenatal Genetic Screening," *The Lancet* 338 (1991): 869.

36. Lainie Friedman Ross, "Disclosing Misattributed Paternity," *Bioethics* 10, no. 2 (1996): 115.

that another child would not be afflicted and leave it at that to protect the secret of the mother. This course withholds information from a couple who has paid for a set of test results. It also sets up the counselor for escalation from withholding information to a direct lie if the parents ask *why* it could not happen again and the counselor persists by giving a false reason. It would be better if the counselor would first inform the woman of the test results and suggest that she reveal it to her spouse before the next appointment. If she prefers not to reveal the information before the appointment, the counselor will explain the test results to the couple then.

The Christian tradition of recognition of the reality of sin and the model of grace and reconciliation would be helpful to a Christian couple here. Also the Christian tradition of lauding adoption as characteristic of God's relationship with his people and with one another is applicable. It would be a misnomer to call the sperm source "the father" or even "the biological father." At a purely biological level, the father who is raising the child has probably done far more to sustain and nurture the biology of the child than the contribution of one cell. The raising father is still the father.

Marriage thrives on transparency. This potentially difficult information is more likely to be confirming than a surprise. It is actually freeing in regard to the presenting problem of genetic risk to future children. In such a case the tests will probably show that future children of the couple will have no risk of the threatened disease at all. However, couples should be informed as part of the consent process that the tests they are requesting may reveal misattributed paternity since that information is directly relevant to the health of their future children. Some couples may decide not to pursue testing when they realize that or other potential implications.

Gamete Selection and Changing the Natural Course

The first option that the couple carrying a risk of Tay-Sachs has that involves more genetic testing is gamete selection. A genetic test could be used to detect the Tay-Sachs allele in ova before they are allowed to be fertilized. The method could include detailed analysis of genetic code of each ovum by polar body biopsy. In contrast, testing a sperm would destroy it, although rough sperm sorting by weight to increase the likelihood of beginning a male or a female has been done for some time. Ova always carry an X chromosome. Sperm carry either a Y chromosome or an X. If a conception occurs with a Y sperm the result is XY, a boy. If an X-carrying sperm unites with an ovum the result is XX, a girl. X-carrying sperm weigh more than Y-carrying sperm. Sperm selection by weight might be pursued to try to assure that no boys would be conceived, hence avoiding a condition such as hemophilia that is only experienced by males. Sperm sorting can also be pursued for gender balance. For example, parents could be grateful for their two sons but would also like to raise a daughter.[37]

Would such intervention cause us to think of our children as things we select instead of people we unquestionably welcome? This important question will be discussed in the section entitled "tentative pregnancy" later in the chapter. The question we will address here is whether such intervention is unacceptable because it changes the natural course of sexual intercourse and procreation. That specific question is a subset of the larger one of whether it is ever good for human beings to change nature. This question and its ramifications are so influential for this study that I will pursue them for several pages. Of course if the reader has already worked out if it can be natural and right for human beings to intervene in nature, including procreation, one could skip ahead to the next section on zygote selection.

Nature as God-Given, Hence Set

Jeremy Rifkin has been effective in publicizing his objection to changing nature. Ted Peters summarizes Rifkin's advocacy this way. "He appeals to a vague naturalism, according to which nature itself claims sacred status. He issues his own missionary's call: 'the sacralization of nature stands before us as the great mission of the coming age.' "[38]

In contrast, the Christian tradition worships God, not his creation. The creation is to be received with appreciation and cared for as God's handiwork entrusted to us, but it is not God. Yet some theologians have argued that because the physical world is God's creation, intervening in its natural course is rebelling against God's intent for it and us. Changing the natural course is arrogantly trying to out-design God. Since God has created the world as it is, it should remain as it is.[39] Its current state is the stan-

---

37. There are also cultures where boys are so much more prized than girls that sex selection has been practiced extensively by ultrasound and abortion. In parts of India birth ratios of male to female are running at 135 to 75. Allen Buchanan, Dan W. Brock, Norman Daniels, and Daniel Wikler, *From Chance to Choice: Genetics and Justice* (Cambridge: Cambridge University Press, 2000), p. 183.

38. Ted Peters, *Playing God: Genetic Determinism and Human Freedom* (New York: Routledge, 1997), p. 13, quoting Jeremy Rifkin, *Algeny* (New York: Viking, 1983), p. 252.

39. For a recent description and critique of this view in regard to genetics see Thomas A. Shannon, "Genetics, Ethics, and Theology: The Roman Catholic Discussion," in Peters, ed., *Genetics: Issues of Social Justice*, pp. 144-79.

dard for the good and simple reason that this is how God created the world to be.[40]

Cardinal Karol Wojtyla, who would later become Pope John Paul II, wrote the theme this way: "This understanding and rational acceptance of the order of nature is at the same time recognition of the rights of the Creator. Elementary justice on the part of man towards God is founded on it. Man is just towards God the Creator when he recognizes the order of nature and conforms to it in his actions."[41]

While fully recognizing God as the Creator, conforming to nature raises at least three major problems. As we describe these three problems, note that they are not a critique of the philosophical tradition of natural law. The natural-law tradition reasons from the recognition of intrinsic goods without appealing to God or nature as authoritative. It is the appeal to present nature as a standard that I am addressing here.

1. The cardinal's statement is that the standard is not simply the surface of nature. It is to understand and accept "the order of nature." One is to see the underlying, God-given intent of the natural order, but how does one recognize the intent built into nature?[42] Nature is varied and needs interpretation to discern purpose. How does one read the natural intent God has designed into earthquakes, hurricanes, the black plague, and malaria? God's intent in nature is not immediately obvious. Proposed interpretations have been myriad. The convictions and commitments one uses to recognize what is "the natural order" may be the actual standard.[43]

2. Even if one could clearly discern a natural order, why consider that a pure expression of God's will? Sin has warped our perceptions and tainted our physical world. The natural order is not currently all that it should be. We cannot assume that what we currently observe is the original intention. The 1984 Panel for Bioethical Concerns of the National Council of Churches asks

---

40. Paul Ramsey, *Fabricated Man: The Ethics of Genetic Control* (New Haven: Yale University Press, 1978), p. 160.

41. Cardinal Karol Wojtyla as quoted by Thomas A. Shannon, *Made in Whose Image? Genetic Engineering and Christian Ethics* (Amherst, N.Y.: Humanity, 2000), p. 36.

42. This problem is noted as well in Hans-Martin Sass, "A Critique of the Enquete Commission's Report on Gene Technology," *Bioethics* 2, no. 3 (1988): 270, and Jean Porter, "What Is Morally Distinctive about Genetic Engineering?" *Human Gene Therapy* 1 (1990): 421.

43. Michael Banner recognizes this in the midst of his appeals to the authority of the created order of nature. "While Christian sexual ethics must assert the existence and authority of the created order, knowledge of the form and character of that order may be reckoned to be irreducibly theological." *Christian Ethics and Contemporary Problems* (Cambridge: Cambridge University Press, 1999), p. 286.

rhetorical questions to affirm that technology should be used "in harmony with the natural environment and responsible care for it."[44] They describe this as reflecting the Christian responsibility to "eco-ethical stewardship." Yet in the same document the panel also states that "all creatures great and small thus share the total phenomenon of life, with its endless complexity, its tragic wastefulness, and its often dubious purposes."[45] If they are correct that the natural order includes "tragic wastefulness" and "dubious purpose," why consider it an authoritative standard to be maintained as it is?

In 1848 Matthew Arnold wrote a poem "to an independent preacher who preached that we should be in harmony with nature."

> "In harmony with Nature?" Restless fool,
> Who with such heat dost preach what were to thee,
> When true, the last impossibility. . . .
> Know, man hath all which Nature hath, but more,
> And in that *more* lies all his hopes of good. . . .
> Man must begin, know this, where Nature ends. . . .
> If thou canst not pass her,
> rest her slave![46]

3. What then is God's intent for nature and our lives in it? Craig Gay articulates what seems to be a widely felt concern,

> The point is not simply that we have lost a kind of natural simplicity or innocence. Rather, what has been lost, in abstracting away from immediate nature, is the possibility of encountering something outside of ourselves which might discipline — and thus give order to — human making and willing. In the absence of such discipline, modern machine technology seems destined to be destructive of nature, of living human cultures, and indeed of living human beings.[47]

We do need guidance, but nature of itself is not adequate to give it. Why follow the lead of tornadoes and parasites? Such a view assumes that the natural order as we find it was created by God as the pinnacle of all that it could

44. National Council of Churches of Christ/USA, Panel on Bioethical Concerns, *Genetic Engineering: Social and Ethical Consequences* (New York: Pilgrim, 1984), p. 25.

45. National Council of Churches, *Genetic Engineering*, p. 23.

46. Matthew Arnold, *The Essential Matthew Arnold: Poems, Critical Essays, Political Writings, Letters* (London: Chatto and Windus, 1949), pp. 52-53.

47. Craig Gay, *The Way of the (Modern) World; or, Why It's Tempting to Live As If God Doesn't Exist* (Grand Rapids: Eerdmans, 1998), p. 99.

be. Any change would be a detraction. Accepting the premise that God created the natural world, it might still be a right and good starting point, not intended as yet a complete fulfillment. Just because it is natural does not mean that it is good. It may be indifferent or bad. If either, it may be changed for the better. Better is whatever supports what God has revealed of God's purposes and our calling. It may be that the standard is not where we start but where we are supposed to be going.

Irenaeus and other early church fathers such as Clement of Alexandria and Saint Gregory of Nazianzus reasoned that God intentionally created human beings and our natural world as less than they should eventually be.[48] This is not a lack of ability on God's part any more than it was when at the end of the first day of the Genesis creation story the earth was formless and empty and God declared it good. It was not God's intent to complete it yet. This would not be a lack of perfection any more than Jesus was imperfect as "he grew in wisdom and stature."[49] The point is that God sovereignly chooses to create over time and has designed us and our world to do the same. Where we started was good. Where we are to end up will be better.

Part of the growth planned for us is to change the natural course. As discussed in chapter 2, we could not survive if we did not. Far beyond survival, we change the natural course of extreme heat in the summer months by air conditioning. We eat with knife and fork when it would be more natural to tear the meat directly with our incisors made for that purpose. The knife and fork make eating less messy and some tougher foods more accessible, yet in switching to knife and fork we lose some spontaneity and direct sensual contact with our food. This is sometimes described as a significant loss in countries where one directly takes food in hand. In changing the natural course there are wins and losses.

The issue is not that the natural course has been changed. We are part of nature and change it naturally. Beavers build a dam across a stream to create a lake to protect their lodges and make trees more accessible for food. They transform the ecosystem in which they live. That is their nature. Human beings build a dam across a river to protect homes and provide easier access to water and irrigated food crops. That is part of our nature. Our actions can be destructive of ourselves and the rest of God's world, but they are not inherently destructive or unnatural simply because they bring about change.

Nature is to be appreciated and responsibly cared for, yet it is not to be

48. John Hick, *Evil and the God of Love* (San Francisco: Harper & Row, 1977), pp. 211-18.
49. Luke 2:52.

189

unchangeable and final. I am not suggesting this as a radical new concept for the Christian tradition. It has been articulated by the early church fathers as cited above and is found in Judaism from the beginning. As Laurie Zoloth-Dorfman points out, "Judaism is not a nature-based religion; the very assertion of circumcision rests on the notion that the body is neither sacred nor immutable."[50] While there is widespread concern about genetic intervention in the ongoing Jewish community that vividly remembers the connection of eugenics and the Holocaust, the long-held commitment of the tradition is to "*tikkun olam,* the mandate to be an active partner in the world's repair and perfection."[51] God is the maker of nature, not its product or captive.

As described in chapter 3, we should sustain, restore, and improve ourselves and the rest of the physical world. We have learned from the study of ecology that nature is intricately interrelated. Vigilant care is needed in changing the finite world we depend upon. What has worked in delicate and intricate balance should only be changed with caution, yet there should be change. The physical world for which we are grateful can be better. For example, the world is a better place without smallpox, which was eradicated only after years of systematic human effort.

Is Procreation a Special Case?

While it is natural and our mandate for us to intervene in nature, is procreation a special case? Procreation is the place where we welcome into the world new human beings, people with potential to know God and live with God forever. Nurturing these new lives is one of the most important things we can do. Is it essential that these new lives only begin physically with a conjugal union?

Human beings were created to naturally change nature. We already without hesitation intervene in the natural process of procreation by giving painkillers to mothers during the often arduous process of birth. We resort to the surgery of a Caesarean section if needed to save the lives of mother or baby. Neither intervention detracts from God's purpose in our procreation. Would gamete selection detract from what God is doing? It is not contrary to the possibility of God sovereignly choosing to shape a particular genetic inheritance. God's will could not be so easily thwarted. If one's view of providence includes God choosing the genetic start for each human being, God can do that through the intentional acts of human beings as well as through apparently random acts of nature. As discussed in chapter 3, God can write

50. Zoloth-Dorfman, "Mapping the Normal Human Self," p. 190.
51. Zoloth-Dorfman, "Mapping the Normal Human Self," p. 190.

straight with our crooked lines. Are there other reasons why procreation would always be outside our mandate to sustain, restore, and improve?

For Paul Ramsey, altering human biological development would be making human beings less human.[52] Anything that would intervene in the natural process of procreation would violate part of what it is to be human.[53] People should be begotten, not made. Unless natural human parentage is kept inviolate, technological civilization will fragment the personal and biological dimensions of human procreation.[54] To separate procreation in any way from the act of love is to reduce it to reproduction, a manufacturing process rather than embodied personhood.[55] In contrast, Joseph Fletcher argues that it is artificial reproduction that is particularly personal and human in that "it is rationally willed, chosen, purposed and controlled."[56] For Fletcher the very definition of "civilized" is to be artificial.[57]

The magisterium of the Roman Catholic Church instructs that each conjugal act must always be open to procreation as its natural course.[58] Since sexual intimacy is where procreation naturally begins, the unitive and procreative aspects are God-designed and inseparable. One cannot, for example, use barrier methods of birth control. Now it is fascinating that the Roman Catholic Church does encourage the rhythm method of birth control. One may intentionally time intercourse to avoid conception. This is understood as working with God-given nature, not contrary to it. Even for the Roman Catholic tradition one can consciously shape the overall process, but to what degree?

It makes sense that rightly lived sexual intimacy is an ideal place to welcome new life. Sexual intimacy in the Christian tradition is a gift of God to cel-

52. Ramsey, *Fabricated Man*, p. 137. Also see Leon Kass, *Toward a More Natural Science: Biology and Human Affairs* (New York: Free Press, 1985), p. 109, and more recently Leon Kass, "The Moral Meaning of Genetic Technology," *Commentary* (Sept. 1999): 36, and Dennis P. Hollinger, "Sexual Ethics and Reproductive Technologies," in *The Reproductive Revolution*, ed. John F. Kilner, Paige C. Cunningham, and W. David Hager (Grand Rapids: Eerdmans, 2000), pp. 79-91.

53. Ramsey, *Fabricated Man*, pp. 89, 132.

54. Ramsey, *Fabricated Man*, p. 136.

55. Ramsay, *Fabricated Man*, p. 89.

56. Joseph Fletcher, "New Beginnings in Life," in *The New Genetics and the Future of Man*, ed. Michael P. Hamilton (Grand Rapids: Eerdmans, 1972), p. 87.

57. Joseph Fletcher, *The Ethics of Genetic Control* (Garden City, N.Y.: Doubleday, 1974), p. 15.

58. *The Gift of Life (Donum Vitae): Instruction on Respect for Human Life in Its Origin and on the Dignity of Procreation* was published by the Congregation for the Doctrine of the Faith of the Roman Catholic Church.

ebrate a lifelong commitment between a woman and a man.[59] The husband and wife choose to love and share intimately at every level. Their marriage is to be a place of acceptance and enjoyment, a place where they develop the habits of self-giving and forgiveness needed to enjoy any long-term commitment. Their relationship skills grow and mature over time. This creates an ideal place to start a new human life. The newborn enters a family where care and enjoyment of one another is modeled and consistent. In contrast, when sex is merely sport between temporary sequential playmates, it is no longer a welcoming place for new life. This is one of the reasons for the high demand for abortion in cultures that have demeaned the role of sexual intimacy. Sexual intimacy can be a committed, unitive celebration of lifelong commitment without bearing children, and a couple who cannot conceive children on their own can still provide the intended welcome and place for new life while receiving physical help in conception or pregnancy.

According to the Christian tradition God chose to be incarnate in a family apart from sexual intimacy.[60] The incarnation is such a unique event that it cannot be referenced as a standard model for imitation. But it is relevant in that God would not bring about the incarnation through such a process, procreation without sexual intercourse, if doing so was inherently evil. The unitive and procreative aspects of sexual intercourse fit well together, but they are not in every case inseparable. Sexual union can be a celebration of life intimacy and commitment even if one does not bear children. A wife and husband can provide a welcoming home for children even if their sexual union by itself does not achieve procreation. Intervening in the biological course of procreation is not automatically harmful to the child or the parents. What is central is how the child is welcomed and raised by her mother and father. A related question of whether intervention to enable or shape procreation encourages harmful attitudes toward children will be addressed later in this chapter and in chapter 13.

Zygote Selection

The couple described earlier in this chapter is faced with a one in four chance of any child they conceive dying a painful death early in infancy. As described so far, they could decide not to marry. If they do marry or already have, they could choose not to have children. If they do want to have children, they

59. Hollinger describes this gift as ideally one of consummation, procreation, love, and pleasure; see his "Sexual Ethics and Reproductive Technologies," pp. 80-86.
60. Matthew 1:18-25 and Luke 1:26-38.

could adopt. If they want to bear children, they could risk the one in four devastating result that might never occur or might happen to every child they conceive. One alternative would be gamete selection. The couple could use polar biopsy of each ovum to assure that the mother's contribution would not lead to Tay-Sachs. The child could still be a carrier but would not be afflicted with the lethal condition.

Another possibility that has been available for some time now is for the couple to use IVF and test each zygote for the disease. A less invasive version of the technique is to locate and remove a normally conceived zygote to read its code and decide whether to return the zygote to the reproductive tract for natural implantation.[61] By either means the couple could choose to implant only the zygotes that do not have Tay-Sachs disease. If they only fertilized the number of zygotes that they would be willing to implant and carry to term, no zygote with a future would be lost.

Zygote selection is already being used for several lethal conditions. However, once one has developed a technique for one purpose, it can be used for others. One could use zygote biopsy and selective implantation to choose any measurable characteristic that naturally occurs. Genotype is not always phenotype, but one could select for predisposition toward a number of traits. Sperm and ova can be brought together to form multiple zygotes. Then three or four zygotes with desired characteristics could be selected and implanted. Typically only one in three or four naturally conceived zygotes survive to birth. The least invasive way to obtain multiple eggs would be by removing a small piece of ovarian tissue containing hundreds of ova. One hundred or so ova could then be fertilized and available for selection. To increase the chances of successful implantation of the selected zygote, it would be possible to clone multiple copies of the zygote to implant until success. However, if a genetic defect is keeping the zygote from implanting, a genetic copy would have the same result.

Zygote biopsy can be done at the six- to ten-cell stage which is two or three days post fertilization and several days before implantation begins.[62] Removing one cell from an eight-cell zygote for testing does destroy a cell that could have been reimplanted and developed into an identical twin. With cloning capabilities that is a potential of all our nucleated cells. That will be discussed in chapter 13. Prevention of Tay-Sachs disease by this means is

61. S. J. Fasouliotis and J. G. Schenker, "Preimplantation Genetic Diagnosis Principles and Ethics," *Human Reproduction* 13, no. 8 (1998): 2238-2245.

62. Richard J. Tasca and Michael E. McClure, "The Emerging Technology and Application of Preimplantation Genetic Diagnosis," *Journal of Law, Medicine and Ethics* 26 (1998): 7-16.

only prevention if there is no person to protect at this point. If a zygote is a person, diagnosis and discard is a search-and-destroy mission, not prevention. Is the zygote selection process deciding whether to adopt a particular child of many that are present or is it selecting a genetic heritage for a child who does not exist yet? It would be abhorrent to reject and kill a person, but not problematic to choose the best available genetic endowment for a child yet to be.

If the latter is the case, zygote selection is deciding whether to support a particular set of genes likely to develop particular physical characteristics. One is already making a decision like this when, for example, deciding whether to have children that are likely to share your potentially lethal or debilitating genes. If an HD adult chooses not to have genetically related children, is that significantly different from having genetically related zygotes and only implanting the ones that do not have the disease? The answer to that question depends on whether a fellow human being is yet present at the zygote stage. When a person is present has been carefully addressed in chapter 5. Any person present should be welcomed and cared for.

It has been argued that even if a person is not yet present, any selecting of characteristics is bad practice for parenting. As a parent one needs to accept whatever child one has regardless of physical characteristics. While that is true, parenting is also about giving the best available support to one's child. For example, one should do the best one can to provide medical care to correct a genetic handicap. While fully accepting a child born with a harelip, one should do one's best to provide surgery to correct that genetic condition so that one's child can more freely speak and eat. We assume that disabilities are to be corrected if possible, not that they are a God-planned design to be embraced. If the disability cannot be changed, then one looks to see if by grace one can make it a teacher and not just an enemy. But the first goal is to restore the function.

Would it also be consistent with that responsibility for parents to select a genetic heritage by which the facial bones and teeth would grow normally in the first place? Such would avoid later suffering and repeated invasive intervention. It is good parenting to accept one's child and to help your child. The pivotal distinction for this intervention is whether a child is yet present. If a child is present, one accepts and works with the one here. If a child is not yet present, one can select from possibilities to give the best possible start. Giving one's child the best start one can is part of good parenting.

We have always had some control over both the timing of children and their later development. Preimplantation zygote diagnosis makes possible some guidance of the child's genetic heritage as well. Our responsibility remains whether we intervene or not. With the availability of the new technology comes responsibility for selecting the child's genetic endowment or for not doing so. Genetic heritage becomes increasingly less fate and more parental choice. There is less surprise and more opportunity to help. The availability of the technology does not leave us the option of not choosing.

In the near future people may be less supportive of those who have children with avoidable conditions. Since they could have avoided it and did not, the parents will be seen as responsible for their own predicament. Thus the possibility of intervention becomes the obligation to intervene. This may be most strongly felt when parents are already using IVF due to infertility. If one is proceeding with IVF, why implant a zygote that will lead to a child's disability when one can implant one that has the potential to develop into a healthy body? It is the same reasoning that is used when choosing a sperm or egg donor. The source of gametes is screened for health. Some day a child may castigate parents for passing on the family nose or not giving them a better genetic foundation when it was in their power to do so.

Even if one intervenes to stop the development of a body before a person is present, any intervention may still have serious consequences. Leon Kass is concerned with the effect of manipulating genetics on those who carry it out, particularly that routine intervention in the physical nature of human beings would lessen the manipulator's sense of human mystery, hence respect for human life.

> Finally, there may well be a dehumanizing effect on the scientist himself, and through him on all of us. On the one hand his power of mastery increases, but on the other hand his power of mystery decreases. . . . The sense of mystery and awe I am speaking of is demonstrated by most medical students on their first encounter with a cadaver in the gross anatomy laboratory. Their uncomfortable feeling is more than squeamishness. It is a deep recognition, no matter how inarticulate, that it is the mortal remains of a human being in which they are to be digging, ultimately a recognition of the mystery of life and death. The loss of this sense of awe occurs in a matter of days or weeks; mastery drives out mystery in all but a very few.[63]

Kass's concern at this point focuses on the one who is intervening, not the recipient of the intervention. He is concerned that the scientist and others

---

63. Leon R. Kass, "New Beginnings in Life," in *The New Genetics and the Future of Man*, pp. 56-57.

might eventually view human beings who have received intervention as less mysterious, hence less worthy of awe. Without that sense of awe toward human beings, respect for human freedom and dignity may be lost.[64]

However, Kass does acknowledge in the next paragraph that the increase of knowledge can increase one's sense of awe. What may have appeared to be a simple phenomenon may turn out to be wondrous in its intricacy. Our language has two meanings for "wonder." It may mean to be puzzled as to how something is or it may also mean to be amazed at what we do understand. Increasing knowledge has the potential to add to that sense of wonder, not always lessen it. In his reference to the "practice" of "most ordinary men of science," Kass's concern may be rooted more in the familiarity of routine exposure than in lack of understanding. Sheer repetition of involvement and responsibility does tend to dull people to their power in shaping others, whether they are genetic scientists or day-care workers. If one regularly shapes developing human beings, one might start to see those persons more as objects than as fellow human beings.

Paul Ramsey also emphasizes this concern that claiming the authority to intervene encourages disdain for the recipients as mere objects of manipulation. Where there is no destiny toward which human beings are moving and which moves in them, self-modification becomes the human being-God. This is to subjugate everything and everyone to the freedom of those who wield it. "Man as a manipulator is too much of a god; as object, too much of a machine."[65] Such manipulation by human beings of other human beings renders the recipients "thingified."[66] It makes human beings objects rather than fellow children of God. Gilbert Meilaender writes, "One whom we beget shares in our being, is equal in dignity to us. One whom we make has been distanced from us, become the product of our will."[67]

Genetic intervention could be done in such a way that it treats the other as merely an object, but that is not a unique problem. The often decried tendency in modern specialized medicine is to think of patients as cases rather than human beings. "Have you seen the liver yet in room 431?" That is not surprising when a physician is faced with briefly advising tens of patients a day. It may be difficult to show that clinical cure of disease through genetic means is more potentially dehumanizing in this regard than

64. Leon R. Kass, "Making Babies — the New Biology and the Old Morality," *Public Interest* 26 (Winter 1972): 53-54.

65. Ramsey, *Fabricated Man*, p. 103.

66. Ramsey, *Fabricated Man*, p. 151.

67. Gilbert Meilaender, "Mastering Our Gen(i)es: When Do We Say No?" *The Christian Century* 107, no. 27 (1990): 875.

other treatments now widely practiced and accepted.[68] Mass inoculations against tetanus or polio can look like assembly lines yet are not necessarily dehumanizing. One could argue that the risk of perceiving and treating recipients as objects rather than persons is heightened by the addition of yet another point of intervention.[69] On the other hand initial widespread genetic intervention may actually decrease the need for later medical intervention, hence reducing rather than increasing the frequency of manipulation and its attendant risks.

Why do zygote selection rather than treat a child after birth? Some conditions have no treatment. A child born with Tay-Sachs will inexorably die. Zygote selection avoids that painful process and end. If treatment is available for a condition, the course of treatment might still be protracted and painful. A zygote with TP53 will develop a fairly normal body but be racked with almost continuous multiple cases of cancer. That is not a condition one would choose for a loved one. Zygote selection can eliminate a health threat from an entire line of descendants so that there would be less overall medical intervention for the individual and those who follow. One can give the child the best available genetic start, choosing a probably healthy constitution and particular characteristics. Selecting superior resistance to disease, for example, would be a valued enhancement. Of course, as stated before, as attractive as these results would be for the resulting children, they should not be pursued via zygote selection if zygotes are human persons. Achieving benefits for some people by killing other people is not acceptable. If a zygote is a developing human life but not yet a person, such intervention is less problematic. Again the status of zygotes and other stages of life in the womb was discussed in detail in chapter 5.

IVF for enhancement is an expensive intervention. Yet many parents make great sacrifices to obtain the best available college education for their children. That investment is made to enhance the child's capability and options. Will clinics be willing to offer this service? Of course. Surface changes of cosmetic surgery are widely available. The second most common surgical procedure in the United States is liposuction. Medical training and technology are often devoted to surface enhancements. Zygote selection would be far more formative. Many of the wealthy will pursue zygote selection as an advantage for their children. That may be important to them as proud parents

---

68. Immanuel Jakobovits, "Some Letters on Jewish Medical Ethics," *Journal of Medicine and Philosophy* 8 (1983): 217-24.

69. James F. Keenan, "What Is Morally New in Genetic Manipulation?" *Human Gene Therapy* 1 (1990): 293.

or as a gift to their children.[70] Once the procedures are clearly safe and efficacious for the resulting children, others will want such intervention for their children lest they start at a disadvantage. Government may eventually provide a decent minimum of such intervention as it does now for education and medical care for the sake of the child and society. We will discuss that more in chapter 12.

C. S. Lewis is often quoted that "what we call Man's power over Nature turns out to be a power exercised by some men over all other men with Nature as its instrument."[71] Selection would best increase the capabilities and choices for children, not decrease them. Children should have an open future.[72] Parents cannot avoid shaping their children, but they can offer more possibilities rather than fewer whether out of beneficence or respect for the child's future choices.[73]

Kirsten Finn Schwandt reports on a couple who desired to be parents and who both had achondroplasia. Inheriting this gene from both parents is lethal, but one normal copy from one parent and the achondroplasia allele from the other results in dwarfism. Such parents have a one in four chance of a child inheriting the lethal form, a 50 percent chance that the child will receive a mix resulting in dwarfism, and a one in four chance that the child will inherit the normal alleles from both parents resulting in more typical height. The couple specifically requested that zygotes be tested so that they might select only the ones who would be dwarfs. They believed that achondroplasia is not a defect but a way of life that they wished to share fully with their children.[74] Tom Murray cites a parallel case of deaf parents wanting to implant only zygotes that would lead to a deaf child.[75] While people with achondroplasia or hearing loss are full and welcome citizens, it is difficult to picture how deliberately seeking that condi-

70. "[U]nder the conditions of the mobile society, child rearing and pedagogical enrichment are part of status maintenance work. . . ." Elisabeth Beck-Gernsheim, *The Social Implications of Bioengineering*, trans. from German by Laimdota Mazzarins (Atlantic Highlands, N.J.: Humanities, 1995), p. 45.

71. C. S. Lewis, *The Abolition of Man* (New York: Macmillan, 1957), p. 34.

72. Joel Feinberg, "The Child's Right to an Open Future," in *Whose Child: Children's Rights, Parental Authority, and State Power*, ed. William Aiken and Hugh LaFollette (Totowa, N.J.: Rowman & Littlefield, 1980).

73. Dena S. Davis, "Genetic Dilemmas and the Child's Right to an Open Future," *Hastings Center Report* 27, no. 5 (1997): 5.

74. Kirsten Finn Schwandt, "Personal Stories: Cases from Genetic Counseling," in *Genetic Testing and Screening*, pp. 45-46.

75. Thomas H. Murray, *The Worth of a Child* (Berkeley: University of California Press, 1996), pp. 115-16.

tion for children would increase the opportunities their children would have in life.[76]

Tentative Pregnancy

A later intervention that our couple facing Tay-Sachs could pursue is beginning pregnancy the usual way and then aborting if a genetic test showed that the fetus was afflicted with the disease. Current case law in the United States allows abortion until the baby's head has left the birth canal. The professor of law John Robertson argues for such complete freedom to abort in that having or not having children has a profound effect on one's life. For Robertson procreative liberty should be strongly presumed.[77] People should have the right to use or not use contraception or selective abortion. What Robertson is advocating here should not be confused with a moral argument as to what one should do. Robertson is articulating a case for a negative right to government noninterference. The fact that the law does not require or forbid something by threat does not address whether the action is a good one. Having a right to do something does not mean that it is the right thing to do. The law does not forbid manipulating a trusted friend, but morality does. The law is written in regard to a bare minimum of actions so egregious that the state will intervene by force. Such intervention is designed to be indiscriminate in an attempt to treat people equally even though they are rarely equal in many regards. Although legal toleration is often interpreted as if it revealed societal acceptance or indeed approval, that something is legal does not confirm that it is moral. How abortion should be treated by law is an important subject for many other books.

What I am pursuing in the rest of this chapter is the moral question of whether it is good to use genetic testing to establish if a fetus has unwanted characteristics and then abort the fetus when the rejected traits are present. To be honest about what we are discussing, it takes considerable effort to end the life of a normal fetus. Current techniques include cutting the fetus into pieces for extraction, at later stages of pregnancy injecting a saline solution to

---

76. A parallel discussion is whether cochlear implants should be provided for the children of hearing impaired parents. A number of deaf parents have refused permission for the treatment, preferring that their children be fully part of their deaf world and culture. See L. Swanson, "Cochlear Implants: The Head-on Collision Between Medical Technology and the Right to Be Deaf," *Canadian Medical Association Journal* 157, no. 7 (1997): 929-32.

77. John A. Robertson, *Children of Choice: Freedom and the New Reproductive Technologies* (Princeton: Princeton University Press, 1994).

burn the fetus to death so that the womb will then expel the now-dead body, and chemically triggering the normal birth process early enough that the fetus is expelled from the womb before viability.

Current Diagnostic Techniques

The common AFP test can be done quite early from a sample of a woman's blood to see if her early pregnancy shows signs of possible spina bifida. Since some fetal cells enter the mother's blood stream, it is hoped that more extensive tests may also be possible in the same manner. Fetal cells are few and far between, but there are ongoing investigations to see if they can be found for such testing. That would make genetic testing available without directly risking the embryo or fetus. Current techniques can cause miscarriage. Chorionic villus sampling is available at ten to twelve weeks gestation. The process itself seems to cause miscarriage about 1 percent of the time, depending on the number of passes and gestational age. Early amniocentesis can begin at thirteen to fourteen weeks, but amniocentesis is more commonly done about sixteen to twenty weeks. With amniocentesis there is about half a percent chance of causing miscarriage.

Amniocentesis usually proceeds to chromosome analysis. About half of the chromosomal anomalies found are Down syndrome. Among women who have an amniocentesis, the report of Down syndrome leads to a high abortion rate. One study in New York City hospitals registered a rate of 90-95 percent.[78] Another in the Midwest measured 93 percent,[79] and by the National Health Service of England 92 percent.[80] In contrast, in another study less than half the women who received a diagnosis of Klinefelter syndrome aborted.[81] Klinefelter syndrome involves sterility and an increased chance of having a learning disability. In either case pregnancies become tentative for many months of their duration. Women hesitate to announce they are pregnant until they have the test results and decide to continue the pregnancy. Post-amniocentesis abortion falls at five to six months. The fetus has probably

78. Shane Palmer et al., "Follow-up Survey of Pregnancies with Diagnosis of Chromosome Abnormality," *Journal of Genetic Counseling* 2, no. 3 (1993): 139-52.

79. Arie Drugan et al., "Determinants of Parental Decisions to Abort for Chromosome Abnormalities," *Prenatal Diagnosis* 10, no. 8 (1990): 483-90.

80. E. Alberman et al., "Down's Syndrome Births and Pregnancy Terminations in 1989 to 1993: Preliminary Findings," *British Journal of Obstetrics and Gynaecology* 102, no. 6 (1995): 445-47.

81. Arthur Robinson et al., "Decisions Following the Intrauterine Diagnosis of Sex Chromosome Aneuploidy," *American Journal of Medical Genetics* 34, no. 4 (1989): 552-54.

been felt independently moving and kicking in the womb for a month or two by that point. Such late-term abortions are difficult for all involved. Genetic testing could move screening to just after fertilization or gametes before fertilization. A crucial question in each case is still at what point there is someone present to nurture. That was discussed in detail in chapter 5. Here we will note that whenever testing occurs, the very process of testing implies that termination is an option.

Nondirective Counseling and Genetic Pregnancy Testing

Pregnancy testing bears all the problem of probabilities that were discussed in chapter 7. How sure is the prognosis? How serious is the predicted condition in a particular case? Down syndrome varies dramatically in its severity. The presence of a third chromosome 21 does not reveal how serious its effects will be. Offering and pursuing tests to establish this ambiguous and highly charged information is directive. Amniocentesis depends on the choice that at the maternal age of thirty-five it is better to accidentally kill one in two hundred healthy fetuses (one half of one percent miscarriage rate) than to allow the birth of a Down syndrome child.[82] Actually low birth weight associated with maternal diet or smoking is statistically a much greater threat than the likelihood of genetic problems included in current standard counseling.[83] Angus Clarke writes emphatically that

> ostensibly non-directive counseling in connection with prenatal diagnosis is inevitably a sham, not because of a personal failure on the part of the genetic counselor but as a direct result of the structure of the encounter between counselor and client . . . an offer of prenatal diagnosis implies a recommendation to accept that offer, which in turn entails a tacit recommendation to terminate a pregnancy if it is found to show any abnormality.[84]

The process of testing itself is directive. Further, the woman is deciding within a powerful social context. Barbara Rothman writes

> I find the language of individual choice untenable in this situation: women are asked to "choose" whether to bring a child with certain genetic predis-

82. D. Alexander et al., "Risks of Amniocentesis," *Lancet* 15, no. 2 (Sept. 1979): 577-78.

83. Rayna Rapp, *Testing Women, Testing the Fetus: The Social Impact of Amniocentesis in America* (New York: Routledge, 1999), p. 70.

84. Angus Clarke, "Is Non-directive Genetic Counseling Possible?" *The Lancet* 338 (19 Oct. 1991): 998, 1000.

positions into the world, but they are not given choices about the environment in which that child would live. When a woman "chooses" abortion rather than bringing to birth a child with a particular condition or predisposition, she is doing that in a world that sets the parameters of that child's life just as surely as genes do. Abortion . . . is always a choice in context.[85]

Part of that context is that increasing options can eliminate options.

Our choices to regard certain things as choices form our selves and our societies. Consider, for example, the life of a night clerk at a convenience store. One determinate feature of her life is frequently identified on the front door: "The night clerk cannot open the safe." Now, suppose that in order to maximize her freedom we give her the option of opening the safe. But to increase her options in this way is to change the determinate features of her life, and not happily or innocently. Not happily because, given the vulnerability of a night clerk, to change the determinate features in this way is to minimize her security. And not innocently — because under the cover of maximizing options we train ourselves to regard the vulnerability of others as a matter of moral indifference.[86]

The choice from genetic pregnancy testing is for the fetus to have the condition or no life at all. If a person is present at the stage of pregnancy when abortion is considered, it is no more licit to kill the individual than at any other point of life. At the point where a fellow human being is present there is a powerful obligation to care for that person as any other neighbor. Once one recognizes that a person is present, a central question becomes what is best for that child. Calling abortion "therapeutic" at that point is disingenuous. Killing a fetus is not therapy for the fetus. If therapy is referring to the woman bearing the fetus, it assumes that an unwanted pregnancy is a disease. Again chapter 5 lays out the discussion of how and when the presence of another might be recognized.

On Not Measuring Up

Children need a great deal of unconditional acceptance. We expect a mother to love her baby wholeheartedly, even if the child "has a face only a mother

85. Barbara Katz Rothman, *The Tentative Pregnancy: How Amniocentesis Changes the Experience of Motherhood* (New York: W. W. Norton, 1993), p. xii.

86. Allen Verhey, "Luther's 'Freedom of a Christian' and a Patient's Autonomy," in *Bioethics and the Future of Medicine*, ed. John F. Kilner, Nigel M. de S. Cameron, and David L. Schiedermayer (Grand Rapids: Eerdmans, 1995), p. 89.

could love."[87] Prenatal testing assumes that the parents can withhold acceptance until the fetus is shown to be acceptable and then fully engage that care. The reality of parenthood is that if we abort for spina bifida, we might then have a healthy child who soon after birth has her back broken in a car accident. The child with no sign of Down syndrome could have just as impaired intelligence from oxygen deprivation during delivery. Having children is a risky commitment to another person whatever his or her condition.

Those who bear genetic disease that others have avoided will still need proper care and for that matter may be a special source of God's self-revelation and grace. This later point is particularly prominent for Bouma et al., who write from a Calvinist perspective that traditionally emphasizes God's pervasive providence. Those bearing handicaps have a needed role to play, for God's glory and provision are often revealed through dealing with suffering or weakness whether from a genetic handicap or the cross.[88] As Gilbert Meilaender writes, "We need the virtue of love. Love — that can say, without qualification, to another: 'It's good that you exist.' "[89] Part of Meilaender's concern is aimed at the practice of aborting fetuses that appear to have genetic disease. Meilaender argues that such abortions are not welcoming these new human lives with the "open hearted acceptance" that they should receive.

All human beings are of inestimable worth, but not all human characteristics are equally desirable.[90] It is better to be well than sick, but that does not make people who are well more important than people who are sick. Healing people of a disease does not disvalue people who cannot be healed. Preferring sight should not cast aspersions on people who happen to be blind. As soon in development as a person is present, killing them because they are blind says people who are blind would be better off dead or have no standing to live. If one is killing people at an early age because they have a disability, that casts aspersions on the life of older people who have the same disability.[91] However, avoiding people getting started with a disability is another

---

87. Rothman, *The Tentative Pregnancy,* pp. 6-7.

88. Hessel Bouma III, Douglas Diekema, Edward Langerak, Theodore Rottman, and Allen Verhey, *Christian Faith, Health, and Medical Practice* (Grand Rapids: Eerdmans, 1989), pp. 266-67.

89. Meilaender, "Mastering our Gen(i)es," p. 874.

90. Bruce R. Reichenbach and V. Elving Anderson, *On Behalf of God: A Christian Ethic for Biology* (Grand Rapids: Eerdmans, 1995), pp. 140-41.

91. Adrienne Asch, "Can Aborting 'Imperfect' Children Be Immoral?" in *Ethical Issues in Modern Medicine,* ed. John Arras and Bonnie Steinbock, 5th ed. (Mountain View, Calif.: Mayfield, 1999), pp. 384-88.

matter. A person who already lives should not be disregarded just because differently abled. However, before a person is present, if one is able to do so, one should provide sight. It is better to see than not to see. That is no aspersion on the blind who are already here.

Genetic tests can reveal conditions so severe that bringing to term is only extending the dying process. An example is that of dystrophic epidermolysis bullosa. In that case skin lacks an essential property of connective tissue so that gentle contact such as a caress or simply lying prone produces large blisters which then burst to open sores. It is difficult to stave off massive infection. The internal blistering in respiratory and digestive tracts adds to the pain. Even with aggressive treatment death comes within a few months. Physicians could induce labor and comfort the child in every way possible in its last hours or do the same later at full term. Trisomy 18 is another example of a lethal condition. Trying to extend the dying process for a child in such extremity requires aggressive technological support. If the child is dying, there is an obligation not to make that more difficult. Even when cure is not possible, care continues. It is not praiseworthy to extend pain or a body that will never have consciousness. But once a child is born, even if dying, we should not deliberately hasten death. Making a person comfortable while not deliberately extending the dying process is the model of the hospice movement. One could induce labor and care for the child until the dying process is complete.

In contrast, Down syndrome children, the most often aborted, have some associated physical difficulties yet are in no sense dying or absent. They vary in capability. Most Down syndrome children learn, just not as quickly as average. They can live fulfilling lives. Theresia Degener reminds us that "disability must no longer be automatically equated with suffering, and non-disability must no longer be seen as the precondition for happiness."[92] That is true for child and parents.

Chapter Summary

While our genetic heritage is quintessentially intimate to each of us, we are inextricably part of a family. When one investigates an individual's genes, one is learning a great deal about that person's relatives. Testing an individual tests a family. The test is not just of genes. Family cohesion and openness are

92. Theresia Degener, "Female Self-Determination Between Feminist Claims and 'Voluntary' Eugenics, Between 'Rights' and Ethics," *Issues in Reproductive and Genetic Engineering* 3 (1990): 98.

stressed as one discovers information that affects others or needs information from others for effective personal diagnosis. Children are a special case in that they are not yet adults competent to decide how they should participate. If the genetic testing can help their current health, it is not controversial to proceed. If the testing offers no current service, it should probably be left until children become competent to decide if they wish to know. Parents are in the best position to make that judgment when it is not clear.

Genetic testing also offers decisive procreation. Prospective mates or parents can know some degree of their genetic risk before pregnancy and are responsible for how they take those risks into account. If one has a one in two or four risk of a child dying or being severely handicapped, it would be wise to consider alternatives. We discussed eight possibilities, including the acceptability of changing the usual course of nature. We then considered reasons to pursue or not to pursue zygote screening and what has become tentative pregnancy. Genetic testing in these settings can be helpful to human health but at too high a cost if it means the destruction of a fellow person who is already present and should be welcomed.

CHAPTER 9

Testing Genes and Community

Currently having a genetic test can mean risking the loss of your medical in-surance or your job. It can also lead to pressure to have an abortion. Testing genes increasingly tests how we live with each other. In this chapter we begin with how genetic testing challenges the medical insurance industry and em-ployment in the United States. The more internationally applicable questions of government identification and cost containment follow. Then we will think about what role the church can best play as genetic testing becomes pervasive.

Insurance

Genetic Information and the Current System of Medical Insurance in the United States

The United States currently has a mixed system of financing medical care. Some people pay for even their catastrophic medical needs out of their own discretionary income. Potential medical interventions have become so exten-sive and costly that few can self-finance such care for a severe illness. Most people depend on some sort of insurance or other guaranteed provision. Government uses tax revenues to provide care for many. That includes, for example, retired people covered by Medicare, some of the indigent covered by Medicaid, the military, and the prison population. Millions of Americans who do not have medical coverage provided by government depend on emer-gency rooms, which by federal regulation must accept them if they have a

206

potentially lethal condition. Once an emergency-room patient is stabilized, private hospitals are no longer legally obligated to continue care. Public hospitals usually treat patients unable to pay for their care on a first come first served basis, rationing their resources by availability. Some people are able to afford their own medical insurance premiums and yet more are covered by company group insurance. In the latter case the employee pays a percentage of the cost by payroll deduction. Insurance that is partly financed by the employer puts considerable financial pressure on a company to avoid hiring or retaining employees who have high medical costs or dependents with high medical costs. We will address that situation in the next section. Here we will begin with individual medical insurance. These policies face some of the same challenges from genetic testing that are substantially duplicated on a larger scale by company group insurance.

The traditional insurance system in the United States is based on actuarial fairness. If you choose to operate a business in a high-risk environment, the rewards may be great but the cost of insurance will be too. The insurance company must charge higher premiums because of the greater likelihood that it will have to pay a claim. Higher insurance costs are part of what one factors into voluntarily providing a particular product or service in a high-risk manner or place. The customer with a lower risk pays a lower premium. If an insurance company does not charge a lower premium to the lower-risk customer, another company will offer that lower premium, gain the business, and turn a welcome profit. The free-market system requires the insurance company to tie premium level to actual risk so that it can compete for customers yet not be overwhelmed by expensive claims.

If both the buyer and seller know that the insured cost will be incurred in the future, the premiums will be set at least high enough to cover the planned claim. Insurance is then a prepayment plan, a type of saving toward a known expense. What insurance companies have usually offered is not a prepayment plan but rather a sharing of risk for costs that are statistically likely to fall to someone but are not yet assigned to anyone in particular. In other words, people buy insurance to cover needs that they might never have but will need help to deal with if they do. Those who never receive any payment back have still gained from the assurance that the money was there if they needed it. By putting in more money than they took out, they have provided care for those who did turn out to need it.

Personal medical insurance may have some characteristics of a savings plan when it is used to cover expected and regular services such as an annual physical, but it is usually focused on providing for the catastrophic surprise expense of severe illness. It depends then on an equality of ignorance between the insurance company and the insurance buyer. Neither knows what the buyer's future medical needs will be. If a medical insurer has indications from age, blood tests, family history, and personal habits such as not smoking that a person is not likely to need medical care, the company will offer a lower premium. Not as much money has to come in from this type of individual to cover payouts. If the company does not offer the individual a low premium, someone else will and the insurer loses that business.

If the insurer has reason to believe that the person seeking a policy is likely to need medical care, the company will either raise the premium, exclude the costly condition from coverage, or not offer a policy. Smokers pay a higher premium than nonsmokers because those who smoke are likely to need more medical care. If the insurer does not charge a higher premium for higher risk, people at high risk will notice that the company's premium is lower for them than that of competitors. They will flock to the insurer that is less expensive for them, and more money goes out to them in payouts than their premiums bring in. This is called adverse selection and can quickly bankrupt the insurer, at which point no one who has been paying premiums to the company has any functional insurance. Insurance companies have to tie premiums to risk to survive in a free-market system.

There are already millions of Americans who find themselves without insurance under the current system. They are fine as long as nothing goes wrong with their health. If they do suffer serious illness, they either go without care, lose work to days in line in possibly available free clinics, or wait until the condition is so critical that emergency rooms are required to treat them. Most of these people without medical coverage are working full time. They make too much money to qualify for Medicaid but not enough to cover even the basic premiums for medical insurance. For example, if one earns seven dollars an hour, that is a full-time salary of fourteen thousand dollars a year. In the year 2000 a standard family medical policy with large deductibles began at about eight thousand dollars a year. Just the monthly premiums have already accounted for more than half of annual income before even considering medical-care deductibles, food, clothing, shelter, and taxes. Our society has ignored this situation because even though it involves millions of people, their percentage of the general population is relatively small, and as a group they have not been savvy about how to gain public attention. Our current system rations medical care by ability to pay.

The number of Americans without medical coverage is increased by those with high-risk indications. They may have sufficient funds to pay typical monthly premiums but not enough to afford the premiums their risk dictates. They may have substantial funds to cover high premiums but no insurance company is interested in offering them a policy, or at least not one that will cover the likely future need. While our current knowledge of future maladies is relatively minimal, there are people now who cannot obtain insurance because of high-risk indications such as a positive blood test for HIV or a family health history riddled with a particular disease. Their numbers are substantial but relatively small compared to the size of the general population. This is the group that is likely to expand rapidly with the increasing use of genetic tests.

While genetic tests do not always indicate an assured future affliction, they can substantially add to knowledge of likely future conditions. That is immensely helpful to medical care. As discussed in chapter 7, such information can lead to treatments in an increasing number of cases, heightened vigilance to lessen the effects of some disease, and better planning when one knows what one faces. These are compelling reasons to offer genetic testing. Physicians will order the tests to improve care and to defend themselves from liability for failing to offer available care. The problem for individual medical insurance is that this newly available information dispels the old foundation of an equality of ignorance between the insurance company and the insured. Under the current system the increased knowledge of likely outcomes leads either to bankrupting the insurers if they *do not* have access to the information or to people being excluded from medical coverage for the very conditions they are most likely to have if insurance companies *do* have access to the information. If the latter route is chosen, as genetic tests multiply each year more Americans will find themselves in the quandary of either losing their coverage because the test results show that they will probably need care or maintaining their coverage by not finding out what care they need.

If insurance companies do not have access to the results of client genetic testing, they fear adverse selection. If the report from one's tests is that one is quite likely to need extensive medical care in the future, the prudent person would buy the most comprehensive plan available. It would be expensive but more than likely to pay for itself. If the report is that one has hit the genetic lottery and is quite unlikely to need medical care, one will naturally put more of one's limited resources into other desirables such as housing or education. A more limited policy for accident or other surprise would seem adequate. Now if insurance companies cannot tell who is likely to be sick, but insurance buyers can, the likely to be sick will sign up in greater numbers and for more coverage than those who are likely to be well. Pay-outs will overwhelm pay-ins and the insurance company will look for new avenues of business or go bankrupt. Then no one has insurance. This process of buyers having more information than insurers and naturally choosing accordingly is the adverse selection feared by insurance companies.

209

The insurance companies need a level playing field to survive in the current risk-based system. Knowing as much predictive genetic information as the buyer, they will offer lower-cost policies to those who do well in the genetic lottery. The competitive necessity is to refine risk analysis. Low-risk buyers are offered low premiums because they will probably need less in return. If a company does not offer the lower-risk buyers lower premiums, another company will, and will make a handsome profit with the added business. High-risk customers will be charged higher premiums or not be insured at all for likely conditions. "The industry has announced its intention to use genetic information the same way it uses other predictive medical information, and defends the economic necessity and fairness of doing so."[1] Note again that this is not rapacious company management. This is the required result of the traditional risk-based system in a new context of greater knowledge.

Medical insurance companies do not need to directly require genetic testing to avoid adverse selection. A great deal of genetic information can be deduced from other tests. Also those with good results from genetic tests will probably be glad to volunteer their records to obtain lower premiums. They would win the genetic lottery *and* the lifetime medical insurance discount. They would be less likely to be sick or disabled and would put less money into the medical care system. The pool of those unwilling to volunteer their results would be recognized as high risk, hence calling for higher, indeed maybe unobtainable, premiums. Those who have lost the genetic lottery will find it difficult to find anyone who will insure them for the very condition that is most likely to put them in need of help. This already happens to many people, but their overall numbers have been small enough to be ignored by the general public. With the development of widespread genetic testing, this category will expand to most of us. We may still be able to buy medical insurance, just not for what we are likely to need.

Already 7 percent of those who do not currently have medical insurance in the United States are considered "medically uninsurable."[2] An example is that of children who have a parent with Huntington disease. Most insurance companies will not insure such a child until he or she has been tested for the

---

1. Deborah A. Stone, "The Implications of the Human Genome Project for Access to Health Insurance," in *The Human Genome Project and the Future of Health Care*, ed. Thomas H. Murray, Mark A. Rothstein, and Robert F. Murray Jr. (Bloomington: Indiana University Press, 1996), p. 134.

2. Nancy E. Kass, "The Implications of Genetic Testing for Health and Life Insurance," in *Genetic Secrets: Protecting Privacy and Confidentiality in the Genetics Era*, ed. Mark A. Rothstein (New Haven: Yale University Press, 1997), p. 305.

gene and found free of it.[3] There has already been a documented case of an insurance company telling prospective parents that it would not cover a baby with cystic fibrosis if the positive test was ignored and the baby brought to term.[4]

Even if genetic intervention becomes highly cost effective for a lifetime of care, that may not be compelling for insurance companies experiencing rapid turnover in whom they insure. Their return on preventative treatment has to be almost immediate to be financially attractive. Otherwise they are spending money to lessen payout for a future insurance company that may not be them. Diagnosis and abortion for expensive conditions have an immediate financial benefit. That will be attractive to insurance providers. The economic pressure is to test and abort any pregnancy that is likely to produce a child with an expensive condition. The insurance company may say you are free to reject testing, but if you have an expensive and avoidable child, that expense is yours. Since caring lifelong for a handicapped child is even more expensive than for a healthy one, such an economic threat would often be daunting.

It is likely that with relatively inexpensive bundled tests and analysis of the entire genome available, it will become evident that almost every individual will be carrying at least some sequences that place them and their children at risk. Probably everyone carries deleterious genes of one sort or another. The University of Minnesota geneticist V. Elving Anderson estimates "that every person carries five to ten seriously defective genes."[5] These are usually hidden by the function of the normal gene that is naturally paired with the defective one. But their presence heightens risk for children who may be part of the proposed policy as dependents. The logic of a risk-based insurance market requires excluding coverage of the very conditions for which people will need coverage. There are multiple reports that such discrimination from genetic test results has already begun.[6]

3. Kass, "The Implications of Genetic Testing for Health and Life Insurance," p. 306.

4. Glenn McGee, *The Perfect Baby: A Pragmatic Approach to Genetics* (Lanham: Rowman & Littlefield, 1997), p. 11.

5. Bruce R. Reichenbach and V. Elving Anderson, *On Behalf of God: A Christian Ethic for Biology* (Grand Rapids: Eerdmans, 1995), p. 199.

6. Robert H. Binstock and Thomas H. Murray, "Genetics and Long-Term-Care Insurance: Ethical and Policy Issues," in *Genetic Testing for Alzheimer Disease: Ethical and Clinical Issues,* ed. Stephen G. Post and Peter J. Whitehouse (Baltimore: Johns Hopkins University Press, 1998), p. 166; Ted Peters, *For the Love of Children: Genetic Technology and the Future of the Family* (Louisville: Westminster/John Knox, 1996), p. 87.

Protecting Genetic Information as a Special Case

Federal and state legislation has sought to lessen the threat of genetic discrimination. For example, in 1997 North Carolina passed Senate Bill 254 that prohibits discrimination against any individual due to the results of a genetic test in regard to health insurance and employment. While this offers some protection, the North Carolina statute and many others do not address premiums. Premiums can still be raised to unobtainable levels to meet the now-known significant risk. The policy then becomes an unaffordable prepayment plan, rather than a pooling of risk.

Many states have passed laws that prohibit denying coverage for someone with a positive genetic test that remains asymptomatic. "Asymptomatic" implies that one has the condition but not visible symptoms. But in the case of genetic propensity, one may never have the condition. This is a point of confusion considering the definition of "preexisting condition." Does that include having a genetic predisposition even if it may never manifest itself in actual symptoms? Also current law does not seem to protect the individual once the symptoms are otherwise apparent. Mark Rothstein asks why we should protect those without symptoms but not those with the disease visibly in progress. He then asks another question. Why single out genetic information for antidiscrimination? Why protect the 5 percent of women with breast cancer whose cancer is genetically inherited and not the 95 percent without a clear genetic cause? If they are protected, why not people with equally devastating diseases that do not happen to be primarily genetic? Nongenetic testing can lead to similar results of people excluded from the very coverage that they most need.[7]

Compared to other revealing information, genetic testing results might be a special case in some ways. With genetic testing entire families can be implicated, the results can affect choices about having children, eugenics is part of our recent history, and both stigma and discrimination are associated with its labels.[8] But treating genetic information as distinct data would require a separate set of medical records for patients and distinguishing entangled etiologies. When is genetic involvement sufficient to classify the disease as genetic? Since most diseases have both genetic and environmental com-

7. Thomas Murray, "Genetic Privacy and Discrimination," Ethical Boundaries in Cancer Genetics conference, St. Jude Children's Research Hospital, Memphis, 27 May 1999. Also Jon Beckwith and Joseph S. Alper, "Reconsidering Genetic Antidiscrimination Legislation," *Journal of Law, Medicine and Ethics* 26 (1998): 205-10.

8. NIH-DOE Working Group, *Genetic Information and Health Insurance: Report of the Task Force on Genetic Information and Insurance* (Bethesda, Md.: National Center for Human Genome Research, 1993).

ponents, it is difficult to separate out genetic information as if it were a special category. Genetic information is so continuous with other medical information such as the health history of the parents that the practical issue quickly becomes one of the privacy of all medical information, not just genetic test results.

Insurance companies generally do not need to do tests themselves to assess risk, since they have access to all the medical records of those they insure. Most insurance companies are part of the eight-hundred-member Medical Information Bureau (MIB). Insurance buyers have to waive confidentiality of medical records to get medical and life insurance. This bureau serves as a central clearing house for applicant information. What one company knows from paying claims or considering a policy is available to all of the others. Withholding promised information is fraud. Further, physicians conscientiously write down testing information and the gist of conversations to better care for each patient. Physicians need free access to all information for the best care. The resulting charts and computer databases that aid the physician are available to third-party payers. Physicians must sign off that they have given the insurance company all the relevant information that they have.

When insurance companies ask for information, physician offices rarely prepare special reports. The usual procedure is to simply send a copy of the entire chart. "Insurers inevitably acquire information about individuals' genetic make-up and family history, even if insurers do no testing of their own and ask no specific questions about genetic disease."[9] They have easy access to genetic test results from the physician records already available to them. Family history, for example, has always served as a crude genetic test. If insurers want a particular genetic test that has been forbidden by law, they can require or check a test result that is not as precisely targeted but is still informative. Deborah Stone writes that "when California banned the use of HIV antibody tests in health and life insurance underwriting in 1985-86, insurers tested applicants with the T-cell test instead."[10]

Anonymous genetic testing has been suggested as a solution.[11] In regard to AIDS testing it was successful at freeing more people to be tested, and that experience could be repeated for occasional concerns about single gene disorders. For example, one could anonymously test for Huntington disease. Of

9. Stone, "The Implications of the Human Genome Project for Access to Health Insurance," p. 144.

10. Stone, "The Implications of the Human Genome Project for Access to Health Insurance," p. 137.

11. Maxwell J. Mehlman et al., "The Need for Anonymous Genetic Counseling and Testing," *American Journal of Human Genetics* 58, no. 2 (1996): 393-97.

course this would isolate the tested individual from follow-up information such as newly available treatment. Granted, one could maintain a post-office box under a pseudonym.[12] However, such a method would be unwieldy. Most importantly, testing anonymously does not help current or future treatment. The point of most testing would not be simply to alert one to a possible condition, but rather to reveal a whole range of propensities and characteristics that would aid prevention and treatment. An example would be recognizing which pharmaceutical would help an individual's particular heart condition. This kind of information is immensely valuable and physicians will want to see it directly recorded in a patient's chart. If it is in one's chart, insurance companies have full access to it. If it is not directly given to one's physician, but one discusses it with one's physician merely for interpretation, such will still probably be recorded in one's chart, which is again accessible to insurance companies. Further, as a protection against off-the-record testing, insurance application forms ask a standard question of whether you have had any other tests done. If you say no when you have, fraud has been committed and the insurance company can demand back all paid-out coverage. This right of recision is standard in insurance contracts. If you answer yes to the standard question that you have had a test, the company will probably want the results of the test or not issue insurance.

If one is already within a plan, one could obtain anonymous test results without reporting it to the insurance company, but the test and its results will probably be revealed if you seek counsel from your physician about it. To achieve anonymous testing one would need to pay for both test and anonymous counseling out of pocket. Any treatment then pursued and charged to your insurance company would of course alert the insurance company to your condition. Also if one changed insurance plans, the standard question reappears asking if you have had any tests, and so you would have to stay with one policy. At best anonymous testing could be useful for assurance that one does not have a genetic condition or for lifestyle choices that can proceed without your physician's counsel or treatment. Widespread genetic testing in this manner would be a great loss to people and society of data that could have been more helpful. One's primary-care provider would not have access to information that could aid one's care.

12. Mehlman, "The Need for Anonymous Genetic Counseling and Testing," p. 396.

Providing a Decent Minimum of Health Care for All

Medical costs are so high that most people need insurance to have access to care. If people know that their insurance is at risk when they obtain genetic information, they are less likely to use genetic testing. They and society will lose the benefit of more effective treatment and planning. If people are to gain from the results of genetic tests, they will need a guarantee of provided or affordable insurance. Such is not possible in the traditional system where people with poor results can be priced out or completely excluded.

American society struggles with how to finance medical care. Now we provide medical care for many of our citizens through programs such as Medicare. For the rest we ration by ability to pay. While one can make poor choices such as smoking that increase risks to one's health, most health problems are not of one's choosing or under one's control. People cannot escape many of their risks. This will be dramatically evident as genetic testing reveals to more people that they were born with heightened genetic vulnerability to some illnesses or disabilities. While one can make a strong case for the fairness of actuarial risk for voluntary activity, it may not be as appropriate for medical care. Neither illness nor medical insurance is optional. If a community values its members, as in the Christian tradition of love for one's neighbors, pricing out or explicit exclusion of care for the very conditions for which people most need care would be unacceptable. One might also make the case from fairness. People will vary in what they are able to achieve, but each citizen should have the basic education and health care needed to have a chance at participating. Health care that includes prevention warrants a different system of provision and payment from television sets. What is needed is a system of financing care that everyone automatically joins and maintains at the same contribution level.[13]

This is not automatically a call for government-run medical care with all the clumsiness of government bureaucracy. The best means of implementation deserves a shelf of other books. The point is that people need health care. They will not be able to get it under the current system that economically requires insurance companies to limit exposure to likely care needs. With genetic testing, increasingly most of our medical needs will be in that category of likely care needs. Health care is an important part of access to education and effective employment. If people matter we should find a way to provide a decent minimum to all. We already do this for a large portion of our population through Medicare, Medicaid, military, and prison programs.

13. Mark A. Rothstein, "Genetic Privacy and Confidentiality: Why They Are So Hard to Protect," *Journal of Law, Medicine and Ethics* 26 (1998): 199-200.

Until then, as more people find themselves at risk of losing coverage over the very conditions for which they most need help, a wider sense of concern may develop. If for preservation of self or society, or out of concern for others, one is convinced that all citizens should have access to a decent minimum of medical care, the medical insurance system will have to change to one that guarantees basic access. When the number of affected Americans is large enough to politically demand it, a decent minimum of basic medical care will probably be guaranteed to all. The delivery system is likely to be some variation of either a national medical insurance or federal requirements that insurance companies offer at least the same premium and basic coverage to any applicant. Again the best way to provide access is a subject for many other books. The point here is that with the rapidly growing availability of genetic tests, the current system increasingly will no longer provide care for many of the people who need it, and that will be most of us.

Until such guarantees are in place, it is prudent for many people to avoid genetic testing that will become part of their record. If it affects their risk, the current system will often exclude the condition from coverage as pre-existing or raise premiums out of reach. This is even more the case in regard to disability insurance. The payout tends to be larger since the treatment time is longer. Life insurance is affected by the results of genetic tests as well.[14] Not using genetic tests is a significant loss to the patient who lacks information that might be helpful to understand, plan, prevent, or treat a genetically related condition. It is also a loss to the community in that often a condition that could have been treated more cost effectively by catching it early is now allowed to run its course into more expensive acute care. Yet in the current context, before individuals agree to a genetic test they would do well to consider how it may affect their access to insurance.[15]

Employers

Avoiding Costs

Popular fears have been expressed in movies and fiction that employers will choose employees not for their ability but for their genetic potential. If they did, they would be misunderstanding human genetics. Genotype is not

14. Rothstein, "Genetic Privacy and Confidentiality," p. 200.
15. Kass, "The Implications of Genetic Testing for Health and Life Insurance," p. 314.

phenotype. Human beings are so complex that a rich genetic endowment can be unfulfilled and a relatively poor one can be substantially transcended. Companies seeking to predict future performance would find better indicators in past performance and current ability-based tests than in genetic heritage.

Where genetic tests are likely to be used by employers is in controlling medical costs. Businesses depend on producing more than they consume to survive long-term. That can be pursued by raising their revenue or reducing their costs. Since the current system of private medical insurance in the United States is usually employer based, medical care is often a major factor in company costs.

Fifty-five to 65 percent of employees in the United States are covered by company self-insurance.[16] Many of the other employees are under experience-based policies where if a company's employees incur greater expenses the premium cost to the company and employees increases. In either case, medical care is a significant part of the employer's costs. Between 1980 and 1990 alone the cost to employers of employee medical and dental insurance rose 150 percent.[17] What employers pay for workers compensation insurance also reflects what they have paid out for such claims. Hiring an employee who is likely to later need workers compensation adds financial risk. The link between medical insurance and employment gives a powerful incentive to companies to find ways to keep the presently or potentially disabled or ill out of their work force. This is heightened by the concentration of medical care costs in one subset of employees.

> In any given year, 5 percent of health care claimants consume 50 percent of health care resources, and 10 percent of claimants consume 70 percent of resources. It quickly becomes clear to health benefits managers that if they can eliminate a class of very high cost users, they are going to save the company a lot of money. And those high-cost users do not even have to be employees; they can be the dependents of employees.[18]

The employer has a powerful incentive to keep out expensive employees. That is devastating both to the employee who cannot find work and to society, which will often still pay for the medical care but lose the worker's

16. Kass, "The Implications of Genetic Testing for Health and Life Insurance," p. 302.

17. Kass, "The Implications of Genetic Testing for Health and Life Insurance," p. 309.

18. Rothstein, "Genetic Privacy and Confidentiality," p. 201.

contribution. As of 1998 there have already been documented cases of genetic discrimination in employment despite relatively few tests in use.[19]

Genetic discrimination is expanded by misunderstanding the difference between having a propensity for a disease and actually having a disease. For example, confusion was rife when the genetic basis of sickle-cell disease was first recognized.[20] As described in chapter 1, a person can be genetically more vulnerable than average to a particular disease without ever manifesting the disease. Also, being a genetic carrier for a disease does not mean that one will have the disease. Being a carrier does heighten the possibility, however, of having dependents with the disease and all attendant expenses.

Laws against genetic discrimination in employment continue to change rapidly, so I will not attempt to parse them out here.[21] Whatever the legal terrain, the potential employer has a substantial financial incentive to find ways to uncover likely high-cost care needs. Savings can be significant in avoiding expensive medical care or lost investment in training. Only a system of medical care that finances a decent minimum regardless of where one works will support employers hiring the best person for the job.

Employee and Customer Safety

Kodak, DuPont, and other chemical companies routinely screen employees for inherited sensitivity to the chemicals they work with.[22] This genetic screening is to be sure that workers who are particularly susceptible to a toxin in that workplace work elsewhere. Some of the earliest workplace screens were for acetylation and ataxia telangiectasia. Half of Americans lack the genetic code to make an enzyme needed to detoxify exposure to the chemical arylamine. Their risk of bladder cancer from frequent exposure is significantly higher than that of those who have the fully functioning gene for acetylation. Ataxia telangiectasia is an autosomal recessive genetic disorder that leaves one more vulnerable to a number of cancers. People with this genetic heritage have five times the risk of cancer as that of the average person

19. Paul Steven Miller, "Genetic Discrimination in the Workplace," *The Journal of Law, Medicine and Ethics* 26, no. 3 (1998): 190.

20. Thomas H. Murray, "Genetic Screening for Cancer Susceptibility," Ethical Boundaries in Cancer Genetics conference, St. Jude Children's Research Hospital, Memphis, 27 May 1999.

21. Miller, "Genetic Discrimination in the Workplace," pp. 190-91; Rothstein, "Genetic Privacy and Confidentiality," pp. 201-2.

22. McGee, *The Perfect Baby*, p. 11.

even when they are heterozygote carriers. Six million Americans are carriers.[23] Companies have used tests for these genetic conditions to remove workers from certain chemical work sites. Excluding workers has appeared more cost effective than reducing exposure for all.

But note that these conditions increase susceptibility. They do not mean that those without the alleles are safe from the exposure, just that they are not as likely to be harmed. Risks still remain for any human being regularly exposed to the toxins. It may be initially less expensive to screen out the most susceptible employees than to clean up the work site. But that strategy still leaves a dangerous environment for the workers that remain. It also means that some employees are excluded due to their genes from what is often high-paying work. Testing should not be a distraction from cleaning up the site for every worker.

Where genetic profiling of specific workers may be justified is in the safety of customers. For example, it would be appropriate for airline pilots with a family history of HD to be tested for the gene. HD does not have a sudden onset, so pilots who had the gene could be watched with heightened vigilance but may not have the disease during their flight career and can be productive in the meantime.

Government

Identification

The government of the United States has already moved toward being able to identify all military personnel by genetic markers. The FBI has recently opened a national DNA database to track criminals. We leave DNA traces everywhere. For the World Trade Center bombing investigation, the FBI was able to match one of the bombers to his saliva on a stamp that mailed a letter claiming responsibility. The FBI DNA database is run by fifty states coordinated by common procedures and FBI computer software. Thirteen sites in the human genome are tested for the arbitrary repeat length of certain genetic patterns. That identification can be assured to one in several billion, but there is no medical or appearance information involved.

When a government begins compiling a genetic database for solving crimes, it becomes imperative to understand that there is no inherently crim-

23. Trudo Lemmens, "Genetic Testing in the Workplace," *Politics and Life Sciences* 16, no. 1 (1997): 60.

inal genome. Wilson and Herrnstein have argued that criminals are more likely to be physical descendants of criminals, even when they never knew those relatives. They also tend to be of lower intelligence, impulsive, and have nervous systems that respond more slowly to vigorous stimuli. Their conclusion is that physical factors do not determine what one does but can predispose one to criminal behavior.[24] Such studies are fraught with difficulties. Where does one find criminals for study? Incarcerated in the criminal justice system. The researchers might be measuring the nervous system response time of people incarcerated for long periods or the lower average intelligence not of all criminals but of those who were caught and doing time for it.

If criminals do tend to be temperamentally impulsive, this does not mean that they are criminals just because they are impulsive. Properly channeled this trait could be refreshing spontaneity or the reflexes necessary to the survival of a fighter pilot. The biological tendency is not inherently harmful; it is a matter of how it is channeled by will and chosen environment. Sin is in the will and the environment we create, not in most biological drives in and of themselves.

Over fourteen thousand adoptees in Denmark were compared to their genetic and raising parents for criminal records. Thirteen and a half percent of the adopted sons had a conviction when neither their genetic nor raising parents did. Of the boys whose genetic parents had a criminal conviction and raising parents did not, 20 percent had a criminal conviction. That is an increase of 67.5 percent. There may be inherited tendencies to being impulsive and a high need for external stimulation that could encourage such activity, but note that 75.5 percent of the boys who had both genetic and raising parents convicted of a crime had no convictions themselves.[25] There is still far more to a person's behavior than even genetic heritage plus environment.

Cost Containment

If government moves toward guaranteeing a decent minimum of medical care for all Americans, there could be significant savings in some categories. Currently about 25 percent of every medical care dollar goes to charging and paying bills, not actual care. The Canadian system spends less than half that

24. James Q. Wilson and Richard J. Herrnstein, *Crime and Human Nature* (New York: Touchstone, 1985), p. 66.
25. British Medical Association, *Human Genetics: Choice and Responsibility* (Oxford: Oxford University Press, 1998), p. 200.

with one standard insurer. Even without assuring medical care for all citizens, the United States government at various levels already provides care for many. Genetic tests that could avoid costly conditions are financially attractive. That is a powerful incentive in a system already under economic pressure.

Screening asymptomatic populations is economical when there is a clear early treatment benefit such as in phenylketonuria (PKU) and hypothyroidism. However, screening for PKU has dramatically raised the number of women with PKU who have reached child-bearing age and are having children themselves. Now that PKU cases are treatable, they are quickly becoming more common.[26] Assessing the economy of an intervention can be complicated.

If the government does provide screening, should it be before issuing a marriage license, during pregnancy, or at another time? The high frequency of thalassemia in Cyprus offers a test case. The disease used to be fatal for afflicted children by the time they were ten years old. With the introduction of a drug called Desferal and the availability of blood transfusions, many patients began to survive on into their child bearing years. This presented an economic challenge for the government of Cyprus.

> The existing thalassemia patients were consuming more than 50% of the available blood supplies, while more than 20% of the total drugs budget of the Ministry of Health was used for the purchase of Desferal. Furthermore, with an expected birth rate of 60-70 new patients per year, the number of patients could double in about ten years, thus stretching the limited blood supplies and other resources to the limit and compromising the quality of care not only for the existing thalassemia patients but for other patient groups as well.[27]

In 1978 the Cyprus government made available free diagnostic testing to couples for carrier testing.[28] Three years later the Holy Church of Cyprus began to require that couples be tested for carrier status before marriage. This was an effort to encourage people to marry only if they would not together have thalassemic children. The birth of new thalassemic patients has dropped to 0 to 2 per year since 1985.[29]

---

26. Diane B. Paul, "The History of Newborn Phenylketonuria Screening in the U.S.," in *Promoting Safe and Effective Genetic Testing in the United States: Final Report of the Task Force on Genetic Testing*, ed. Neil A. Holtzman and Michael S. Watson (Baltimore: Johns Hopkins University Press, 1998), p. 146.

27. Panayiotis Ioannou, "Thalassemia Prevention in Cyprus," *The Ethics of Genetic Screening*, ed. Ruth Chadwick et al. (Dordrecht: Kluwer Academic, 1999), p. 58.

28. Ioannou, "Thalassemia Prevention in Cyprus," p. 59.

29. Ioannou, "Thalassemia Prevention in Cyprus," pp. 61-62.

Rogeer Hoedemaekers and Henk ten Have are concerned that over the years there has been a shift from prevention of conception found in pre-marriage testing for carrier status to selective abortion. This allows couples to marry regardless of carrier status but also makes second-trimester abortion an expected policy. Hoedemaekers and ten Have see significant social pressures in Cyprus to the effect that if parents do not have a prenatal test or an abortion for those who test positive for thalassemia, they are responsible for the burdens that then fall on the child, family, and society.[30] With available prenatal testing one is now responsible for not intervening as well as for intervening. Most alarming to Hoedemaekers and ten Have is that these late-term abortions are not encouraged for the sake of the child but rather to save the difficulty and costs for the parents and society. This is a dangerous precedent. Saying that the decision is up to the parents is not enough. The medical system has framed the decision to a particular end, eliminating costly thalassemia patients.[31]

Combining genetic tests with economics will offer significant incentives to countries to encourage the diagnosis and abortion of a fetus that is probably disabled. Robert Edwards, one of the founders of *in vitro* fertilization, stated at an international fertility conference in France that "soon it will be a sin of parents to have a child that carries the heavy burden of genetic disease. We are entering a world where we have to consider the quality of our children."[32] Margery Shaw has written in the *Journal of Legal Medicine* that to give birth to a child with HD should be a criminal offense.[33] It is more economically efficient to require diagnosis and abortion of those likely to be disabled than to care for them after birth. If it is not directly required, government could assign all financial responsibility to "irresponsible" parents for not diagnosing and aborting. The involved finances are so demanding that such might bear almost as much pressure as a legal requirement. Government could require premarital screening so that couples know the involved risks in having children and could make more informed decisions before pregnancy. If government requires genetic testing during pregnancy, abortion is powerfully endorsed.

---

30. Rogeer Hoedemaekers and Henk ten Have, "Geneticization: The Cyprus Paradigm," *Journal of Medicine and Philosophy* 23, no. 3 (1998): 279-80.

31. Hoedemaekers and ten Have, "Geneticization: The Cyprus Paradigm," pp. 284-86.

32. As quoted by *The American Feminist* 6, no. 4 (1999-2000): 22.

33. Margery Shaw, "Conditional Prospective Rights of the Fetus," *Journal of Legal Medicine* 63 (1984): 99.

The Influence of Public Policy

Public policy is not only economic. There is a symbolic aspect to public policy choices. The public application or restraint of technology not only works toward particular goals but also expresses specific values.[34] That expression of values influences perception, particularly when a process is new and people are just beginning to evaluate it. Once a societal response is in place, attitudes are also shaped by the pervasive assumptions of such structures.

Eric Mount suggests that the old saying "seeing is believing" might be reframed as "seeing is behaving" and "behaving is seeing."[35] When society channels a particular behavior, people tend to accommodate their perceptions to their resulting experience. As their experience changes, so do their perception and further choices. What is legal, for example, is not always moral. Many people would condemn adultery as immoral, yet it is not illegal in most states and it is rarely prosecuted where it is illegal. Yet if society says something is legal, within the realm of what will be tolerated, that acceptance carries an implicit degree of approval that influences people's perceptions, hence their choices.

When such influence is combined with a technology that directly intervenes in something as basic as human physical nature, its effect on people's conception of self, others, and community can become particularly important. A significant factor of a technology as potentially pervasive and shaping as genetic intervention is how it might affect the perception people have of themselves and each other. Perception influences choices and action. How would the government's active involvement in genetic testing affect us? Government policy will not only provide or prohibit certain uses of genetic tests, but it will thereby influence the public's voluntary use. All the implications of engendered attitudes discussed in chapter 8 then come into play.

If it becomes a matter of government support and routine to abort any potentially disabled fetus, it will raise the concern of how remaining people with handicaps would be perceived. The Panel on Bioethical Concerns of the National Council of Churches warns against attitudes of condescension, discrimination, or contempt for those whose genetic endowment remains less than ideal.[36] Care would need to be taken that people who continue to have

34. James F. Childress, *Priorities in Biomedical Ethics* (Philadelphia: Westminster, 1981), pp. 113-14.

35. Eric Mount Jr., *Professional Ethics in Context: Institutions, Images, and Empathy* (Louisville: Westminster/John Knox, 1990), pp. 13-27.

36. National Council of Churches of Christ/USA, Panel on Bioethical Concerns, *Ge-

genetic problems would not be ostracized or rejected.[37] For Hessel Bouma et al., eliminating birth defects too easily slips into eliminating the defective.[38] One's perception here will depend on whether aborting a disabled fetus is killing a person in the womb who happens to be handicapped or intervening before someone has a disability. When we should recognize a person's presence is walked through in chapter 5.

Without government restrictions, will gamete or zygote selection lead to a characteristics arms race? Realizing the benefits in our society that go with height, will parents select the tallest child they can get? If such was a widespread pursuit, nobody would be better off relative to one another among those with access to diagnosis and selection. The intervention would either be self-defeating for competitive advantage or heighten class differences between those who could afford intervention and those who could not. We will discuss that effect more in chapter 12.

A central danger in all the above government concerns is distraction from other change that might be more effective. Genetic intervention should not overshadow alternative use of limited resources that might sometimes be more effective.

Church

The church is a formative community by God's grace for followers of Jesus Christ. As a gathering of people, churches have all human foibles, yet remarkable things have been and continue to be achieved through church communities. There worship, teaching, and caring can help the church's people best use genetic testing.

---

netic Engineering: Social and Ethical Consequences (New York: Pilgrim, 1984), p. 28. Also Allen Verhey, "The Morality of Genetic Engineering," Christian Scholar's Review 14, no. 2 (1985): 137-38, and Kenneth Vaux, Birth Ethics: Religious and Cultural Values in the Genesis of Life (New York: Crossroad, 1989), p. 129.

37. Roger L. Shinn, "Gene Therapy: Ethical Issues," in Encyclopedia of Bioethics, vol. 2, ed. Warren T. Reich (New York: Free Press, 1978), p. 525.

38. Hessel Bouma III, Douglas Diekema, Edward Langerak, Theodore Rottman, and Allen Verhey, Christian Faith, Health, and Medical Practice (Grand Rapids: Eerdmans, 1989), pp. 266-67.

Worship

The church is a community of worship where God is at the center. That keeps the body and genes in perspective. Life depends on the physical body, but the physical body is not what life is about. Worship should be a place where the body is appreciated, like all other good gifts of God, but also recognized as diminutive compared to who God is and God's plan for our lives. This perspective is described in more detail in chapter 3 and in the book conclusion.

Teaching

The teaching function of the church takes place in formal venues such as preaching and Sunday school. It can also be quite effective in home groups that meet regularly for fellowship, study, and accountability. Churches should consciously help parishioners to think through genetic testing so that it is not suddenly a surprise and spur-of-the-moment decision. There is no need for each member of a congregation to confront these choices as if no one has ever faced them before. My own experience is that there is already a great hunger present in churches for help in framing and starting to think through the questions of human genetics from a Christian perspective. That will only increase as more congregation members find themselves making these decisions. They should not be blindsided by them.

Another teachable moment can be in premarriage counseling. Marriage is under such stress in our culture that it warrants careful preparation more than ever. Establishing if genetic testing might be important for a couple and referring them well could be a great service. Pastors cannot assume expertise in genetics but should be ready to counsel parishioners on a Christian perspective of pursuing and receiving such information. If members of the congregation have been prepared in the broader channels of preaching and teaching, counseling can settle on particular needs of the person or couple present.

Just as medical schools have been dramatically increasing the genetics portion of their already-stretched curriculum, seminaries will need to help students think through the theology and ethics of human genetics. The churches they lead will need their understanding and insight in what is becoming the genetics century.

A number of denominations have seen these developing concerns and have published resources on genetic testing to inform and spur discussion in congregational life. Two were written specifically to help thinking about genetic tests. *Genetic Testing and Screening: Critical Engagement at the Intersec-*

tion of Faith and Science holds nine thoughtful essays by members of the Evangelical Lutheran Church in America.[39] The essays represent a broad range of relevant expertise, including articles by a clinical geneticist, a genetic counselor, the chief financial officer of a genetic technology company, and several theologians. In the same year the Episcopal Diocese of Washington, D.C., presented *Wrestling with the Future: Our Genes and Our Choices*.[40] This booklet is organized as a series of questions with one- or two-paragraph responses. It raises many of the practical issues involved with choosing whether to avoid or do genetic testing. Some denominations have also developed brief position statements. Audrey Chapman has surveyed the range of denominational statements in *Unprecedented Choices: Religious Ethics at the Frontiers of Genetic Science*.[41]

Formal teaching is influential but also benefits from effective modeling. In a healthy church the people know and love one another. There is an openness of victories to celebrate and struggles to share that enables the members to learn from each other's experiences, good and bad. Each member is responsible not only for herself or himself, but for the influence that he or she has on others. Openness about experiences with genetic testing can create teachable moments as people see how others deal with the involved challenges.

Caring

Congregations should develop centers of expertise. There will be godly members of the congregation who have knowledge and experience in dealing with genetic testing. They would not replace the professional counsel of physician or geneticist, but could add a rich resource for people as they confront genetic testing for the first time, a particularly vexing aspect, or begin to assimilate results. The full emotional impact of bad news from a genetic test, as for any bad news, is usually well after the news is first heard.[42] Even good news has its

39. Roger A. Willer, ed., and Division for Church in Society, Evangelical Lutheran Church in America, *Genetic Testing and Screening: Critical Engagement at the Intersection of Faith and Science* (Minneapolis: Kirk House, 1998).

40. Committee on Medical Ethics, Episcopal Diocese of Washington, D.C., *Wrestling with the Future: Our Genes and Our Choices* (Harrisburg, Pa.: Morehouse, 1998).

41. Audrey R. Chapman, *Unprecedented Choices: Religious Ethics at the Frontiers of Genetic Science* (Minneapolis: Fortress, 1999).

42. Harry Karlinsky, "Genetic Testing and Counseling for Early-Onset Autosomal Dominant Alzheimer Disease," in *Genetic Testing for Alzheimer Disease: Ethical and Clinical Issues*, pp. 112-13.

stresses. This is particularly true if the recipient hoped good news would have more impact than it did, had made major decisions that are now recognized as mistaken and irreversible,[43] or feels guilt at surviving what has afflicted family members. Congregations would do well to prepare and refer to people within the congregation for needed support.[44]

The members of the church are to share in one another's joys and help bear one another's burdens. When one suffers, all suffer.[45] Healthy children are often stressful. Many marriages have strained to the point where they did break during the challenge of raising a disabled child. One's healthy spouse and other children still need care. Families with genetic afflictions need help. The church should rally around them. Sometimes even a volunteer to organize and schedule the needed volunteers is crucial. In some ways it is safer to help someone with an acute affliction. The needed help is temporary. Prayer lists at churches tend to be long on hospitalizations and surgeries. Chronic needs can be more daunting. They require commitment, a frightening idea for the general culture. Here the concept of serving in teams might make the difference. An individual church member might be overwhelmed reaching out to a difficult and extended situation. If a group of church members works together, taking turns, significant long-term needs can be handled graciously.

Chapter Summary

Testing genes tests community. The formative influence of genes on individuals and families ripples through our communities. Genetic testing does as well. In this chapter we have looked at four primary areas of community concern.

1. If we value our neighbors sufficiently to assure a decent minimum of medical care for all, genetic testing will require substantial changes in how such care is financed in the United States.

2. Guaranteed access to medical care will be essential to approach full employment.

3. Government is already interested in genetic testing for identification and probably will be increasingly so for cost containment. More than economics is at stake in government policy in that the genetic testing that government provides or requires will influence public perception of its best use.

43. Karlinsky, *Genetic Testing and Counseling*, pp. 112-13.
44. Further help is often available from support groups for people with genetic disorders. A directory of such support groups is available at www.geneticalliance.org.
45. 1 Corinthians 12:26.

4. Churches can play an important role. Through worship, teaching, and support they can help people deal with genetic testing and other genetic interventions that will be discussed in the following chapters.

PART III

GENETIC DRUGS: ADDING GENE PRODUCTS TO THE BODY

CHAPTER 10

Genetic Drugs and the Individual

In the 1700s British sailors began to be called "Limeys" after the vitamin C-rich limes they took on long sea voyages to avoid scurvy. We have long been intentional about providing diet supplements that the body needs to sustain itself. Reading the genetic code reveals the "recipes," often complex, for many substances that the body needs. With that information these substances can be assembled and introduced into the body of a person who previously lacked them. The first such genetic pharmaceutical to be widely used was Humulin, approved by the Food and Drug Administration in 1982. Diabetics are unable to make sufficient insulin to regulate the level of sugar in their blood stream. This can result in severe harm such as blindness and eventually death. Insulin from slaughtered pigs was sufficiently similar to human insulin to sustain human diabetics but was not a perfect match to human insulin. With the discovery of the genetic code for human insulin it became possible to place those instructions in bacteria. The single-cell bacteria then follows the genetic instructions as it does the rest of its own genes and churns out human insulin. Human insulin which has never been in a human body is now the treatment of choice for millions of diabetics. The human body is being sustained in its proper functioning by a genetically engineered product made outside the body but to the body's specifications. The body's ability to make its own insulin is not restored, but the body is sustained by making available the insulin it needs.

In chapter 3, I described the mandate to sustain, restore, and improve what has been entrusted to us, including our physical bodies. Part I discussed issues involved in the process of genetic research. Part II worked through ethical questions raised by genetic testing. Here in part III we are looking at genetically synthesized pharmaceuticals. In current discussions genetic pharmaceuticals are often included under the category of gene transfer because that is the method used to make them. The genetic sequence that guides the making of a particular substance is transplanted from one organism to a recipient. The new host begins to synthesize the desired substance, which is then available for a person who lacked it. Since this book is focused on the best use of genetic technology rather than the description of laboratory techniques, I have divided what is usually discussed as one category of gene transfer into the two applied categories of genetic drugs and genetic surgery. Gene transfer is often technically important to achieving each, but in application genetic drugs and surgery are significantly distinct. Genetic pharmaceuticals involve no direct alteration of a person's genes. Instead the human being is receiving a genetically synthesized product. In contrast, genetic surgery to some degree establishes or changes a person's genetic endowment. Genetic surgery is also a broader term than gene transfer. It includes gene transfer, as well as gene removal, inactivation, or duplication (as in cloning).

Considering genetic pharmaceuticals before genetic surgery has at least three advantages. First, our discussion proceeds in the order of widespread availability of the intervention. As described above, Humulin is a genetic drug that already has a two-decade history of effective use. Genetic surgery is only beginning. Second, addressing genetic drugs as a separate category helps to keep our complicated task of analysis more manageable by raising issues such as genetic intervention for cure or enhancement before adding the further complication of directly changing a person's genes. Third, focusing on genetic pharmaceuticals at this point emphasizes continuity with questions already raised by nongenetic drugs.

The example of genetically manufactured human insulin helping diabetics has already been given as an example of sustaining the body. Genetic pharmaceuticals can also be useful in restoring the human body. For example, genetic drugs in development "can stimulate the growth of collateral vessels around severely obstructed arteries in the leg and result in dramatic and lasting clinical improvement."[1] Such treatment might well become the preferred long-term therapy when heart muscle is starved by blocked arteries.

Genetic drugs also have the potential to be used for the improvement of one's body. When one receives a vaccine, one is improving the body's immune system by teaching it to quickly recognize an enemy organism it has not encountered before. New vaccines to improve one's immune system may

---

1. William B. Schwartz, *Life Without Disease: The Pursuit of Medical Utopia* (Berkeley: University of California Press, 1998), p. 129.

be genetically designed and targeted. Hepatitis B vaccine is now produced in vats of genetically modified yeast. The technique is safer and less expensive than the former method of using human blood. Also a genetic pharmaceutical that restores function will often be capable of improving it. A gene product to increase muscle mass of a patient with degenerative muscle disease might be able to provide the increase in muscle mass desired by a competitive weight lifter.[2] Ritalin can be used to help a child with attention deficit disorder to focus better, but it can be used by an adult as well to better concentrate at work or to enjoy a contemplative experience. Pharmaceuticals developed for one purpose are sometimes quite effective for another. Once a substance is available, it is difficult to control off-label uses.

Many thoughtful people argue passionately that any human genetic intervention should only be used to cure disease or maintain health. Sustaining and restoring the body are welcome, but not attempts at improvement. In 1985 W. French Anderson, a leading scholar in molecular hematology then at the National Institutes of Health, may have been the first to use the now standard term *enhancement* for genetic attempts at improvement.[3] He stated that "on medical and ethical grounds we should draw a line excluding any form of enhancement engineering. We should not step over the line that delineates treatment from enhancement."[4]

The Medical Ethics Code of the American Medical Association affirms that genetic intervention should "be utilized only for therapeutic purposes in the treatment of human disorders — not for the enhancement or eugenic development of patients or their offspring."[5] European medical research councils have affirmed correcting specific genetic defects while stating that at-

---

2. Example from Erik Parens, "Is Better Always Good? The Enhancement Project," in *Enhancing Human Traits: Ethical and Social Implications,* ed. Erik Parens (Washington, D.C.: Georgetown University Press, 1998), p. 2.

3. In congressional hearings and "Human Gene Therapy: Scientific and Ethical Considerations," *The Journal of Medicine and Philosophy* 10 (1985): 275-91. Also in regard to human genetics, but with quite a different process in mind, Hermann J. Muller called for "*enhancing* genetic selection" (Muller's italics) in a 1963 article advocating voluntary artificial insemination by selected donors: "Genetic Progress by Voluntarily Conducted Germinal Choice," in *Man and His Future,* ed. Gordon Wolstenholme (Boston: Little and Brown, 1963), pp. 247-62.

4. W. French Anderson, "Genetics and Human Malleability," *Hastings Center Report* 20 (Jan./Feb. 1990): 24; W. French Anderson, "Human Gene Therapy: Why Draw a Line?" *The Journal of Medicine and Philosophy* 14 (1989): 681-93.

5. As quoted by Arash Kimyai-Asadi and Peter B. Terry, "Ethical Considerations in Pulmonary Genetic Testing and Gene Therapy," *American Journal of Respiratory and Critical Care Medicine* 155, no. 1 (1997): 7.

tempts at enhancement should not even be contemplated.[6] At the Council for International Organizations of Medical Sciences Twenty-fourth Round Table Conference in Inuyama, Japan, the "Working Group B: Genetic Screening and Testing" concurred that "the paramount guiding principle in the proper use of genetic services must be the concern about an actual or possible health problem."[7]

In contrast, at the same conference the "Working Group C: Human Gene Therapy" acknowledged the distinction between cure and enhancement as important, but called for more discussion, not a ban, on whether enhancing normal capacities might be appropriate.[8] Paulina Taboada writes that they are not. "To draw a distinction between the use of gene transfer techniques to enhance traits in relation to the treatment of health problems and their use to enhance or improve human traits *per se* seems to be of the utmost importance."[9]

Distinguishing between intervention to cure disease and enhancement is so commonly cited as the line between right and wrong use that I will devote the rest of this chapter to thinking through what the differences are between cure and enhancement. That will be relevant to the best use of genetic pharmaceuticals and later in the book to genetic surgery. The distinction depends on how one defines disease and health, so that will be our focus. Three definitions of disease and seven definitions of health will be described. One observation will be that the divide between cure and enhancement is often difficult to distinguish at any given time, but, more important to this discussion, all but one of the definitions welcomes physical improvement over time. The definition that rejects change is not persuasive. Since the more compelling definitions accommodate physical improvement, over time there is actually little difference between the ultimate goals of cure and enhancement. More clear and relevant criteria are needed to guide the use of genetic pharmaceuticals and any other genetic intervention. Such an alternative standard is offered in chapter 11.

6. H. Danielson, "Gene Therapy in Man: Recommendation of European Medical Research Councils," *The Lancet* 1 (1988): 1271.

7. Council for International Organizations of Medical Sciences, "The Declaration of Inuyama and Reports of the Working Groups," *Human Gene Therapy* 2 (1991): 126-27.

8. "Declaration of Inuyama," pp. 128-29.

9. Paulina Taboada, "Human Genetic Enhancement: Is It Really a Matter of Perfection?" *Christian Bioethics* 5, no. 2 (1999): 192.

Intervention Only to Cure Disease

Some ethical systems emphasize choice guides other than lines of demarcation. For example, virtues can take central place. Virtues are "a kind of second nature that dispose us not only to do the right thing rightly but also to gain pleasure from what we do."[10] Even from this perspective emphasizing character, descriptive lines may be used to help recognize what is indeed virtuous. Another focal point for ethical evaluation can be the situation itself met by responsive flexibility, yet here one still typically uses some sort of standard to judge the particular situation. Joseph Fletcher, famous for his "situation ethics," could be described accurately not so much as a total situationist as a rule-monist.[11] While emphasizing flexibility in context, he still appealed consistently to one rule, to always do the loving thing. Some sort of line-drawing is typical of moral reflection. It is also characteristic of law or other public policy where consistent expectations and results are highly valued. "Line-drawing is the ordinary business of moralists and lawmakers. It says that up to a certain point such-and-such a value will be preserved, but after that point another value will have play."[12]

Some scholars have accepted the possibility of conceptual distinctions but despaired of lines actually functioning in society, yet historically there have been some ethical lines effectively drawn and honored. What can be done is not necessarily what is done. Our society's ban on using prisoners for medical research is a case in point.[13] While the potential utility of a controlled population with a debt to society invited experimentation in the past, a line has been drawn not to use prisoners as research subjects. The coercive environment in prison makes freedom of choice too problematic. That line has been honored in the practice of our society.

Is there a line that should be honored between ethically acceptable and unacceptable genetic intervention?[14] LeRoy Walters has created an influential classification of genetic intervention into four types.[15]

10. Stanley Hauerwas, "Virtue," in *The Westminster Dictionary of Christian Ethics,* ed. James F. Childress and John Macquarrie (Philadelphia: Westminster, 1986), p. 648.

11. Joseph Fletcher, *Situation Ethics: The New Morality* (Philadelphia: Westminster, 1966).

12. John T. Noonan Jr., "An Almost Absolute Value in History," in *The Morality of Abortion: Legal and Historical Perspectives* (Cambridge: Harvard University Press, 1970), p. 50.

13. John Fletcher, "Evolution of Ethical Debate about Human Gene Therapy," *Human Gene Therapy* 1 (Spring 1990): 65.

14. Paul J. M. Van Tongeren, "Ethical Manipulations: An Ethical Evaluation of the Debate Surrounding Genetic Engineering," *Human Gene Therapy* 2 (1991): 73.

15. LeRoy Walters, "Genetics and Reproductive Technologies," in *Medical Ethics,* ed. Robert M. Veatch (Boston: Jones and Bartlett Publishers, 1989), pp. 220-21.

	cure of disease	enhancement of capacity
somatic cells	1	3
germ-line cells	2	4

The classification depends on two lines of distinction.[16] One is the difference between somatic and germline therapy, which will be discussed in chapter 14. The other is a distinction between interventions to cure disease and those to enhance capacity. We will try to discern that line here.

The difficulty of setting an exact line between cure of disease and enhancement of capacity is freely admitted by those who advocate the distinction.[17] Others reject the distinction for this very reason. When a disease is cured, the capacity of the recipient has been enhanced. Cure of disease includes increase in functional capability. Someone cured of blindness has gained the capacity to see. People cured of a fever have regained the capacity to regulate their body temperature and energy to pursue tasks of their choice. Cure of disease can be described as enhancement of capacity, yet all enhancement of capacity may not be cure of disease. Enhancement could be a broader category that includes cure of disease as a subset. The proposed line could be where enhancement achieves more than a specific cure.

Juan Torres offers a variation on this view that what is important is not whether there is an actual enhancement, but rather what the goal of the enhancement is.[18] An enhancement of the body's immune system in general or in its ability to survive a disease treatment is an enhancement beyond the cure

16. Others have developed four-part typologies, although as in the case of W. French Anderson the distinguishing terms are not always the same. In 1985 W. French Anderson termed his four divisions "somatic-cell therapy, germline cell therapy, enhancement genetic engineering, and eugenic genetic engineering" ("Human Gene Therapy: Scientific and Ethical Considerations," *Journal of Medicine and Philosophy* 10 [1985]: 275-91). Richard A. McCormick, for example, uses the same description in *The Critical Calling: Reflections on Moral Dilemmas since Vatican II* (Washington, D.C.: Georgetown University Press, 1989), p. 265.

17. Clifford Grobstein and Michael Flower, "Gene Therapy: Proceed with Caution," *Hastings Center Report* 14, no. 2 (1984): 15; Thomas H. Murray, "Ethical Issues in Genetic Engineering," *Social Research* 52, no. 3 (1985): 488-89; Gregory Fowler, Eric T. Juengst, and Burke K. Zimmerman, "Germ-line Gene Therapy and the Clinical Ethos of Medical Genetics," *Theoretical Medicine* 10, no. 2 (June 1989): 151-65; John Lantos, Mark Siegler, and Leona Cuttler, "Ethical Issues in Growth Hormone Therapy," *Journal of the American Medical Association* 261, no. 7 (1989): 1020-24; Martin Benjamin, James L. Muyskens, and Paul Saenger, "Short Children, Anxious Parents: Is Growth Hormone the Answer?" *Hastings Center Report* 14, no. 2 (1984): 5-9.

18. Juan Manuel Torres, "On the Limits of Enhancement in Human Gene Transfer: Drawing the Line," *Journal of Medicine and Philosophy* 22, no. 1 (1997): 43-53.

of a single disease, yet appropriate since the intent is related to the cure of disease.[19] Enhancement of traits other than those related to avoiding or curing disease would not be acceptable. Torres' distinction, as the others between cure and enhancement, still depends on effectively defining *disease*.

The words *cure* and *enhance* both call for development from the person's present state to an ideal standard. Enhancement works toward a point where improvement is no longer possible in that the optimum state has been reached, while cure of disease works toward being cured, disease free, as an optimal state. Almost any desired change can be described as relief from a negative, hence a cure or attainment of a positive, hence an enhancement. *Cured* and *enhanced* do not of themselves differentiate one end state from the other. Distinguishing between them rests on stipulation of their end goals. What is this "disease" that cure seeks to eliminate?

It will not help us to focus on specifically *genetic* disease. Defining a genetic subcategory of disease raises unique questions such as whether to include recessive heterozygotes or only phenotypes.[20] The problem before us is that of defining the more general term *disease*. The distinction between cure of disease and enhancement of capacity does not depend on the cure of specifically genetic disease, but simply the cure of disease through genetic means. Any disease that could be treated through genetic intervention would be included without needing to define the disease as exclusively or primarily a *genetic* disease. For example, cardiovascular disease is not generally categorized as a genetic disease, yet genetic drugs have been projected as a way to prevent or ameliorate it.[21]

Definitions of disease are many.[22] Since they often set general expectations of health-care givers and others, they have received a great deal of atten-

19. Eric T. Juengst argues a similar tack specifically for the profession of medicine in "Can Enhancement Be Distinguished from Prevention in Genetic Medicine?" *Journal of Medicine and Philosophy* 22 (1997): 125-42.

20. For a detailed analysis of the concept of genetic disease in particular see Eric Thomas Juengst, *The Concept of Genetic Disease and Theories of Medical Progress*, vols. 1 and 2 (Ann Arbor: University Microfilms International, Dissertation Service, 1985).

21. Schwartz, *Life Without Disease*, p. 129.

22. Whether disease is in some sense an entity in itself as in the medieval realism of the Platonic tradition or descriptive of a recurring pattern with no ontological reality outside its particular manifestation, as in nominalism, does not have to be resolved here. If one wishes to pursue the venerable discussion of the metaphysics of disease, a good place to become oriented would be with Lester S. King, "What Is Disease?" in *Concepts of Health and Disease: Interdisciplinary Perspectives*, ed. Arthur L. Caplan, H. Tristram Engelhardt Jr., and James J. McCartney (Reading, Mass.: Addison-Wesley, 1981), pp. 114-18.

tion.[23] This is said with full recognition that individual treatment choices are often influenced more in practice by highly specific guidelines for particular situations than by broad general theories of disease.[24] Theories of disease that are broad enough to account for treatment programs from emergency surgery to counseling for clinical depression are generally not specific enough to guide direct treatment decisions. A physician faced with a child's bulging ear drum and pain is more likely to think in terms of ampicillin dosage than whether "disease" is present. On the other hand, clinical judgments do not occur in a vacuum. In the case of growth hormone therapy, to be discussed more in chapters 11 and 12, one group of writers approves of such therapy only if there is a measurable disease of growth-hormone deficiency or height projection that could be labeled a "deformity."[25] Here basic conceptions of disease and deformity are setting the perimeter for intervention. The concepts of disease brought to bear by the physician, patient, and society in a clinical encounter will set perspective and expectations that influence the considerable latitude of negotiated treatment choices.

I am particularly interested for this study in whether definitions of disease lead to a steady state of maintaining function as we now know it or if over time to improvement in physical capacity. If the latter is the case, cure of disease and enhancement might differ more in speed of change than in goal. Most definitions of disease can be grouped roughly into two sets. These two sets receive various titles, including complaint / functionalist,[26] normativist / neutralist,[27] and evaluatory / explanatory.[28] The labels carry different connotations and emphasis, but in each pair the first term refers to disease as rooted in the preferences of the evaluator and the second to disease as a particular measurable standard of human biology. In other words, disease is either a matter of human preference as to what is undesirable or it is a matter of human biology in failing to meet a physical standard.

23. Mervyn Susser, "Ethical Components in the Definition of Health," in *Concepts of Health and Disease,* p. 94. See also H. Tristram Engelhardt Jr. and Stuart F. Spicker, eds., *Evaluation and Explanation in the Biomedical Sciences* (Dordrecht: D. Reidel, 1975).

24. Mark Siegler, "The Doctor-Patient Encounter and Its Relationship to Theories of Health and Disease," in *Concepts of Health and Disease,* p. 631.

25. Lantos, Siegler, and Cuttler, "Ethical Issues in Growth Hormone Therapy," pp. 1020-24.

26. Siegler, "The Doctor-Patient Encounter and Its Relationship to Theories of Health and Disease," p. 629.

27. H. Tristram Engelhardt Jr., "Health and Disease, Values in Defining," in *The Westminster Dictionary of Christian Ethics,* pp. 261-62.

28. H. Tristram Engelhardt Jr., "The Concepts of Health and Disease," in *Concepts of Health and Disease,* p. 31.

If disease is defined by human preference, eliminating it calls for improvement toward a physical form that best meets human goals. In contrast, definitions of disease based on human biology try to be value-free by building from standard human biology. However, the biology-based definitions face a new situation with the potential for genetic intervention. Human biology can be genetically changed. Definitions of disease based on human biology are built on potentially moving ground. There is only one influential definition of disease and health that has an unchanging application over time. Its standard is not in human preference or biology. We will address that one last. The other definitions offer standards that welcome improvement over time.

Disease as a Value-Free Description

The case for disease as a value-free observation of human biology is made by Christopher Boorse. For Boorse "diseases are internal states that depress a functional ability below species-typical levels."[29] The attempt is to make disease judgments value-neutral. Disease can be recognized as a matter of the natural sciences without evaluation of whether the observed state is desirable or not.

In "Health as a Theoretical Concept," Boorse critiques several alternative definitions of disease.[30] To be disease free is desirable, but many undesirable shortcomings such as clumsiness are not necessarily diseases and some diseases such as cowpox before an outbreak of smallpox are in fact desirable, so value judgments are not definitive. He rejects the definition of medical positivism, that disease is whatever undesirable condition physicians treat, since doctors often recognize diseases that they are not able to treat and do other procedures such as circumcision or cosmetic surgery that are not considered disease treatment. Statistical normality alone would not be a necessary or sufficient condition in that red hair or type AB blood are unusual conditions but not diseases. Pain and suffering are not sufficient in that teething and childbirth are both extremely painful yet not generally considered diseases. Disability is more favorably received by Boorse as a possible definition but requires careful qualification so as not to include, for example, the inability to swim. Adaptation is rejected as too dependent on a given environment

29. Christopher Boorse, "Health as a Theoretical Concept," *Philosophy of Science* 44 (1977): 542.
30. Boorse, "Health as a Theoretical Concept," pp. 544-50.

and homeostasis as too narrow to account for deafness, limb paralysis or other diseases that are not specifically loss of homeostasis.

What Boorse advocates as an alternative to the above is to begin with a reference class consisting of an age group of one gender of a species. Disease is then a type of internal state which reduces one or more functional abilities below the usual efficiency of one's class. By *function* he intends "contributions to individual survival and reproduction."[31] By *normal efficiency* he includes within or above typical functioning. Diseases then are inferences from empirically discoverable species design, requiring no value judgment about what forms of life are desirable. Boorse wants to keep an empirical, value-neutral recognition of physical freedom from disease distinct from "the most controversial of all prescriptions — the recipe for an ideal human being."[32] Note, however, that if all individuals are cured to the "usual efficiency of one's class," some individuals will naturally be better off than average and the average will rise. Over time what was once considered enhancement beyond species typical functioning would become cure to reach the new typical level.

Disease as a Rejected Physical State

In contrast, the role of values is emphasized by Caroline Whitbeck. "Diseases are, first of all, psychophysiological processes; second, they compromise the ability to do what people commonly want and expect to be able to do; third, they are not necessary in order to do what people commonly want to be able to do; fourth, they are either statistically abnormal in those at risk or there is some other basis for a reasonable hope of finding means to effectively treat or prevent them."[33] Here the emphasis is on desired capability with no firm tie to statistical norms or current form. Disease is a normative concept that designates certain states as unwanted.[34] While a particular disease often refers to a set of physical phenomena, the judgment that a particular set of phenomena constitutes a disease is value laden, hence flexible according to the values of the definers. In an 1851 edition of *The New Orleans Medical and Surgical Journal* the desire of a slave to run away was labeled a disease, "Drapeto-

31. Boorse, "Health as a Theoretical Concept," p. 556.
32. Boorse, "Health as a Theoretical Concept," p. 572.
33. Caroline Whitbeck, "A Theory of Health," in *Concepts of Health and Disease,* p. 615.
34. William K. Goosens, "Values, Health, and Medicine," *Philosophy of Science* 47 (1979): 102.

mania."[35] Nearsightedness could be deemed a disease, as could lack of resistance to heart failure. The terms *"disease* and *health* appear to involve evaluation as well as description."[36]

The thesis is not that disease is in every case a net loss, but that every disease involves something unwanted. "Either medicine is blindly bent on curing disease, oblivious to other consequences, or medicine serves the best interests of its patients, but not both."[37] "What defines diseases are true benefits and harms."[38] "There is then no theoretical difference between beneficial abilities never had by a species and those lost by individuals. . . . The moment a newly created beneficial ability became available, persons without the 'treatment' would be considered as lacking something."[39] "Disease increasingly means whatever we have a reimbursable treatment for."[40] By this view medical care should be guided by whatever is perceived as a benefit, not by a concept of proper functioning. Disease is that harm, inherent functioning or not, which the individual would benefit from being without.

One response to this perception of the terms or application as changing would be to further stipulate that by cure of disease one means relief of "suffering, morbidity, and mortality."[41] "Suffering" is still an evaluative bracket that changes in definition as much as the concept of disease that it is meant to specify. "Morbidity" is usually defined as that which is related to disease and so does not help us to define disease. Even relief from "mortality," as the most clear of the three qualifiers, might shift over time in the sense of expectations of appropriate forestalling. Human evaluation remains central even with these three descriptors.

Lester King brings statistical norms back into the formulation of what is disease, but qualifies that requirement to an almost entirely value-determined definition. "Disease is the aggregate of those conditions which, judged by the

35. Samuel A. Cartwright, "Report on the Diseases and Physical Peculiarities of the Negro Race," *The New Orleans Medical and Surgical Journal* 7 (May 1851): 707-9. This article was first brought to my attention by F. C. Redlich, "The Concept of Health in Psychiatry," in *Concepts of Health and Disease*, p. 381.

36. Hessel Bouma III, Douglas Diekema, Edward Langerak, Theodore Rottman, and Allen Verhey, *Christian Faith, Health, and Medical Practice* (Grand Rapids: Eerdmans, 1989), p. 266.

37. Bouma et al., *Christian Faith*, p. 104.

38. Bouma et al., *Christian Faith*, p. 108.

39. Bouma et al., *Christian Faith*, p. 109.

40. David Healy, "Good Science or Good Business?" *Hastings Center Report* 30, no. 2 (2000): 19.

41. John C. Fletcher, "Evolution of Ethical Debate about Human Gene Therapy," *Human Gene Therapy* 1 (Spring 1990): 64.

prevailing culture, are deemed painful, or disabling, and which, at the same time, deviate from either the statistical norm or from some idealized status."[42] When labeling a particular pattern a disease depends on the judgment of the prevailing culture and includes the option of appeal to ideals as well as statistical norms, values again predominate. Cure of disease and enhancement could be indistinguishable by this definition.

Disease as a Hindrance to Basic Function

H. Tristram Engelhardt emphasizes the ambiguity of the concept of disease and traces this characteristic of the term to the presence of both explanatory and evaluatory notions.[43] Descriptive and normative, *disease* describes factual conditions and judges them good or bad. With acknowledged indebtedness to Hegel and Kant, Engelhardt argues that the perception and portrayal of reality is a cultural product.[44] The application of medical knowledge, defining and treating disease, is as pervasively shaped by human goals as any other human activity. Even if one could define, as Boorse does, what is typical of a species, why should that be of interest? Boorse may be defining typical biological function, but for Engelhardt disease has to do with how human beings evaluate that biological function. The definition of disease is not only an observation of a physical pattern but also a judgment of how well the current level of function enables or hinders the individual's goals and then an individual and social construct of what will be treated medically. As a social agreement, what is disease is subject to negotiation that cannot be resolved by appeal to empirical observation. Even if all could agree on the empirical observation that something is typical or atypical, whether that state is desirable or not and the appropriate response would not thereby be automatically resolved.

"What is considered a disease condition in one biomedical tradition may not exist at all in others, as appears to be the case with AD [Alzheimer's disease]. Other diseases only labeled as such in certain cultures include posttraumatic stress disorder (PTSD) and premenstrual syndrome (PMS) in the United States, dropped stomach and *shinkieshitsu* in Japan, neurasthenia in China, *triste tout le temps* and *crise de fois* (liver crisis) in France, and heart in-

---

42. King, "What Is Disease?" p. 112.
43. Engelhardt, "The Concepts of Health and Disease," p. 31.
44. H. Tristram Engelhardt Jr., *The Foundations of Bioethics* (Oxford: Oxford University Press, 1986), pp. 157-59, 194-95.

sufficiency in Germany."[45] There is no universally recognized definition of health, disease, or the goals of medicine. Defining disease is extremely difficult. Lyme disease could have been described as "erythema chronicum migrans," a disease already present and described in Europe. Diagnosis as a new disease, despite similar if not identical pathological basis, affected how it was investigated and the proposal of therapies.[46]

Since for Engelhardt the perception of disease will reflect evaluative goals, response should reflect the personal goals of the recipient and only more broadly that of the community. Such an understanding will protect individual choice but could allow limitless permutations of what is or is not disease. Edward Berger and Bernard Gert have attempted to avoid such idiosyncrasy by describing disease as characterized by "universal evils" that all human beings avoid.[47] Such a definition still reflects social values but seeks some objectivity in wide consensus. However, universal consensus would be difficult to prove, and consensus may vary in degree at many points. Berger and Gert recognize such borderline cases and suggest that they may be ignored, since they are advocating that only the most serious maladies, which by their definition are those that are recognized universally, should be treated.

Joseph Margolis has made an interesting proposal that could be helpful in light of all the above. Disease could be described as any disorder of the body relative to basic prudential function.[48] Basic prudential function would be those capabilities that enable basic necessities of human life. Such a definition which recognizes a strong value component would be more inclusive and goal-oriented than a definition tied to species-typical capacity. Disease would not be merely a statistical abnormality. It would be that which hinders prudential function. Prudential function is desired by virtually all human beings since it is basic to life. By limiting disease to lack of prudential function a recognizable standard is possible.

Would such a definition be objective? Yes. A measurable standard of

45. Atwood D. Gaines, "Culture and Values at the Intersection of Science and Suffering," in *Genetic Testing for Alzheimer Disease: Ethical and Clinical Issues,* ed. Stephen G. Post and Peter J. Whitehouse (Baltimore: Johns Hopkins University Press, 1998), p. 258.

46. Robert A. Aronowitz, *Making Sense of Illness: Science, Society, and Disease* (Cambridge: Cambridge University Press, 1998), pp. 16-17.

47. Edward M. Berger and Bernard M. Gert, "Genetic Disorders and the Ethical Status of Germ-Line Gene Therapy," *The Journal of Medicine and Philosophy* 16 (1991): 671-72, 675.

48. Joseph Margolis, "The Concept of Disease," in *Concepts of Health and Disease,* p. 575.

what was generally necessary to support basic life choices could be recognized and pursued. However, application of such a standard could change over time. It would reflect the environment that challenges and constrains survival, as well as what the society considers appropriate levels of basic life. At a time when almost everyone has eyesight genetically corrected to the best naturally occurring 20/15, signs and public events would assume such, and having 20/20 vision would be perceived as in need of cure up to basic levels. What could be described at one point in time as enhancement could later be categorized as cure of disease.

Each of the above definitions of disease is open to physical improvement that raises the standard expectation of physical capacity. By these definitions the goals of curing disease and enhancing capacity are not distinguishable over time.

Intervention Only to Health

When discussing the point of curing disease, the summary term that is often used is *health*. In the last section we saw that over time the goal of curing disease is not easily distinguishable from the attempt to enhance capacity. Both welcome improvement. Does the definition of health offer a clearer distinction? How is health as the pursued goal of cure of disease different from an ideal pursued by enhancement of capacity? We will look at several definitions.

At the start it is notable that three definitions of health are too ambiguous to guide us on this particular question. The World Health Organization stated in 1958 that "health is a state of complete physical, mental, and social well being and not merely the absence of disease or infirmity." This grand definition of health might call for fixed or changing application, depending on the definition of "complete physical, mental, and social well being." The question has merely moved from what is "health" to what is "well being."

Pope John Paul II gives a similarly expansive definition. "From a Christian perspective, then, health envisions optimal functioning of the human person to meet physiological, psychological, social, and spiritual needs in an integrated manner."[49] It is easy to picture improvements beyond current forms that would more clearly approximate "optimal functioning."

---

49. Pope John Paul II, "The Ethics of Genetic Manipulation," in *Origins* 13, no. 23 (1983): 385, cited by Kevin D. O'Rourke and Philip Boyle, *Medical Ethics: Sources of Catholic Teachings,* 2nd ed. (Washington, D.C.: Georgetown University Press, 1993), p. 8.

A third definition, health as "intuitively self evident," may also be unclear on this point in that whether pursuing health leads to the current status quo or change over time would depend on the individual's intuitive apprehension of that self-evidence. While appeals to self-evidence have a long and august history, including functioning as the opening warrant of the founding document of the United States,[50] it is often difficult to gain consensus by appeals to intuition, and such consensus has not so far occurred on a definition of health.

Definitions That Welcome Physical Improvement

Health as a Statistical Norm

Health could be defined as a statistical norm.[51] Anything outside of a small average range would be unhealthy. This standard has the advantages of being grounded in objective data but raises at least three problems. The first is that few people advocate that health includes eight teeth cavities per person, a life expectancy of thirty-five years as in Uganda, or osteoporosis, which in some populations is a common condition for elderly women. These are statistical norms.[52] Defining health as the statistical norm would affirm as health what many would object to as unhealthy. Second, what is typical changes from society to society and from day to day.[53] Such a definition would not be consistent across time or place. Third, such a standard could produce the oddity of declaring someone with unusual longevity or strength unhealthy, since he or she would be deviating from the small average range. The definition can be qualified that health is the statistical norm plus desirable abnormality, but that qualification would introduce the role of values which leads to a definition the statistical norm definition is often cited to avoid.

Normal health could also be defined as the greatest capacity naturally

50. "We hold these truths to be self evident, that all men are created equal . . . ." Thomas Jefferson, "The Declaration of Independence," Philadelphia, Continental Congress, 1776, p. 1.

51. E. Murphy, *The Logic of Medicine* (Baltimore: Johns Hopkins University Press, 1976).

52. The noting of this ambiguity can also be found in H. Tristram Engelhardt Jr., "Persons and Humans: Refashioning Ourselves in a Better Image and Likeness," *Zygon* 19 (Sept. 1984): 282.

53. Robert M. Veatch, "Ethical Issues in Genetics," in *Progress in Medical Genetics*, vol. 10, ed. Arthur Steinberg and Alexander Bearn (New York: Grune and Stratton, 1974), p. 257.

occurring, without the limit of what is naturally occurring *now*. This standard would be likely to increase with time if one corrects any heritage below average up to the average. Some people will naturally have genetic endowments that combine or mutate a little better. If negatives are corrected, but positive changes are left intact, over time the statistical norm will rise. What was average becomes below average. Correcting to average would lead to incremental and continuing enhancement of physical capacity. Change would be relatively slow, but change would occur nonetheless. While enhancement of capacity is associated with an ideal that may be considerably different from the present state of humanity, and cure of disease is associated more closely with whatever the current state, the two would converge over time.

Health as Whatever Evolution Produces

Christopher Boorse argues for defining health as whatever maintains the evolutionary goals of survival and reproduction.[54] Enhancement might then coincide with cure of disease as a powerful adaptive and sustaining strategy. In contrast it has been argued that we should submit to the cumulative wisdom of evolution by remaining just as we are, but it would actually be difficult to derive that commitment from the "wisdom of evolution." Such an argument would be for prudence, not commitment to the present. If we assume that evolution has adapted humanity so far, its very nature would be to continue to change and adapt to an environment quite different from that in the past. An appeal to the wisdom of evolution would call for respectful care to insure that changes would work as well as ones already long proven but would not offer positive guidance as to what is desirable.

Health as a Means to What Is Valued

Health may be the maximization of those physical capabilities which are valued for the goals they render attainable.[55] This would include the definition presented by Talcott Parsons, "Somatic health is, sociologically defined, the state of optimum capacity for the effective performance of valued tasks."[56] Health is essentially goal oriented, hence varying with people and circum-

54. Christopher Boorse, "On the Distinction Between Disease and Illness," *Philosophy and Public Affairs* 5 (1975): 49-68.

55. Lennart Nordenfelt, *On the Nature of Health: An Action-Theoretic Approach* (Hingham, Mass.: Kluwer Academic Publishing, 1987).

56. Talcott Parsons, "Definition of Health and Illness in the Light of American Values and Social Structure," in *Concepts of Health and Disease*, p. 60.

stances. For Bernard Häring health is "the fullest possible capacity to develop relationships with God, with one's neighbor, and within the community."[57]

These explicitly value-oriented definitions of health are sometimes called "normativist."[58] Feet crippled by binding would be unhealthy in our society, while such was the epitome of health and desirability for upper-class women during part of China's history. Some forms of hallucination that would now be labeled schizophrenia have at times been honored as the highest form of health and special blessing. If health is whatever physical capacity we value, that leaves open the restriction or encouragement of enhancement of capacity according to those values. The definition of health as the maximization of the physical capabilities that we value could be limitless.[59] Whitbeck has argued that "the absence of an upper limit on health does not make that concept any more obscure than concepts such as wealth, which also have no upper limit."[60] Such a definition would coincide with enhancement of capacity.

There has been considerable continuity both geographically and temporally in what people physically value. If conceptions of health are as fluid as human values, why has there not been more variety in how health is perceived? It could be argued in response that despite the great variety of human cultures, survival needs are consistent due to the consistent physical nature of human beings. While kinds of shelter vary from igloos to palm-frond huts, human beings generally require shelter. While food varies from blubber to grubs, all human beings require food. Such common survival issues remain whether living in the Arctic or tropical jungles. Strategies to meet them would vary with circumstances but maintain considerable commonalities. Health could reflect the diversity of human values while still maintaining considerable consensus if it reflects common needs basic to human function.

Health as Tied to Our Current Condition

The Catholic Hospital Association defines health as "a functional whole, in which all necessary functions are present and acting cooperatively and har-

57. Bernard Häring, *Manipulation: Ethical Boundaries of Medical, Behavioral, and Genetic Manipulation* (Slough, U.K.: St. Paul Publications, 1975), p. 56.

58. Arthur L. Caplan, "The Concepts of Health and Disease," in *Medical Ethics*, ed. Robert M. Veatch (Boston: Jones and Bartlett, 1989), p. 57.

59. Elisabeth Beck-Gernsheim, ed., *Welche Gesundheit wollen wir?* (What Kind of Health Do We Want?) (Frankfurt: Suhrkamp, 1995).

60. Whitbeck, "A Theory of Health," p. 616.

moniously."[61] This functional focus is common to a number of definitions. Here it is left unclear as to what is cooperative, harmonious, and necessary. For our inquiry the crucial ambiguity is in the word *necessary*. It may imply basic prudential function, as discussed earlier in Margolis's work,[62] or an appeal to the *current* natural order as a standard may be intended. As discussed in chapter 8, a theme in one part of the varied Christian tradition has been that the current natural order should be supported in that God designed and ordained its present form. Created by God, the natural course as we have received it bears divine authority. This is the one definition of disease and health that is permanently tied to our present state. Health is not defined here by human choice or biology, but rather by God's prerogative as Creator.

We have discussed in chapter 8 the difficulty of discerning God's purpose merely by observing nature. The Christian tradition has often spoken of the corruption of God's good creation due to the human fall into sin. Which natural patterns are the God-given ones? There have been numerous mutually exclusive interpretations. The standard one uses to distinguish what of nature is God-given and what is not or what is the underlying God-given intent, may then be the actual standard. Also, as described in chapter 3, the debate between Irenaeus and Augustine remains on whether our received physical form was the pinnacle of God's intent for human beings or the starting point. If the natural order is perceived as ordained and set, efforts to enhance would be contrary to the God-established pattern. If the natural order is perceived as one which God is not only restoring but also improving over time, enhancement may be part of the fulfillment of human responsibility.

Holding the Line at Cure of Disease

Some of those who reject the distinction between cure of disease and enhancement of capacity because they find it imprecise are concerned about attempts at the cure of disease beginning a slippery slope to attempts at enhancement.[63] By *slippery slope* I am referring to an argument also described by metaphors such as *the thin edge of the wedge*. The slippery-slope argument depends on two assertions. The first is that the projected end result of crossing a particular line is undesirable. The second is that once one crosses the

---

61. Catholic Hospital Association, *Health Care Ethics: A Theological Analysis* (St. Louis: Catholic Hospital Association, 1978), p. 28.

62. Margolis, "The Concept of Disease," pp. 561-78.

63. Bouma et al., *Christian Faith*, p. 266.

line one is on a slope where there is no place to stop. This is due to the lack of a conceptually clear moral distinction at which to stop along the slope or that even if there is a conceptually clear stopping point it would probably not be honored in practice.[64] Bernard Williams calls these two versions of the slippery-slope argument the "arbitrary result" argument and the "horrible result" argument.[65]

C. Keith Boone sees no slippery slope between cure of disease and enhancement of capacity. For Boone the distinction between cure of disease and enhancement of capacity is "seismic," so that a slippery-slope argument at that point would be a "bad axiom," unhelpful and misleading.[66] In contrast, the World Council of Churches paper on genetic engineering claims a conceptual (or arbitrary) slippery slope between cure of disease and enhancement of capacity. "There is no absolute distinction between eliminating 'defects' and 'improving' heredity. Correction of mental deficiency can move imperceptibly into enhancement of intelligence, and remedies of severe physical disabilities into enhancement of prowess."[67]

Willard Gaylin observes a social pattern that he projects will create a practical slippery slope for human genetic intervention. "This technology can and will be used to reduce genetic faults and increase the opportunity for a normal, healthy child. It will also be used to effect changes whose merits are not so cut-and-dried, changes that will be seen as making a more nearly ideal or optimal child . . . we inevitably turn from the replacement of deficiencies to additions for enhancement and ennoblement."[68] This is already our pattern from dental braces to correct a bite problem to braces for that perfect smile.

Sheldon Krimsky seems to argue *both* conceptual and practical slippery slopes at the line between cure of disease and enhancement of capacity.

> Moral rules based upon nebulous distinctions are most vulnerable to slippery-slope outcomes. . . . The distinction between enhancement and medical therapy is a socially constructed category influenced by many factors

64. James F. Childress, "Wedge Argument, Slippery Slope Argument, etc.," in *The Westminster Dictionary of Christian Ethics,* p. 657.

65. Bernard Williams, "Which Slopes Are Slippery?" in *Moral Dilemmas in Modern Medicine,* ed. Michael Lockwood (Oxford: Oxford University Press, 1985), pp. 126-37.

66. C. Keith Boone, "Bad Axioms in Genetic Engineering," *Hastings Center Report* 18, no. 4 (1988): 11.

67. World Council of Churches, Church and Society, *Manipulating Life: Ethical Issues in Genetic Engineering* (Geneva: World Council of Churches, 1982), p. 7.

68. Willard Gaylin, *Adam and Eve and Pinocchio: On Being and Becoming Human* (New York: Viking Penguin, 1990), p. 9.

that contribute to the current taxonomy of clinical disorders. Moreover, once the right of somatic cell therapy becomes established, it is doubtful that its use can be restricted to "medical therapy." Consider all the surgical techniques that are used for cosmetic purposes. . . . No satisfactory moral rule has been advanced that sets boundaries on somatic cell human genetic engineering.[69]

For Krimsky, there is no clear division between cure of disease and enhancement of capacity, and the distinction would probably not be honored in practice even if it was conceptually clear. Before one can hope for a distinction to be honored in practice by society, there needs to be a clear conceptual distinction. The distinction is lacking between cure of disease and enhancement of capacity. The color spectrum shades from red to orange to yellow without precise moments of transition, yet we still refer to distinct recognizable colors. But genetic intervention is not merely a matter of taste. In this case the stakes are much higher. Conceptual clarity is essential if the distinction is to be actually applied.

If disease or health is fixed to the current average and not any future average, there is a clear and permanent distinction between cure of disease and enhancement of capacity. Almost any restoration to health could be characterized as enhancing the recipient's capacity, but one could stipulate that there should be no improvement beyond the present average. One could cure to today's typical level of health and no further. There would then be no conceptual slippery slope. On the other hand, the question of relevance would remain. If curing disease is enhancing an individual's capacity to an average level typical today, why stop there? Why not use the technique to prevent diseases from occurring in the first place or to further augment desired capability? Why limit health and capacity to today's average, which as cures are achieved will become statistically less than average?

If the definitions of health and disease that have changing application are more convincing, there may still be a rough distinction between cure of disease and enhancement of capacity at any given time, but its application would vary over time. The line of division would move. What would be ruled enhancement of capacity now might well fall under cure of disease later. By these definitions the applied result of the distinction would be conceptually consistent but not fixed in application for all time.

Would a changing application of cure of disease necessarily be a dangerous slippery slope? The distinction could still be serviceable as a method

69. Sheldon Krimsky, "Human Gene Therapy: Must We Know Where to Stop Before We Start?" *Human Gene Therapy* 1 (1990): 173.

of prioritization if the projected end state of increased capacity is not rejected. In such a case there would be change, but it would not be uncontrolled or negative as implied by the metaphor of the slippery slope. W. French Anderson, who advocates the distinction between cure of disease and enhancement of capacity, consistently states his affirmation of genetic cure as cure of *serious* disease. This affirmation is often qualified with words such as "initial"[70] or "at least for the time being."[71] For Anderson the diseases chosen for therapy "should and undoubtedly will" expand to less serious diseases as the techniques improve. Early therapy, entailing more risk than when refined, would be more appropriate for those at greater risk from the severity of their disease.[72] The earliest interventions then would be for the most severe diseases. When techniques are proven to be safe and effective, less serious disease could be treated.

Anderson ends one of his articles with the statement, "But until we have acquired considerable experience with regard to the safety of somatic cell gene therapy for severe disease, and society has resolved at least some of the ethical dilemmas that this procedure would produce, non-therapeutic use of genetic engineering should not occur."[73] The implication from the word *until* seems to be that just as less serious disease may eventually be treated, so nontherapeutic intervention, enhancement, might begin at a future date. Since according to Anderson the main current limitation is our ignorance, as our knowledge increases the range of appropriate interventions may increase as well.[74] The distinction between serious and less serious disease, as well as between cure of disease and enhancement of capacity, could be a way to set research and treatment priorities at any given time, even as its precise application shifted over time. Once risk and expense are incurred to intervene, it will probably not be much more difficult to enhance than to restore to a current average.

Genetic pharmaceuticals may offer many such opportunities. In the

70. W. French Anderson, "Human Gene Therapy: Why Draw a Line?" *The Journal of Medicine and Philosophy* 14 (1989): 688, 690, and "Genetics and Human Malleability," *Hastings Center Report* 20 (Jan./Feb. 1990): 24.

71. Grobstein and Flower, "Gene Therapy: Proceed with Caution," p. 15.

72. W. French Anderson and John C. Fletcher, "Gene Therapy in Human Beings: When Is It Ethical to Begin?" *New England Journal of Medicine* 303 (Nov. 1980): 1293.

73. Anderson, "Human Gene Therapy: Why Draw a Line?" p. 691.

74. This position may be a development from his earlier statement in 1985 that interventions should not be for other than therapeutic reasons "regardless of how fast our technological abilities increase"; Anderson, "Human Gene Therapy: Scientific and Ethical Considerations," pp. 289-90.

next chapter I will offer four standards for genetic intervention that may better recognize best use than the often physically and morally ambiguous distinction between cure and enhancement.

Chapter Summary

Adding gene products to the body has already well served many people in the case of Humulin. Genetic pharmaceuticals show great promise for sustaining, restoring, and improving the bodies entrusted to us. Many have argued that uses should be limited to the cure of disease. Enhancements beyond this line would be unacceptable. This distinction depends on the definitions of disease and health. As we studied these definitions, it has become clear that one could use them as a shaded division like colors on a spectrum, yet many border cases would be difficult to categorize and even the family groupings would shift with time. What is sufficient to be called health now could be counted as a disease in the future. One could retain the distinction between cure and enhancement for general research funding priorities. The more clearly an intervention is to cure disease, the more important it is that it should be pursued, but pursuing health while avoiding enhancement is problematic. The problem is both with the indistinctness of the line and with the rationale for holding it. There is no adequate conceptual distinction between cure of disease and enhancement of capacity that would allow us to make a principled argument for cure of disease that would not over time also allow genetic intervention for enhancement. In chapter 11 I will offer an alternative in four standards that do not rely on the popular but problematic distinction between cure and enhancement.

Genetic Drugs and a Family's Children: Four Standards

Genetic intervention should only proceed when the intervention is (1) safe and (2) a genuine improvement for the recipient. It is best for adults, and imperative when deciding on behalf of children, that the intervention also (3) increase capacity for a more open future and (4) be the best available use of limited resources.[1] As discussed in chapter 5, particularly high standards are essential when deciding on behalf of others such as children, who depend on us so completely.

To remind us in part of how this work has come to this point, chapter 3 explained the responsibility of human beings to sustain, restore, and improve ourselves and our world. Chapter 10 described some of the possible uses of genetic pharmaceuticals to carry out that mandate for the human body. It then asked if the frequently cited distinction between cure of disease and enhancement of capacity was the best standard for discerning between genetic interventions that we should pursue and those we should not. My conclusion

---

1. In some ways these four standards parallel the influential principles proposed by Tom L. Beauchamp and James F. Childress in *Principles of Biomedical Ethics*, 4th ed. (New York: Oxford University Press, 1994). Those are autonomy, nonmaleficence, beneficence, and justice. Beauchamp and Childress write that they have abstracted these clusters of principles from our common morality and medical tradition. The four standards proposed in this chapter may receive wide affirmation as well but are presented as grounded and shaped by a particular tradition. For example, the Beauchamp and Childress principle of beneficence, that one should generally contribute to the welfare of others, is more "thickly" described here (under the standard of genuine improvement) as to help others better grow and develop in a way that pleases God, particularly as known in the Christian tradition.

was that when the involved definitions are carefully studied, it becomes clear that over time the pursuit of both cure and enhancement lead to the same improvements. Another frequently drawn line between acceptable and unacceptable genetic intervention is that between interventions that affect one recipient and those that can be inherited by descendants. In chapter 14 I will explain that these are roughly distinguishable physically but largely irrelevant morally. More people are potentially affected by an intervention that is inheritable. In that sense the stakes are higher, but intervention should not be done even for the first patient until its success and safety is substantially assured. Once it is clear that the procedure is helpful, provision for descendants is an advantage. The following four standards are offered as an alternative to help distinguish appropriate from inappropriate human genetic intervention.

Safe

As with the introduction of any new drug, carefully observed and phased introduction embodies the maxim to first do no harm. While virtually no action is completely risk free, risk should be at a minimum proportional to the intended benefit. Standard drug testing begins with gradually increasing amounts of the drug in animal models. When the drug is shown to be safe for animals, then it is introduced in small amounts in just a few human beings. Then larger amounts are tested in larger numbers of human beings. The process is characterized by gradual implementation that puts each individual at minimal incremental risk and the number of people at gradually increasing exposure. Unfortunately some rare untoward effects will not be manifested until large numbers of people are involved over time, but such harms can be minimized by careful testing before introduction to the wider public. This process of gradual and watchful introduction was described in chapter 4.

Because children often respond differently to drugs, safety testing needs to occur specifically in them, but with even greater care since they are a vulnerable group unable to defend themselves. Children's involvement in research was discussed in chapter 5. Before widespread use, a genetic drug should be proven to be safe and effective.

Improvement for the Recipient

Generally an improvement is an improvement. We do not usually question whether it is good to improve or grow. If we are consistent, how we define or recognize improvement will stem directly from what we understand to be the purpose of our lives. Education, athletic training, and proper diet are examples of long-lauded attempts in the Western tradition to help us fulfill that purpose. As described in chapter 3, the Christian tradition describes the spiritual life as most important. One is consciously to train oneself, to be transformed from one glory to another closer and closer to the image of Jesus Christ.[2] Growth is part of the mandate to sustain, restore, and improve what has been entrusted to us. That includes our physical selves and the rest of the physical world. Whether physical improvement should specifically include changing our genes is addressed in chapter 13. For this chapter the question is whether we should improve our physical form by the introduction of genetic products.

What is crucially at issue for this standard of improvement is whether a given genetic change actually *is* an improvement and whether it is a net improvement when all of its implications are considered.[3] On the one hand, injecting a genetically produced vaccine to increase the body's immunity against a debilitating or even lethal disease is easy to recognize as a genuine improvement. On the other hand, because genetics are so basic to human structure and function, apparently small changes can have surprisingly pervasive and varied effects. These can quickly complicate the evaluation of an intervention.

The stakes are particularly high in psychopharmacology in that it can alter the brain and in so doing alter the person.[4] The physical brain and the genes that found its structure deeply shape our perception and behavior.[5] For exam-

---

2. James Keenan describes this process in the Roman Catholic tradition as one of pursuing "perfection," the fulfillment of what one is called to be. For Keenan, properly used genetic enhancement might aid that goal: "A Virtuous Consideration of Enhancement," *Christian Bioethics* 5, no. 2 (1999): 104-20.

3. In the case of research, as discussed in chapter 4, a competent individual can volunteer to define improvement more broadly than just benefit to her own body.

4. The book edited by Warren S. Brown, Nancey Murphy, and H. Newton Malony, *Whatever Happened to the Soul? Scientific and Theological Portraits of Human Nature* (Minneapolis: Fortress, 1998), has been influential in describing the human soul as a distinct yet emergent property of the physical body, while J. P. Moreland and Scott B. Rae have recently responded in *Body and Soul: Human Nature and the Crisis in Ethics* (Downers Grove, Ill.: InterVarsity Press, 2000) that the body and soul are two distinct entities intimately related to each other. For our discussion here, both books agree that the physical body affects the soul and that the soul affects the physical body.

5. For example, Donald W. Pfaff et al., eds., *Genetic Influences on Neural and Behavioral Functions* (Boca Raton, Fla.: CRC Press, 1999); Kenneth Blum and Ernest P. Noble, eds., *Handbook of Psychiatric Genetics* (Boca Raton, Fla.: CRC Press, 1997).

ple, genes may contribute a tendency toward some forms of bipolar disorder or schizophrenia,[6] as well as toward personality traits such as novelty seeking.[7] *The Hastings Center Report* cited a 1997 study that researchers had found one gene on chromosome 17 that seemed to be linked to bearers being more anxious, fearful, or pessimistic. The article reported that this genetic variance accounts for about 3 to 4 percent of the variation in neurotic behavior.[8]

Characteristics are often double-edged. If one uses genetic pharmaceuticals to emphasize one particular temperament, one will lose another that may have attendant advantages. Compulsiveness can be an annoying trait and in sufficient degree a psychiatric disorder, yet a deep sense of responsibility, vigilance, and attention to detail can be virtues.[9] They are necessary traits for earning a Ph.D. Different cultures reward different personality constellations. Drugs that alter personal styles may be not so much improving persons as helping them better fit a particular society's expectations.[10] The compulsive industry that one culture may admire may be spurned as hyperactivity by another. What one person considers laudable acting on principle can be seen by another as stubbornness, cooperativeness as collaboration, or rugged individualism as feckless egoism.[11] "There are no genes specific for criminal behavior. Instead, there are genes that predispose children to impulsive, compulsive, hyperactive, and aggressive behaviors, and when these behaviors persist for a lifetime, the affected individuals are at a greater risk of becoming involved in various illegal behaviors. Of course, some of these same individuals may also become congressmen, senators, entrepreneurs, and CEOs."[12]

One can take Prozac (flurocine) to relieve psychic pain. Flurocine limits serotonin uptake, hence leaving more serotonin available in the system. It is not manufactured by gene transfer techniques but exemplifies one of the important questions that multiplying genetic pharmaceuticals will raise more emphatically. Such treatment could be quite helpful or it may sometimes be missing the point. It could mask needed attention to what is causing the pain. If we are in a

6. Office of Technology Assessment (US Congress), *Mental Disorders and Genetics: Bridging the Gap Between Research and Society,* OTA-BP-H-133 (Washington, D.C.: U.S. Government Printing Office, 1994).

7. C. Robert Cloniger, Rolf Adolfsson, and Nenad M. Svrakic, "Mapping Genes for Human Personality," *Nature Genetics* 12 (1996): 3-4.

8. *Hastings Center Report* (Mar./Apr. 1997): 48.

9. Peter D. Kramer, *Listening to Prozac: A Psychiatrist Explores Antidepressant Drugs and the Remaking of the Self* (New York: Penguin, 1993), p. 35.

10. Kramer, *Listening to Prozac,* p. 41.

11. Charles Frankel, "The Specter of Eugenics," *Commentary* 57, no. 3 (1974): 31.

12. David E. Comings, "Both Genes and Environment Play a Role in Antisocial Behavior," *Politics and the Life Sciences* (Mar. 1996): 84.

school for souls as described in chapter 3, we would do well to recognize if the pain is a warning sign of something that needs correction before the damage is worse. One of the most feared aspects of leprosy is the inability to feel pain. While not being able to feel pain sounds wonderful to anyone experiencing it chronically, losing the ability to feel any pain makes it difficult to defend the body from injury. Physical pain usually warns that something is wrong and needs immediate attention before the damage is worse. Not feeling the warning can result in the loss of body parts as they are injured.

Now psychic pain can be purely physical, in which case a physical response is quite appropriate. It might also be part of a destructive pattern embedded by years of abuse or bad choices. Prozac might restore brain chemistry back to a relatively fresh start where positive choices are in reach.[13] In that case the pharmaceutical is not to elevate mood, but rather to restore the normal affective range. This would be parallel to properly dosed insulin that frees a diabetic for a normal range of physical activity. However, like physical pain, emotional pain can also be a sign that something is wrong and emphatically needs our attention in our physical, psychological, social, or spiritual life. To turn off the warning light without dealing with the actual problem might be even more devastating over time.

Granted, if psychic pain is signaling an overwhelming underlying need, there may be times when it is helpful to temporarily postpone that need chemically until one is able to handle it better. But simply because drugs can provide relief does not mean that the problem was merely physical. The drug may mask a call to deal with something deeper. Solving the problem and taking Prozac might both increase levels of serotonin in the brain, but the more important change might be in priorities, convictions, and personhood, not just the brain chemistry that one addresses and the other does not. We need to be careful that we do not settle for counterfeit fixes that turn off the warning light instead of addressing the reason it is calling for our attention. We are here to learn, not to pass the years in maximum comfort.

In some places Ritalin has become the drug of choice for many young children. It may be helpful for a genuine condition of attention deficit disorder (ADD). It may also be a convenient way to pacify large classes or to avoid the results of the widespread choice to use the rapid images and passive entertainment of television as a baby-sitter from the earliest ages. A genetically designed pill might release the same neurotransmitters in the brain as prayer, so

13. Michael J. Boivin, "Finding God in Prozac or Finding Prozac in God: Preserving a Christian View of the Person Amidst a Biopsychological Revolution," manuscript, Indiana Wesleyan University, 2000, p. 19.

that there is a parallel feeling, but for the Christian tradition prayer is not about pursuing a physical sensation. Prayer is about a relationship with God. A pill might duplicate one physical side effect of prayer but not its purpose. If the drug comes to replace prayer, it has been a distraction from what most mattered.

The point is a good life, not just a good time. Thoughtful hedonists realized that the best net effect of maximum pleasure and minimal pain over a lifetime might require foregoing some immediate pleasure. A hedonist might give up the pleasure of rich desserts to extend life for other pleasures that are less life shortening. But from a Christian perspective, a good life is not the pursuit of passing pleasure at all. Pleasure is not the point. This brief life on earth is a school for souls. It is a place of growth, a place of learning and becoming the kind of people who can glorify and enjoy God forever. Part of the joy is found in giving up pleasing self as the central project. Jesus was characterized by abundant joy yet apparently lived under often dismal circumstances and ended in a torturous death. His life is a model for all that human beings can be. The good life does not lie in amassing and securing personal comforts and prizes. It is about abiding joy, not just moments of happiness. It is about becoming and living in a way that pleases God and that is suitable for the life to come. The needed growth in character may be embodied in chemical patterns in the human brain. Our brain begins to embody our choices over time in physical connection patterns that encourage angles of perception and response. Character is about developing mental reflexes to see, do, and value what is right.

In the light of the above, what would be an example of an improvement? For example, would a genetic pharmaceutical that increases quickness and facility of memory be an enhancement? Yes, in the sense that it would be immensely helpful to learning languages and other memory-intensive endeavors that are crucial to communication and community across national borders or for tasks as simple and important as a teacher learning student names. Would memory enhancement make a painful memory difficult to forget? Improved ability to recall does not require one to recall. One still has a choice in what one chooses to revisit. What of capacity overload? Like a computer's memory chip or a full cup, is there simply no room for more information if one's brain is already filled to capacity? There may be the possibility of more storage than most people have tapped, or that could be expanded as well. Increasing information does require increasing search capabilities to use it well. So for some enhancements to be helpful, they would have to progress in tandem with others. Whether the effect a genetic pharmaceutical offers is a genuine improvement will often be a complex judgment.

Open Future

People often make choices that limit their future options. Some seem trivial and are never noticed. An arbitrary turn down one store aisle can lead to meeting a future mate or not. There are consciously chosen life commitments such as to marriage or a monastic order that make possible great goods, but for most decisions it is an advantage to keep options open. Circumstances or one's vision can change. Genetic pharmaceuticals usually lend themselves to this reversible caution. One can stop taking the drug and return to one's initial state. People often practice temporary self-medication to their benefit and no one's harm. For example, caffeine is a common choice. While it is mildly dehydrating, it offers measurable improvement in concentration for focused tasks. It is open, not predestining. One can use the heightened awareness for a task of one's choice and when one stops taking in the caffeine, it leaves the body in a few hours.

Other drugs have irreversible effects. For example, powerful sterilizing drugs may not offer the chance to return to fertility. When a pharmaceutical is irreversible, one should be dramatically more sure of its use, just as one should take care in choosing, say, a tattoo. The "lizard man," in the news of late for filing his teeth, adding implants under his skin, and tattooing his entire body to more closely resemble a lizard, will find it difficult to return to a more normal appearance if he someday finds his earlier choices less interesting.

The value of an open future is especially important for children who have not yet had the opportunity to make their own life-shaping choices.[14] Because genetic pharmaceuticals are likely to be costly as they are first introduced,[15] offers to intervene in children are likely to be pitched to the desires and needs of the parents who choose the intervention and pay the bills.[16] Parents will need to make a conscious effort to be sure that any intervention is to benefit the child, not just themselves. An important part of assuring that fo-

14. Joel Feinberg, "The Child's Right to an Open Future," in *Whose Child: Children's Rights, Parental Authority, and State Power,* ed. William Aiken and Hugh LaFollette (Totowa, N.J.: Rowman & Littlefield, 1980).

15. For example, human growth hormone currently costs ten to thirty thousand dollars a year for daily injections. LeRoy Walters and Julie Gage Palmer, *The Ethics of Human Gene Therapy* (New York: Oxford University Press, 1997), p. 113.

16. Elisabeth Beck-Gernsheim recounts this historical development particularly for the roles of women in *Die Kinderfrage: Frauen zwischen Kinderwunsch und Unabhangigkeit* (The Question of Children: Women Between the Wish for Children and Independence) (Munich: Beck Verlag, 1988) and *Mutterwerden-der Sprung in ein anderes Leben* (Becoming a Mother: The Leap into Another Life) (Frankfurt: Fischer Verlag, 1989).

cus is that any genetic pharmaceutical should increase the child's capacity and hence future choices, not decrease them. Otherwise the child might have reason to feel predestined. Chapter 8 mentioned parents who wanted their children to be deaf or dwarfs like themselves. Such interventions would decrease the choices eventually available to their children. External influences would be so direct and pervasive as to encourage the children to abdicate their responsibility for the lives they then lead.[17]

Bernard Häring responds that the intervention need not be extreme or harmful.[18] The intervention might increase capacities and choice rather than attempt programming, and be publicly known as such. Häring's concern is that "the essential point,"[19] "final concern and criterion,"[20] "key word,"[21] for guiding such intervention is freedom. For Häring such freedom is the unique dignity and capacity of humanity.[22] Human beings find their highest state in self-transforming love relationships that are founded on freedom. These are characterized by a longing for ever growing knowledge of what is good and truthful, the capacity to love what is good, and to put it into practice.[23]

Human beings are not just to preserve freedom, but rather they are to enhance it. Human beings may intentionally shape their environment and themselves as long as the sacredness of each person and the growth of freedom for all is furthered. Consciously shaping ourselves and accepting the influence of others is part of our society of hospitals and schools.[24] Such can serve freedom. Intentional shaping of others takes place best with free consent if the recipient is able to consent, and it should further future freedom of choice for those who have yet no will to exercise. For Häring then the constructive use of genetic interevention to that end should be pursued, cautiously and responsibly, as a God-given and empowered mandate.

In contrast Alasdair MacIntyre has described seven virtues that people would want their children to have. Among these would be a commitment not to manipulate others. He then argues that if parents have those virtues them-

---

17. Hessel Bouma III, Douglas Diekema, Edward Langerak, Theodore Rottman, and Allen Verhey, *Christian Faith, Health, and Medical Practice* (Grand Rapids: Eerdmans, 1989), p. 185.

18. Bernard Häring, *Manipulation: Ethical Boundaries of Medical, Behavioral, and Genetic Manipulation* (Slough, U.K.: St. Paul, 1975).

19. Häring, *Manipulation*, p. 14.

20. Häring, *Manipulation*, p. 50.

21. Häring, *Manipulation*, p. 57.

22. Häring, *Manipulation*, p. 55.

23. Häring, *Manipulation*, p. 57.

24. Roger L. Shinn, "Genetic Decisions: A Case Study in Ethical Method," *Soundings* 52 (Fall 1969): 308-9.

selves, they will not want to genetically manipulate their children to have them.[25] In response it is important first to clarify that genes do not encode virtues.[26] Virtues stem from a higher level of reasoning and commitment than is carried in our genes. Also virtuous parents will seek to encourage virtues in their children through means such as education, modeling, and conversation. Raising a child so that the child has no guidance until she is of age to decide for herself raises hellions that will not have the ability to choose. An open future has to do with helping children build skills of discernment that they are then free to evaluate and apply as persons in their own right.

> [O]ur children are not our children: they are not our property, not our possessions. Neither are they supposed to live our lives for us, or anyone else's life but their own. To be sure, we seek to guide them on their way, imparting to them not just life but nurturing, love, and a way of life; to be sure, they bear our hopes that they will live fine and flourishing lives, enabling us in small measure to transcend our own limitations. They are sprung from a past, but they take an uncharted course into the future.[27]

Best Use of Limited Resources

We should hone our skills and character to better serve. This was the theme of the parable of the stewards described in chapter 3. Adults should eat a diet that supports their health, and a pregnant woman should eat enough folic acid in the form of vitamins or green vegetables to reduce the risk of neural-tube defects such as spina bifida. Vaccinations should be used to help protect ourselves and our children from crippling or even lethal disease. There is much that we can do to support and augment our lives as adults, and as parents we want to help children to the best available start.[28] Investing effort and other limited resources in genetic intervention involves a choice that it is the best use of limited resources.

25. Alasdair MacIntyre, "Seven Traits for the Future," *Hastings Center Report* 9, no. 1 (1979): 7.

26. Bruce R. Reichenbach and V. Elving Anderson, *On Behalf of God: A Christian Ethic for Biology* (Grand Rapids: Eerdmans, 1995), pp. 206-8.

27. Leon Kass, "The Wisdom of Repugnance: Why We Should Ban the Cloning of Humans," reprinted in *The Human Cloning Debate*, ed. Glenn McGee (Berkeley: Berkeley Hills, 1998), pp. 171-72.

28. On the role of the family in the Christian tradition see Stephen G. Post, "Marriage and Family," in *Christian Ethics: Problems and Prospects*, ed. Lisa Sowle Cahill and James F. Childress (Cleveland: Pilgrim, 1996), pp. 265-83, and Rodney Clapp, *Families at the Crossroad: Between Traditional and Modern Options* (Downers Grove, Ill.: InterVarsity Press, 1993).

Just as important as the physical start for children is sharing life with parents. Parents provide the steady base that frees the child to become and explore. My own daughters as toddlers explored most boldly when they could regularly toddle back to hug my knee before the next foray. They would naturally check their base and silently think to the effect that "O.K. Daddy is right here. It is safe to explore out from this point." The way parents treat their children through the opening years of their lives in being with them and caring for them is far more influential than anything we will be able to do genetically. It would be counterproductive to spend little time with one's child because one worked a second shift to pay for a genetic drug to increase her intelligence 10 percent. One would probably have more positive influence on her intelligence by interacting with her during that time. Human beings can augment intellectual ability by training and use. Genetics is only part of physical health, and physical health is only part of what we support for ourselves and our children as we grow and develop, yet genetics can make a genuine difference. Intervention should only proceed when it is the best service available to our always finite resources. When this balance is pursued at the community level, it begins to shade into issues of justice. That will be discussed in part in the next chapter.

Chapter Summary

Whether choosing for oneself or one's children, considering gene product use should include the following four standards. First, one should only proceed when the intervention is safe. While virtually no action is completely risk free, risk should be at a minimum proportional to the intended benefit. That is particularly true when choosing for a dependent. Developing an intervention incrementally so that relatively few people are exposed to it at first and it is proven over time is already standard for introducing other new drugs. Introducing a change in an individual incrementally also makes sense lest one be surprised by unexpected and untoward interactions. Second, intervention should achieve a genuine improvement. That is an obvious goal, but considering the complex interactions of human characteristics and environment it is not always an easy one to assure. Third, interventions should promote an open future, not a predestined one. Adults can change their minds as they develop and change or experience the results of the intervention. When parents decide on behalf of a child, they need to take extra care to increase the child's future options and not limit them. Fourth, human genetic intervention should only proceed when it is the best available use of always limited resources.

CHAPTER 12

Genetic Drugs and Community

Drugs are already a quandary for our communities. Genetically produced drugs add at least two new challenges. First, in the United States the most potent pharmaceuticals such as steroids are controlled by physician prescription. New drugs manufactured with genetic techniques are dramatically increasing the number of drugs that can increase human capacity in one form or another. Their availability under physician control requires physicians to discern more precisely what they are trying to accomplish in medical care. Second, genetic drugs that can reduce or increase physical differences between people test in strikingly new ways the often voiced commitment of government to equality of opportunity. What is this equality of opportunity, and does it apply to physical heritage as well as to institutional access?

Physicians and the Goals of Medicine

How one describes the goals of medicine is more than an intramural debate. Medicine in our society has considerable prestige, institutional power, and financial endowment. How it states and pursues its goals is widely influential. The goals of medicine have been described in a number of ways. Relief of suffering and increase of capacity have been commonly cited, along with the traditional commitment to *primum non nocere* (first do no harm). In all, serving the patient is to be the priority.

Fowler, Juengst, and Zimmerman advocate extending the patient-centered ethic of clinical therapy to decisions about genetic intervention.[1] They argue that

1. Gregory Fowler, Eric T. Juengst, and Burke K. Zimmerman, "Germ-Line Gene

the medical model is already in place and effective. Genetic pharmaceuticals or other genetic interventions would be offered to help presenting patients achieve their own goals. Decisions would be made by patients for themselves or by parents with physician guidance. As characteristic of clinical medicine, the focus would be on compassion for those immediately present, with little if any concern for long-run or wider community implications.[2] Selection of treatment would depend on negotiation between a particular doctor and patient.[3]

Eric Juengst argues further that the involved physician as a professional should be the one who sets intervention limits. Questions of what genetic intervention is medically appropriate "are ultimately questions for professional conscience and vision, not public policy."[4] The apparent conviction is that only fellow professionals can fully understand and guide the practice of the profession, hence they set the appropriate standards. Lantos, Siegler, and Cuttler seem to assume this in their prescription for the use of human growth hormone. They conclude that when faced with new therapies, "If pediatricians do not respond to these dilemmas thoughtfully, carefully, and forcefully, other decision makers, whose decisions may reflect political, social, or market forces, may then respond in ways that do not reflect the best interests of children."[5] Undoubtedly pediatricians will be affected by "political, social, and market forces" as well, but they at least have a professional commitment to put their patients first.

How does one put one's patient first in the case of the genetic product human growth hormone (HGH)? To affect height, HGH must be administered before a child is of legal age to give consent. The attempt is to change end physical endowment, although it is not clear yet whether growth hormone produces a net increase in final height or merely temporary acceleration of growth.[6] Growth hormone therapy requires multiple interventions. Each in-

---

Therapy and the Clinical Ethos of Medical Genetics," *Theoretical Medicine* 10, no. 2 (June 1989): 151-65.

2. Arthur Zucker and David Patriquin, "Moral Issues Arising from Genetics," *Listening: Journal of Religion and Culture* 22, no. 1 (1987): 65-85.

3. Mark Siegler, "The Doctor-Patient Encounter and Its Relationship to Theories of Health and Disease," in *Concepts of Health and Disease: Interdisciplinary Perspectives,* ed. Arthur L. Caplan, H. Tristram Engelhardt Jr., and James J. McCartney (Reading, Mass.: Addison-Wesley, 1981), pp. 628, 643.

4. Eric T. Juengst, "The NIH 'Points to Consider' and the Limits of Human Gene Therapy," *Human Gene Therapy* 1 (1990): 431.

5. John Lantos, Mark Siegler, and Leona Cuttler, "Ethical Issues in Growth Hormone Therapy," *Journal of the American Medical Association* 261, no. 7 (1989): 1024.

6. Ruth Macklin, "Growth Hormones for Short Normal Children: An Ethical Analysis," *ASBH Exchange* 2, no. 4 (1999): 9. Melvin M. Grumbach, "Growth Hormone Therapy and the Short End of the Stick," *The New England Journal of Medicine* 319, no. 4 (1988): 238-41.

tervention is painful and in most cases the intended effect is primarily cosmetic. Twenty thousand children in the United States currently receive HGH. Seventy-five percent of them are HGH deficient. Most of the rest are girls with Turner syndrome that leads to average adult height of 4 foot 8 inches. Some children with short parents and normal levels of human growth hormone are receiving HGH. The drug costs ten to thirty thousand dollars a year[7] and requires daily injection for years at a time. An editorial in *The Lancet* argues concerning growth hormone treatment that it is better to change society's attitudes than the physical endowment of the individual.[8] Here the argument is assuming a case where the intervention does not concretely help the recipient except to greater social acceptance. In such a case it would be more humane for society to welcome and respect all as they are than to insist on aesthetic conformity or an antiquated connection of leadership with physical dominance.

Lantos, Siegler, and Cuttler approve of such therapy only if there is a measurable disease of growth hormone deficiency or as a cosmetic intervention for shortness that is at a level of "deformity."[9] Here basic conceptions of disease and deformity are setting the perimeter for medical intervention.[10] They advocate using HGH only for cure, not enhancement.[11] Chapter 10 addressed the challenges to that distinction. Many of their colleagues already offer cosmetic surgery such as face-lifts and tummy tucks. There is also extensive health care outside the traditional system from herbs to acupuncture. Even if some enhancements are not carried out by the medical establishment, that of itself does not eliminate other people offering those services. The massive alternative health-care industry is a case in point. James Lindemann Nelson has coined the term *schmoctors* for a whole new group of providers who might practice *schmedicine*.[12] These schmoctors might specialize in providing genetic enhancements if they are not offered by more traditional doctors.

7. LeRoy Walters and Julie Gage Palmer, *The Ethics of Human Gene Therapy* (New York: Oxford University Press, 1997), p. 113.

8. Editorial, *The Lancet* 335, no. 8692 (31 Mar. 1990): 764.

9. Lantos, Siegler, and Cuttler, "Ethical Issues in Growth Hormone Therapy," pp. 1021-23.

10. Eric T. Juengst, "Can Enhancement Be Distinguished from Prevention in Genetic Medicine?" *Journal of Medicine and Philosophy* 22 (1997): 125-42.

11. Lantos, Siegler, and Cuttler, "Ethical Issues in Growth Hormone Therapy," pp. 1020-24. Martin Benjamin, James L. Muyskens, and Paul Saenger, "Short Children, Anxious Parents: Is Growth Hormone the Answer?" *Hastings Center Report* 14, no. 2 (1984): 5-9.

12. Erik Parens, "Is Better Always Good? The Enhancement Project," in *Enhancing Human Traits: Ethical and Social Implications,* ed. Erik Parens (Washington, D.C.: Georgetown University Press, 1998), p. 11.

Another longstanding case that will be exacerbated by new genetic drugs is that of athletic competition. Standards and monitoring methods for steroids and other drugs have been developed to enforce fair competition.[13] Genetic pharmaceuticals are likely to become a more difficult case. They are natural body products that occur in varying degrees in different people by genetic heritage. It would be difficult to tell if genetic products were artificially introduced. Athletic commissions already sort out competitors into various classifications such as sex, weight, and age. To assure relatively even competition in the future, new distinctions may be needed that measure what is present in the body, not how it came to be there. Otherwise the competitive edge in athletic competition may become more a matter of chemistry than training and skill.

While physicians cannot as a group control all the uses of their expertise, they can choose what they will endorse as a group. What the medical establishment offers has an imprimatur effect of a prestigious and pervasive system. Physicians cannot unilaterally determine all uses of genetic pharmaceuticals, but they are still responsible for their part. The medical community has been generously endowed by public and private payment and charity and has a rich heritage built from the service of many. If it speaks with a unified voice, its influence will be substantial.

Government Interest in Equality of Opportunity[14]

There are at least three points where equality of opportunity comes into contact with genetic intervention.[15] First, some religious and philosophical tradi-

---

13. George Khushf argues that performance-enhancing drugs are not only unfair but more importantly that they are inherently dehumanizing. However, he states repeatedly that it is difficult to say why they are dehumanizing. "Thinking Theologically about Reproductive and Genetic Enhancements: The Challenge," *Christian Bioethics* 5, no. 2 (1999): 171, 179 (fn 29-30).

14. Equality of opportunity can be seen as a subset of a wider discussion of distributive justice. Many material standards have been argued for the just distribution of resources, including according to equal share, need, merit, past contribution, and voluntary exchange. In the United States we use each of these methods; "equal share" for one person-one vote in civil elections, "need" for emergency-room treatment, "merit" for college admission, "past contribution" for awards such as the Pulitzer Prize, and "voluntary exchange" for consumer goods. Conflict over what is just in a given situation is often at root about which material standard is appropriate in that case.

15. I am not limiting "equality of opportunity" to concerns about institutional discrimination. For a description of different uses of the phrase see John Roemer, *Theories of Distributive Justice* (Cambridge, Mass.: Harvard University Press, 1996).

tions emphasize the priceless worth of each human being. Hence as much as possible an important advantage should be available as much to one person as another. If forms of genetic intervention become important goods that can help pursue basic life plans, as education and medical care do, the opportunity to benefit from them should be as available as possible to all.

Second, seeking equality of opportunity for the citizens within a particular country can be grounded in an affirmation of fairness. Equality of opportunity is often cited to justify large disparities in the distribution of goods. The assumption is that this is fair because in a free-market society everyone has roughly equal opportunity to develop his or her skills and work. That rewards vary is not unjust in that each person had a substantially similar opportunity to earn them. Those who have gained more have in some sense contributed more.

It is certainly true that as one maximizes the opportunities that one has, one will gain more opportunities. There is much to the old saying that the harder one works, the luckier one gets. So equality of opportunity cannot mean lifelong equal results or even options. The person who works hard to master chemistry and then to complete medical school will have opportunities not available to someone who skipped classes for other pursuits that mattered more to him then. The ideal of equality of opportunity is a societal effort to raise up starting access with basic education and medical care. As time passes opportunities will tend to increase or decrease for an individual according to the individual's use of the opportunities that he or she has already had.

A major factor that this theory has traditionally not had to take into account is the natural lottery of starting capacity. Some people have genetic advantages of health and other capabilities far superior to others. If a society is able to redress some of that difference by genetic pharmaceuticals and claims to support equality of opportunity, there will be pressure to implement universal intervention to actually approximate equal opportunity more closely. If two children are quite short, one due to lack of human growth hormone (HGH) and the other due to family trait, both would be equally at risk concerning equality of opportunity.[16] By that standard each may warrant an HGH intervention. If HGH were not equally available, an uneven access to growth hormone could decrease equality of opportunity for those goods associated with height in our society.[17] It is important to note, however, that to

16. David B. Allen and Norman C. Fost, "Growth Hormone Therapy for Short Stature: Panacea or Pandora's Box?" *Journal of Pediatrics* 117 (1990): 16-21.

17. Benjamin, Muyskens, and Saenger, "Short Children, Anxious Parents," pp. 5-9.

the degree the concern with height is for social advantage rather than physical access, it would be better to change such arbitrary culture rather than people's physical form.

Third, if genetic intervention is only available to a few, the recipients could claim a meritocracy of measurably superior abilities. Claims to superiority have been made by various aristocracies without any basis in physical reality.[18] To in fact have a physical distinction between groups would lend itself to one group oppressing the other. Enhancement of capacity, if available to only a few, could increase such stratification. There might be reinforcement of already harmful social evils such as heightism and racism.

Making genetic pharmaceuticals widely available would not homogenize the population if the choice whether to use it and how to use it is left up to the adults who receive it or parents on behalf of their children. There might be a "genetic decent minimum" that is almost universally recognized and provided as helpful to the recipient.[19] Yet beyond that, the multiplicity of decision makers would lead to a variety of choices and results. Variation is healthy for a society in offering complementary traits for complex daily life and more adaptive pathways when circumstances change. Winston Churchill seemed always a bit out of sync before brilliantly leading Great Britain through World War II. Treasured during the war for his unique combination of traits to lead his country through those dark hours, he was quickly retired by the voters when it was over.

Affording Provision

The pharmaceutical executive John Varian estimates that it takes about 280 million dollars to bring an idea for a drug on through development to market availability.[20] Genetic pharmaceuticals could be a societal drain from other more effective strategies.[21] However, genetic drugs are likely to be paid for

---

18. Mark H. Haller, *Eugenics: Hereditarian Attitudes in American Thought* (New Brunswick: Rutgers University Press, 1963), p. 5.

19. Allen Buchanan, Dan W. Brock, Norman Daniels, and Daniel Wikler, *From Chance to Choice: Genetics and Justice* (Cambridge: Cambridge University Press, 2000), pp. 81-82.

20. John Varian, "Genetics in the Marketplace: A Biotech Perspective," in *Genetic Testing and Screening: Critical Engagements at the Intersection of Faith and Science,* ed. Roger A. Willer (Minneapolis: Kirk House, 1998), pp. 64-65.

21. Karen Lebacqz, "Bioethics: Some Challenges from a Liberation Perspective," in *Faith and Science in an Unjust World,* ed. Roger L. Shinn (Geneva: World Council of

initially by the wealthy as a hoped advantage for themselves or their children. This would come not from tax revenue but from their discretionary income. If they can spend their money on college tuition, a personal trainer, or gambling in Las Vegas, why not to increase their physical capacity? This also means that relatively high start-up risks fall on volunteers who pay for the drug's development. When it becomes clear that the intervention is safe and advantageous, government will probably be called upon to provide it for all.

Mehlman and Botkin suggest that the fairest distribution of genetic services would be an access lottery. Every set of parents would have an equal chance at being selected randomly for full genetic services. This would protect not equality of results or capability, but at least equal chance of genetic help. Society would benefit from gene-enhanced abilities in the work force, but no one group could monopolize the technique and take over almost all positions distributed by merit. Genetic intervention would be limited in its share of the medical economy.[22]

A genetic lottery would be superior to the unfairness and social disruption of generations of compounding disparate access. However, it is likely that once it is clear that some genetic pharmaceuticals are safe and efficacious for the improvement of physical life, most parents will want them for their children. Simply a chance at the service would not be any more sufficiently consoling to the losers than everyone having an equal chance at the first twelve years of education. Equality of opportunity calls for all to actually be offered that resource, not just for all to have a chance at it in a lottery. Once it is clear that intervention is safe and works, demand will soar and the costs per intervention will decline, particularly as less labor intensive techniques are developed. Most of the objects and services that we now consider standard were originally luxuries. A primary-school education was unaffordable to many for much of the history of the United States. Now education through the twelfth year is a guaranteed provision and considered a necessity.

While the issue of cost has been raised as a limit to widespread intervention, cost will probably become an incentive for intervention rather than a restraint. The one-time cost of genetic intervention could be considerably less than the cost of repeated hospitalization and lost work.[23] Combining the eco-

---

Churches, 1980), p. 279, and Jean Porter, "What Is Morally Distinctive about Genetic Engineering?" *Human Gene Therapy* 1 (1990): 423.

22. Maxwell J. Mehlman and Jeffrey R. Botkin, *Access to the Genome: The Challenge to Equality* (Washington, D.C.: Georgetown University Press, 1998), pp. 124-28.

23. LeRoy Walters, "The Ethics of Human Gene Therapy," in *Contemporary Issues in Bioethics*, ed. Tom L. Beauchamp and LeRoy Walters, 3rd ed. (Belmont, Calif.: Wadsworth Publishers, 1989), p. 524.

nomic advantage of less needed care along with citizens probably being capable of longer taxable lives, it would be in the government's economic interest to make genetic intervention available to all. Intervention could be an economic boon, freeing resources for other services.

While it would probably be technically intensive and expensive at first, through improving techniques it would be progressively less so. After the initial bulge in start-up costs, reduction in other medical costs would begin immediately. Myopia, diabetes, trisomy 21, and many diseases and handicaps treated at great expense to the sufferer and society are genetically caused or related. A one-time intervention to lessen these diseases and disabilities might become economically desirable. If such a situation were to develop, the economic issue of genetic intervention may become more an autonomy issue of whether a society should require it to avoid the greater expense of not doing it. Economic costs will initially pressure against its government use and then probably eventually for its use.

Provision Across Borders

The sociologist Dorothy Wertz has observed that Japan, China, and India have the technology for enhancement efforts and few internal cultural barriers to it. She was not surprised to hear the president of the East Asian Society of Bioethics argue that we should use recombinant DNA techniques to promote "artificial evolution-positive eugenics."[24] Pre-birth tests are already very popular in China, where many parents know they will only be able to have one child, who carries all their dreams and enjoys every advantage they can provide. The government of Singapore has created policies to encourage well-educated citizens to marry and have many children.[25] In India ultrasound has been so commonly used for sex selection that birth ratios of male to female are running in some areas at 135 to 75.[26] It is doubtful that the more extensive and precise techniques of genetic intervention would be eschewed. Interest in genetic intervention is high around the world.

Concerns of humanity, equality, and economy extend across national borders. While it can be argued that obligations are greatest to those nearest,

24. Dorothy C. Wertz, "Society and the Not-So-New Genetics: What Are We Afraid Of? Some Future Predictions from a Social Scientist," *Journal of Contemporary Health Law and Policy* 13 (1997): 332.

25. Chee Khoon Chan and Heng-Leng Chee, eds., *Designer Genes: I.Q., Ideology, and Biology* (Kuala Lumpur: Institute for Social Analysis, 1984).

26. Buchanan et al., *From Chance to Choice*, p. 183.

that is a tempering factor and not necessarily a set limit to care. The National Council of Churches of Christ/USA has advocated that the benefits of biogenetic innovation should be available regardless of geography, economic ability, or race.[27] A policy statement adopted by the governing board reasons that "the whole of humankind, created by God, living under God, offered salvation by God, is a unity."[28] All people then should have access to the benefits of biogenetic innovations.

New concerns at the international level might include lack of control in how such shared technology is implemented. Other governments might use the technology in ways the originating country considers unconscionable. In such cases, the first government could limit its aid, yet practically at best would be able only to delay the other's use. The strongest impact may be in pressure to keep up. Nations may fear falling behind the most aggressive programs. Nuclear power and weapons proliferation offer a parallel case.

Chapter Summary

Communities face a number of challenges in how genetic drugs should be used. As a profession with the right to prescribe pharmaceuticals, physicians are in a position to deeply influence how our society uses genetic drugs. Government will probably eventually provide some genetic pharmaceuticals to uphold a traditional justification of society in equality of opportunity. Cost will initially work against that widespread use. As genetic drugs become less expensive and prove their effectiveness, the economic pressure will be on governments to provide and even require them.

---

27. National Council of Churches of Christ/USA, Panel on Bioethical Concerns, "Genetic Engineering: Social and Ethical Consequences," pp. 4, 35; also Richard A. McCormick, *The Critical Calling: Reflections on Moral Dilemmas since Vatican II* (Washington, D.C.: Georgetown University Press, 1989), p. 269.

28. National Council of Churches Governing Board, "A Policy Statement of the National Council of Churches in the United States of America: Genetic Science for Human Benefit" (New York: Pilgrim, 1986), p. 9.

PART IV

GENETIC SURGERY: CHANGING GENES IN THE HUMAN BODY

CHAPTER 13

Changing Genes and the Individual

Possibilities

To change our genome is to change us. Human beings are far more than genes, but genes are part of who we are. It is much easier to read the genetic sequence than it is to repair or otherwise change it, yet the fourth wave of developing genetic technology is a kind of surgery that directly alters a person's genes. These changes can be introduced by means such as gene transfer, removal, inactivation, or duplication (such as in cloning).

The first apparent cure by a type of genetic surgery was announced in April 2000. Marina Cavazzana-Calvo and Alain Fischer stated in *Science* that their team had saved the lives of two babies afflicted with SCID-X1 (severe combined immunodeficiency-X1).[1] The bone marrow of the babies lacked part of the genetic instructions needed for a working immune system. The physicians were able to insert the needed genetic material into marrow cells, which then multiplied and displaced cells with the defective gene. At the time of the announcement the babies were continuing to sustain their newly functioning immune systems ten months after treatment. While the gene therapy community has been chastened by the plethora of unfulfilled expectations that has swirled around grant proposals and news stories, it seems that here at last is confirmation that in some cases gene therapy might actually bring about a cure.

Other potential uses for gene surgery are multiplying rapidly. Whole journals such as *Human Gene Therapy* are devoted to publishing research

---

1. Marina Cavazzana-Calvo et al., "Gene Therapy of Human Severe Combined Immunodeficiency (SCID)-X1 Disease," *Science* 288, no. 5466 (28 Apr. 2000): 669-72.

studies that investigate possible applications.[2] William Schwartz catalogs progress toward genetic surgery contributing to treatment for Amyotrophic Lateral Sclerosis (ALS), cancer, coronary artery disease, cystic fibrosis, epilepsy, glaucoma, Huntington disease, organ transplantation, stroke, and viral infection.[3] The ideal is not just gene addition but gene replacement. When an old gene is supplanted by a new gene, there is a better chance that it will express cooperatively without being covered by the defective gene. Such intervention could even correct dominant disorders.[4]

The closest familiar parallel to many genetic surgeries might be tissue or organ transplantation. Everyone has the same number of genes. What varies is the versions (alleles) of each gene. To have the gene for cystic fibrosis is to have a variation of the sodium pump gene that is not able to do its job. The lungs fill with mucus, which makes it difficult to breathe and invites infection. With our best current treatments the disease is still usually fatal by the patient's twenties. One possible treatment is to transplant healthy lungs with the proper code, but it would be far better to transfer the lacking genetic instructions to enable the patient's own lungs to do their life-giving work. Genetic surgery would not be limited by the number of organs available for donation. It would be less invasive and cost less than transplanting whole organs. Also it would avoid a lifelong regime of drugs to suppress rejection of the foreign lung tissue. Genetic surgery could compare to transplantation surgery the way polio vaccine does to the iron lung. One clumsily tries to replace function already lost. The other helps the body to prevent the harm before it occurs.

Currently we place the human recipe for insulin into bacteria so that they will make human insulin. Many diabetics then need to inject the insulin several times a day. Genetic surgery might eventually offer introducing the genetic recipe directly into the body, where it would function normally so that the body could make insulin as needed for itself. One could transplant the entire Isles of Langerhans tissue that makes insulin, but that leads to all the problems of rejecting foreign tissue. If one could introduce just the missing instructions, all the usual problems of organ transplantation are avoided.

---

2. E.g., genetically engineered antibodies for immunization or vectoring the introduction of new genetic material; see Mireia Pelegrin et al., "Genetically Engineered Antibodies in Gene Transfer and Gene Therapy," *Human Gene Therapy* 9 (10 Oct. 1998): 2165-75.

3. William B. Schwartz, *Life Without Disease: The Pursuit of Medical Utopia* (Berkeley: University of California Press, 1998), pp. 131-47.

4. LeRoy Walters and Julie Gage Palmer, *The Ethics of Human Gene Therapy* (New York: Oxford University Press, 1997), pp. 72-74.

Patients who inherit familial hypercholesterolemia (FH) lack the genetically recorded instructions for removing low-density lipoprotein (LDL) cholesterol from the blood stream. The result is that most who are homozygous (with two bad genes) die of heart attacks at an early age. One treatment method is a combined heart and liver transplant. The heart is replaced because of cholesterol damage and a new liver is provided in order to add the missing cholesterol-fighting capability to the body. James M. Wilson has pursued methods to inject only the genetic instructions the body needs to deal with the cholesterol, rather than transfer an entire organ. Livers are in short supply for transplantation, the surgery is massive, and transplantation requires immunosuppression as long as the transplanted liver survives. Simply giving the body the needed instructions it lacks would be far better. Wilson has had success removing liver cells, adding the genetic instructions to them, and then returning them to the body. The patient's LDL levels improved. The next step is to find ways to inject needed genes, so that cells do not need to be removed from the body for treatment.[5] Another promising method of gene introduction might be that of inhaled aerosols.[6] A third might be the introduction of an artificial chromosome.[7] This last method would eliminate the risk of randomly interpolating genetic material where it might disrupt a needed gene's work.

Genetic surgery might extend to behavior issues as well. A 1993 Netherlands study found an extended family mutation on the X chromosome that interferes with production of monoamine oxidase A. The lack of this enzyme affects levels of serotonin and noradrenalin in the brain. This seems to predispose to the quick aggression associated with criminal activity.[8] One of the original project areas of the Hastings Center was the ethics of behavior control through interventions such as psychosurgery.[9] Genetic surgery may raise many of the involved issues again.

Because genetic surgery on behalf of an individual has so much in common with other medical interventions, it is not surprising that "nearly all task forces, ecclesial communities, and individual theologians who have addressed the first of these types, somatic cell therapy [genetic intervention that cures

5. Walters and Palmer, *The Ethics of Human Gene Therapy*, pp. 27-28.

6. Walters and Palmer, *The Ethics of Human Gene Therapy*, p. 29.

7. M. A. Rosenfeld, "Human Artificial Chromosomes Get Real," *Nature Genetics* 15, no. 4 (1997): 333-35.

8. H. G. Brunner et al., "Abnormal Behavior Associated with a Point Mutation in the Structural Gene for Monoamine Oxidase A," *Science* 262, no. 5133 (22 Oct. 1993): 578-80.

9. There are many articles focused on this topic, beginning with the 1971 volume of the *Hastings Center Report*.

disease in one person], have approved of its use once the scientific and technical difficulties have been solved."[10]

Yet several concerns are raised in the midst of the potential goods. One prominent concern is that we not extend somatic therapy to inheritable changes. That will be addressed in chapter 14. Another is the line between cure and enhancement. We studied that distinction in chapter 10, and I concluded that more effective standards would include asking of the proposed changes only if they are safe, a genuine improvement, increase the recipient's open future, and are the best use of limited resources. Genetic surgery would change the natural course. Chapter 8 addressed changing nature as natural to human beings. Considering the impact of sin and the open question of whether nature even before the fall was all it could ever be, our standard is what should be, not what has been. That is why there are generally no qualms about doing a Caesarean section to save the life of a mother and her baby. Babies do not naturally exit the womb through the uterine wall, but that intervention is welcome when needed. Should our regular intervention in nature include improving our own genes, or are they a special case, off limits? The first section of this chapter will address that question. The second section of the chapter will trace the attitudes such intervention might encourage, and the last section the case of human cloning.

Improving Our Genes?

Genetic surgery raises the possibility of changing to some degree human physical attributes. That challenges us with the question of what our physical nature is and what we want it to be. "The question of what constitutes the normatively human is the most important issue that lurks in all the more specific and concrete problems we face when ethical issues are raised about developments in the field of genetics."[11]

The physical nature of human beings is part of what it is to be human. We spend much of our time providing for physical needs such as food and shelter and would quickly cease to exist if such concerns were not actively pursued. Even if one thinks of human beings as primarily mind or soul, or

10. James J. Walter, "Theological Issues in Genetics," *Theological Studies* 60 (1999): 125.

11. James M. Gustafson, *Theology and Christian Ethics* (Philadelphia: Pilgrim, 1974), p. 274. This important point is raised as well by others such as more recently Mark J. Hanson, "Indulging Anxiety: Human Enhancement from a Protestant Perspective," *Christian Bioethics* 5, no. 2 (1999): 121-22.

only temporarily physical while preparing for a transformed eternal life, the current nature of human life is mediated through and dependent upon the physical body. How one perceives the purpose and capacity of that physical nature is crucial to whether changing it is appropriate. If a particular intervention would benefit or harm a human being depends in large degree on what a human being is or should be. "Our fundamental decisions will be made, perhaps they have already been made, at the point where we accept a particular understanding of man."[12]

Does being human include remaining as we are physically or changing physically? Human beings have studied what it is to be human for centuries and have thought about conscious choices that would affect physical heritage. Already 2500 years ago Plato made a program to encourage good physical inheritance part of his ideal republic. But we are now approaching the first time that we can directly and precisely change physical attributes. H. Tristram Engelhardt Jr. has suggested that "human nature, as we know it today, will inevitably — for good secular moral reasons — be technologically reshaped."[13] I will first critique arguments for not doing so and then reasons why we should.

Physical Attributes Should Remain as They Are Now

Static Nature

Leon Kass is concerned that changing the current physical nature of human beings would make us less human. If alteration occurs, "the nonhuman life that may take our place will in some sense be superior — though I personally think it most unlikely, and certainly not demonstrable. In either case, we are ourselves human beings; therefore, it is proper for us to have a proprietary interest in our survival, and in our survival as human beings."[14] For Kass *current* human physical form is essential to what it is to be human. Kass affirms this point repeatedly and strongly but does not explain why it is the case.

---

12. Paul Ramsey, "Genetic Therapy: A Theologian's Response," in *The New Genetics and the Future of Man,* ed. Michael P. Hamilton (Grand Rapids: Eerdmans, 1972), pp. 171-72.

13. H. Tristram Engelhardt Jr., "Human Nature Technologically Revisited," in *Ethics, Politics, and Human Nature,* ed. Ellen Frankel Paul, Fred D. Miller Jr., and Jeffrey Paul (Oxford: Basil Blackwell, 1991), p. 183.

14. Leon R. Kass, "Making Babies — The New Biology and the Old Morality," *Public Interest* 26 (Winter 1972): 54.

Paul Ramsey makes the same affirmation in *Fabricated Man,* where he states that to physically transform human beings is to replace the current species with a life mistakenly thought to be superior. Ramsey calls such an action "species suicide." Our present condition should be the standard. Human beings are the particular bodies that they are now as much as they are souls and minds. To change this flesh is as much a violation of what is human as violating human will or freedom. The human body is not a tool for a given human being. Human beings *are* bodies. The body as it currently is, is part of what it is for a person to be human. To change the current physical form of humans would be to be less human and less than human.

Our bodies are central to our current life, yet in the Christian tradition the promise is claimed that someday God's people will be transformed as Jesus Christ was. "So it will be with the resurrection of the dead. The body that is sown is perishable, it is raised imperishable; it is sown in weakness and raised in power."[15] God's people will still be the same individuals, but in a strikingly new form. There is still a description of embodiment, but in a new way. According to the Christian tradition our current physical form is not required for us to be human. Also our current bodies are not eventually necessary to who each of us is as a person.

As for the present physical condition of humanity, it is common for newborns to have major genetic disabilities. Most people do not feel an allegiance to Tay-Sachs disease or inherited diabetes, yet that is part of what human beings are now. Society invests substantially to attempt to change nature in this regard. Why not in others? Here one might advocate change of current nature to an ideal nature that is disease free. Cure of disease would be acceptable, although not further enhancement. As a goal, maintaining the current form of human beings does not clarify by itself if that includes the elimination of common diseases or handicaps, in that current life for human beings includes such maladies. If it does include correcting those conditions, why not include preventative or strengthening interventions? If preventative or strengthening interventions are included, at what point is such intervention straying from loyalty to our present form?

Roger Shinn argues that part of the essence of our civilization is to modify our environment and ourselves from how we cook food to brain surgery.[16] Our society has considered it particularly beneficial to change what

15. 1 Corinthians 15:42-44.
16. Roger L. Shinn, "Gene Therapy: Ethical Issues," in the *Encyclopedia of Bioethics,* vol. 2, ed. Warren T. Reich (New York: Free Press, 1978), p. 525, and "The Ethics of Human Genetic Engineering," in *The Implications of the Chemical-Biological Revolution* (North Dakota State University, 1967), p. 24.

naturally occurs when the natural is deemed detrimental. Smallpox was a naturally occurring disease until it was eradicated by protracted international immunization. In fact, if one is arguing from what seems to be natural, the natural pattern seems to be one of development and change.

Notice that I am not arguing right now for or from the theory of macroevolution. I am quite aware that I will have readers for whom that theory is foundational and others for whom it is anathema. I am simply observing here that over time descendants within a species can be quite different. We have dachshunds and Great Danes within a few centuries. Medieval armor looks like it was made for children, since the adult men of the day who wore it were quite a bit shorter than adults today. In just the last generation the average height of a Japanese male has increased almost six inches. It is incontrovertible that we see change.

Would descendants of human beings be less human to the degree that they change? Only if "human" is tied to our current physical form. Are current well-nourished generations, generally taller and stronger than those of the past, more or less human due to those evident physical changes? Is "human" a static concept that has already been quintessentially achieved or does it consist in qualities that can be positively developed? While human beings are physical beings, that does not of itself require that their physical nature remain unchanged. Future descendants, in some way physically different from our current state of development, may be more human in what we value as human.

Paul Ramsey adds a further reservation about modifying our physical form. He is concerned that such modification would separate us from the ethics of medical practice and our civilization.[17] Apparently for Ramsey the ethics of medical practice and our civilization are based on the current physical state of human beings. To change our physical state would cut us loose from our ethical moorings. Actually this concern leads not to being adrift but to shifting anchorage. If the point of ethics is to live out appropriately one's physical endowment, as the argument assumes, a new physical endowment would lead to a new ethic, not a state of being ethics free. Granted, the new ethic could eventually be quite different from the old one. Genetic change would shift the stated foundation, not eliminate ethical systems being built upon it.

17. Paul Ramsey, *Fabricated Man: The Ethics of Genetic Control* (New Haven: Yale University Press, 1978), p. 105.

By Divine Command

"Human beings should not play God before they have learned to be human beings and when they are human beings they will not want to play God."[18] Ramsey does not explain the term *playing God*. In fact, concerning a different case addressed in an earlier work, he uses the term positively as an appropriate goal for humans to imitate God's gracious impartiality.[19] The concern that intervention in human genetics is in some sense playing God might find part of its resonant power in its ambiguity. It is used by Ramsey and throughout the literature to refer to many different concerns.

In this case, for Ramsey, God-given dominion reaches over all the animal world but not to human beings themselves.[20] To intervene in human genetics is to make oneself lord and creator of future generations. Human beings are to serve human life, not change it.[21] Humans are in no position to claim the ability to improve on what God has made. To even attempt such would be a hubristic attempt to be what we are not. Only God has the authority and ability to form and change us. Altering human physical form is taking on a role that human beings do not possess and should not usurp. Human beings are to enjoy and work within our God-given design as we have received it. Since the body is human, it is not part of the nonhuman world given to human beings for the dominion of stewardship and potential modification.

Such a position would possibly allow some genetic intervention for the cure of disease to better maintain the God-given design. Further change would be unacceptable. Carl Henry speaks in a similar vein in regard to genetic intervention. "It is the tragic irony of human existence, however, that when fallen man seeks to elevate himself to super humanity, as if he were himself God and the creator of his cosmic destiny, he soon succeeds only in transforming himself into an iniquitous monster who grandiosely rationalizes the aberrations he elevates as sacred and moral."[22]

A foundational assumption in these statements is that God's will is known and that it is to maintain the present physical attributes of human beings just as they are. That such is God's will may not be immediately

---

18. Ramsey, *Fabricated Man*, p. 151.

19. Paul Ramsey, *The Patient as Person: Explorations in Medical Ethics* (New Haven: Yale University Press, 1970), p. 256.

20. Ramsey, *Fabricated Man*, pp. 88, 132.

21. Ramsey, *Fabricated Man*, pp. 95, 138.

22. Carl Henry, "The New Image of Man," in *The Scientist and Ethical Decision*, ed. Charles Hatfield (Downers Grove, Ill.: InterVarsity Press, 1973), p. 170.

clear.[23] Traditions such as Islam, traditional Roman Catholicism, and evangelical Christianity are convinced that God is all powerful and purposeful, the source and active sustainer of the world, indeed the founding and ultimate standard of all that is. A key issue for them is what does God want human beings to do and what does God retain at God's sole initiative?[24] Different understandings of God and God's will lead to different conclusions concerning appropriate genetic intervention.

Physical Attributes Should Change

Dynamic Nature

The point of this perspective is not that human beings should abandon their nature or change it. The argument from this perspective is that human nature *is* to change. In contrast to Ramsey's understanding of human beings as already in the form they are meant to be, James Gustafson sees human form developing. Human beings are not yet what we should be and indeed we find our purpose in development. When considering obligations to future generations, the current physical form of human beings is not a given. In physical form and the rest of what it is to be human there is a central place for change. For Gustafson it is primarily in our experience and scientific observation of the human that we gain a sense of what values preserve and enhance the qualities of life that give a sense of fulfillment.[25] That sense of fulfillment is found in having a vocation to surpass what we are now to a *telos* we have not yet obtained.[26] The *telos* is not a set image of what human beings already are or should be, but rather one to be discovered, one that changes and develops as human beings change and de-

---

23. Richard McCormick, testimony during United States Congress Committee Hearing on Human Genetic Engineering, 1983, Committee Print #170, 334; Charles E. Curran, "Theology and Genetics: A Multi-faceted Dialogue," *Journal of Ecumenical Studies* 7 (Winter 1970): 78; Ronald S. Cole-Turner, "Is Genetic Engineering Co-Creation?" *Theology Today* 44 (Oct. 1987): 340.

24. For an apologetic on behalf of divine command ethics (in this case in the Calvinist tradition), see Richard J. Mouw, *The God Who Commands: A Study in Divine Command Ethics* (Notre Dame: University of Notre Dame Press, 1990). Mouw addresses criticism from moral development theories to feminist critiques of inherent patriarchy.

25. James M. Gustafson, "Basic Ethical Issues in the Bio-Medical Fields," *Soundings* 52 (1970): 178.

26. James M. Gustafson, "Genetic Engineering and the Normative View of the Human," in *Ethical Issues in Biology and Medicine*, ed. Preston N. Williams (Cambridge, Mass.: Schenkman, 1973), pp. 46-58.

velop.[27] The self-creativity of human initiative and freedom is central to humanity itself.[28] For Ramsey human beings are static, for Gustafson dynamic. For Ramsey it is pridefully and foolishly stepping outside of the God-given human role for human beings to modify their God-given selves. For Gustafson, self-creation is part of the essence of being human. Enhancement could be a welcome part of that self-creation.

Gustafson has argued at length that ethics should be theocentric (God-centered), but he means something quite different from Ramsey's appeals to God's person and authority. By theocentric Gustafson does not suggest that God has an intention for the future of humanity. For Gustafson God has no intentionality.[29] Rather, Gustafson describes God as the limiting order closely akin to if not synonymous with the natural order.[30] In any choices, including ethical ones, if human beings hope to survive, they must not choose actions that will destroy their physical existence. God, as the natural order, while not offering positive direction of what is desirable, does make demands as to what must be taken into account if human beings are to survive. Concern to maintain the physical is then a necessary factor to take into account as human beings modify themselves, if they value the continuance of their lives and that of their descendants.[31] Genetic intervention could be pursued within those survival limitations.

By Divine Command

Authors such as the scientist Donald MacKay and theologian Bernard Häring agree with Gustafson that human beings would do well to change, but carry on the Irenaean tradition that God intentionally created this world and is not finished with it yet. It is not yet all that the Creator intends it to be. Human beings are called to be part of God's ongoing creation. While Ramsey asserts that creation is to remain as it stands, for MacKay creation is still properly developing and human beings have an appointed role in guiding its development. MacKay appeals to several biblical passages.

---

27. Gustafson, "Basic Ethical Issues," p. 178.

28. Gustafson, *Theology and Christian Ethics*, p. 285.

29. James M. Gustafson, *Ethics from a Theocentric Perspective*, vol. 1: *Theology and Ethics* (Chicago: University of Chicago Press, 1981), p. 272.

30. See the discussion in "Part One: Focus on the Ethics of James M. Gustafson," ed. James F. Childress and Stanley Hauerwas in *The Journal of Religious Ethics* 13, no. 1 (1985): 1-112.

31. Gustafson, "Basic Ethical Issues," p. 173.

"He that knoweth to do good and doeth it not, to him it is a sin," says the proverb; and this teaching runs throughout the Bible, indicating that although there is indeed a sense in which everybody and everything owes its being to God as Creator, it is simply bad theology to conclude from this that God places an embargo on taking our share of responsibility for the way things turn out next.[32]

Fully recognizing human limitations, human beings are called to pursue the Creator's priorities, not human-generated ones. Human beings are given the world as stewards to fulfill their God-given purpose, not to do with it as they will. Lynn White has argued that often in the Christian tradition dominion has been interpreted as the free use or destruction of the physical world for any human whim.[33] Those arguing as above make every effort to differentiate themselves from such an interpretation.

> The book of Genesis spells out the mysterious dignity of what it means to be human. Encamped in a complex and dangerous world whose every feature, like himself, owes its being to the say-so of his Creator, man is singled out as one who has an explicit task. He is to "have dominion." He alone, it would seem, among his fellow creatures, will be held accountable for the management — or mismanagement — of the resources at his disposal. For the biblical writers the world is not merely something to be enjoyed or admired — though it is both — but something to be explored and developed by man in a spirit of thankful and responsible stewardship, for the benefit of his fellows and the glory of God. With all his limitations, he has enough likeness to his Creator to be called upon to accept and pursue his Creator's priorities. This — no less — is his destiny and his dignity.[34]

MacKay does not argue that human dominion may act out its every whim, but rather that it is to be exercised in a spirit of stewardship responsible to God. Such intervention maintains the world, develops it toward a better form, and mitigates some of the effects of the "fall" of human beings. Hans Schwartz and more recently Ronald Cole-Turner have also emphasized

---

32. Donald M. MacKay, *Human Science and Human Dignity* (Downers Grove, Ill.: InterVarsity Press, 1979), p. 58.

33. Lynn White, "The Historical Roots of Our Ecological Crisis," *Science* 155 (10 Mar. 1967): 1203-7. Contrast, for example, the understanding of dominion as stewardship in Norman Anderson, *Issues of Life and Death* (Downers Grove, Ill.: InterVarsity Press, 1977), p. 37.

34. MacKay, *Human Science*, p. 13.

this later theme of redemption.[35] Current fallen nature is not identical with God-given creation.[36] Without hesitation we try to repair disabilities that come from genetic heritage. When a child is born with a harelip, we immediately do surgery so that the child can eat and speak with greater ease. When a child is born with the genes for retinoblastoma, we heighten vigilance and treatment so that the child will not lose her eyes or die of cancer. We have discussed in chapter 3 how such harms and our responses to them are understood in light of God's providence and our responsibilities.

The common warning against the prideful taking of God's place in genetic intervention assumes that God has forbidden intervention or reserved it for God alone. Those who argue that genetic intervention is part of the God-given mandate for human beings to share in creation, redemption, and transformation of creation, would see the greater danger not in an attitude of pride, but of sloth. Not fulfilling the responsibility to turn genetic intervention to service would reflect a dangerous and destructive attitude of disobedient apathy. James Walters, for example, says that in Roman Catholicism genes have the same honored status as the rest of the human body, not a unique status.[37] If one can do surgery on other parts of the body, one can do surgery on one's genes.

Paul Ramsey adds the further concern that even as human beings gain in knowledge, we will never have sufficient wisdom to properly use that knowledge for any physical change. "The boundless freedom of genetic self determination leads to boundless destruction since no human being or collection of human beings will ever have the wisdom to do it rightly."[38] Only God possesses such wisdom.[39] At this point Ramsey appears to be evaluating human will and discernment more than human knowledge and predicting it to be permanently wanting. Henry Stob concurs: "To tamper with the genes seems to me to 'outrun' God into an unknown future and to exercise an 'elective' discrimination mere men do not possess," nor ever will.[40] In short, for

35. Hans Schwartz, "Theological Implications of Modern Biogenetics," *Zygon* 5 (Sept. 1970): 263; Cole-Turner, "Is Genetic Engineering Co-Creation?" 338-49, and *The New Genesis: Theology and the Genetic Revolution* (Louisville: Westminster/John Knox, 1993).

36. Max L. Stackhouse, *Public Theology and Political Economy: Christian Stewardship in Modern Society* (Lanham, Md.: University Press of America, 1991), p. 144.

37. James J. Walter, "Theological Perspectives on Cancer Genetics and Gene Therapy," Ethical Boundaries in Cancer Genetics conference, St. Jude Children's Research Hospital, Memphis, 28 May 1999.

38. Ramsey, *Fabricated Man*, p. 96.

39. Ramsey, *Fabricated Man*, p. 149.

40. Henry Stob, "Christian Ethics and Scientific Control," in *The Scientist and Ethical Decision*, ed. Charles Hatfield (Downers Grove, Ill.: InterVarsity Press, 1973), p. 20.

these authors such an attempt would claim a wisdom and right that belong to God alone.

Human beings are able to change themselves but in their finitude are always at risk of endangering themselves in those choices. MacKay specifically warns of the potential havoc of even one well-intentioned error but responds that such risk does not of itself necessarily abrogate the responsibility to proceed. The implication for MacKay is not inaction, but rather that human beings need divine guidance.[41] He uses the metaphor that navigating by a landmark tied to your own ship's head is pointless.[42] A proper heading can be set by seeking God's revealed will such as found in Christian Scriptures. MacKay writes as one convinced that God has purposefully created the world and human beings and that God sets standards for their best development. If one is open to listen and obey, God will guide that one to the best service of God and others.

MacKay is hopeful on the basis of biblical texts such as James 1:5 that God will graciously give wisdom to those who ask for it. Sufficient guidance will not come from the current order of nature. The apostle Paul wrote that the whole creation groans in travail waiting for redemption.[43] It is not currently all that it was meant to be. The standard for intervention is rather one of prayerful submission to God and then acting as seems to best express love of God and neighbor. "All human exploitation of natural laws and resources must be an expression of this love, and of nothing else, if it is to be acceptable."[44]

MacKay is not sanguine that people will consistently listen to God or listen well. Granting the presence of sin that obscures judgment and twists motives, MacKay calls as well for "answerability" to God and others in all such decisions. Any intervention must be slow and reversible so that adjustments can be made in the face of the tremendous complexity.[45]

One's understanding of being physically human as fixed or changing is central to whether physical change has the potential to lead toward the positive development of human beings, or in choosing to change, the destruction of humanity. If the description of human beings as physically static beings is

41. MacKay, "Biblical Perspectives in Human Engineering," in *Modifying Man: Implications and Ethics*, ed. Craig Ellison (Washington, D.C.: University Press of America, 1978), p. 72.

42. Donald M. MacKay in *Man and His Future*, ed. G. E. W. Wolstenholme (Boston: Little, Brown and Company, 1963), p. 286, as quoted by Paul Ramsey, *Fabricated Man*, p. 124.

43. Romans 8:19-23.

44. MacKay, "Biblical Perspectives," p. 69.

45. MacKay, *Human Science*, p. 61.

most compelling, no genetic improvement is appropriate. If human beings are called to develop themselves, purposeful and direct enhancement of capacity could be appropriate, or according to some even required. MacKay and Häring, for example, argue that human beings are responsible to God for proper care, correction, and development of creation. This brings us back again to the mandate to sustain, restore, and improve what God has entrusted to us. That includes ourselves and our world. So used in God's service, genetic intervention may yield positive changes to human form.

Attitudes Encouraged by Intervention

That there could be positive results from genetic intervention is not enough reason to pursue it. How they are obtained is also an important consideration. For example, what effects would changing human genetics have on the attitudes of those who receive intervention and those who carry it out?

Making People?

At this time many different methods of gaining access to genes for surgery are being investigated. They warrant careful attention as they enter use. Introducing genes could be by methods such as injection, inhaling an aerosol, or removing cells for treatment and then placing them back in the body. The degree of concern generally increases with the degree of invasiveness, particularly in any departure from the way human beings now come to be. A brief intervention after conception and before extensive development would not interrupt the loving bond and intimacy of intercourse at the start of life nor later development in the womb. *In vitro* fertilization would be a more substantial departure but might still be in support of a couple committed to each other and any children for life.

The resulting families, as all families, would need to be built "on acceptance of people as they are with all their imperfections."[46] Gilbert Meilaender asserts that if one fully accepts other human beings one should not seek to change their genetic endowment. He may be conflating the acceptance of children as fellow human beings with being resigned to whatever happens to be their current physical state. Meilaender himself seeks to avoid this confu-

46. Thomas H. Murray, *The Worth of a Child* (Berkeley: University of California Press, 1996), pp. 5-6.

sion in the same article when he writes that "we should turn against disease, not against those who are diseased."[47] Torbjorn Tannsjo points out what may be a similar move on the part of the European Parliament that on January 26, 1982, affirmed a right to be born free of artificial intervention in one's genes but in the next paragraph clarified that this prohibition did not apply to therapeutic applications "since such therapy promises to make possible the elimination of certain hereditary diseases."[48] It is not a rejection of people crippled by polio to give children a polio vaccine. One can honor people who happen to be crippled without affirming that it is good to be crippled. It is not a rejection of particular persons with spina bifida that we are sure that pregnant women have adequate B12 in their diet. Why should genetic inoculation be perceived as a rejection of people afflicted by the unwanted condition? Genetic intervention need not imply a lack of care for a particular person. It may embody it.

Some groups have argued that any intervention imparts harmful attitudes. The Enquete Commission was established in Germany to inform the public and advise the Bundestag concerning genetic legislation. The commission stated that "everybody must have the possibility of seeing himself or herself, his or her own essence, as the result of a fate separated from human beings — or as created by God — but not as the project and as the more or less successful experiment of other human beings."[49] For the commission, finding one's origin in chance or providence is important to individual worth and independence. While the assertion is evocative of several important matters, the commission does not explain it.

As often happens in committee documents, members may have had different reasons for affirming the final statement. One might be the concern that human beings should not be perceived or treated as manufactured objects, hence disposable or discounted from full respect. Children are our progeny not our creations.[50] "What we beget is like ourselves. What we make is not; it is the product of our free decision, and its destiny is ours to determine . . . it is, in fact, human begetting that expresses our equal dignity, we

47. Gilbert Meilaender, "Human Cloning Would Violate the Dignity of Children," reprinted in *The Human Cloning Debate,* ed. Glenn McGee (Berkeley: Berkeley Hills, 1998), p. 194.

48. Torbjorn Tannsjo, "Should We Change the Human Genome?" *Theoretical Medicine* 14 (1993): 239.

49. Enquete Commission, "Prospects and Risks of Gene Technology: The Report of the Enquete Commission to the Bundestag of the Federal Republic of Germany," *Bioethics* 2, no. 3 (1988): 257-58.

50. Charles Frankel, "The Specter of Eugenics," *Commentary* 57, no. 3 (1974): 33.

should not lightly set it aside."[51] Genetic intervention is "a giant step toward turning begetting into making, procreation into manufacture, making man himself simply another one of the man-made things. . . . As with any other product of our making, no matter how excellent, the artificer stands above it, not as an equal but as a superior, transcending it by his will and creative prowess . . . human children would be their artifacts."[52]

The charge that genetic surgery would turn the recipient into a man-made object for our use is a serious one. But would it? Does surgery after birth for a harelip make the child "made not begotten"? No. Physical intervention can be out of love for the recipient, fully recognizing and motivated by concern for him or her as a person. Early genetic surgery could someday correct the genetic code so that the child's teeth and lips grow straight in the first place. Why would corrective surgery after birth not devalue a child, but before birth it would? Further, it is not the process of apparently random conception that makes us people or things. The Christian tradition describes angelic beings who are persons of significant dignity and worth without human begetting at all. Being a person is not limited to being born of a human being, let alone a particular method of conception. A thought experiment is nicely provided by the popular Star Trek series. One of the most admirably personal characters is Data, an android. The character's origin is beside the point. What matters is the person who is present.

The Enquete Commission's concern may be that the domination of parents and others should not be harmfully increased over their children. Chapter 11 discussed the importance of an open future. Another concern may be that genetic intervention might come to be organized by sales order and purchase. That would make it difficult to keep the all-important attitude of welcome and acceptance for every child. If genetic intervention becomes primarily a contract to purchase a guaranteed product, the attendant attitude of commodification would be damaging.[53] Giving offspring the best possible genetic start is not grounds to treat them as disposable products once they have arrived, any more than financing a tonsillectomy means one can toss out Johnny the toddler if he needs more surgery. People should not be treated as if they were mere objects. However a child comes to be and whatever her condition, she should be welcomed and loved.

---

51. Meilaender, "Human Cloning Would Violate the Dignity of Children," p. 194.

52. Leon Kass, "The Wisdom of Repugnance: Why We Should Ban the Cloning of Humans," reprinted in *The Human Cloning Debate,* pp. 169-70.

53. Murray, *The Worth of a Child,* p. 33.

Staying Connected

Some people seek immortality in passing on their genes. The irony is that such a quest is both impossible and trivial. Genes quickly disperse over generations and the physical form that they embody is not our most important presence or legacy. The design of human genetics is one of dispersal and recombination. A child will carry only half of each parent's genotype, and even that half may not be what is most prominent in the phenotype of the parent. A parent with green eyes could pass on a recessive gene for blue eyes. It would be from the parent's genotype but would not duplicate the parent's phenotype. By the fourth generation a child is equally related genetically to sixteen different people, in the fifth generation to thirty-two, and in the sixth to sixty-four.

For various motivations many societies have put great effort and importance into tracing and maintaining blood lines. For example, the child as genetically connected property dominated Roman law. This may have been to reinforce distinctions of birth between slave and free.[54] Genetic parents could demand the return of a child that had been adopted a decade before and raised completely apart from them. The United States legal system has carried on this tradition in stressing genetic connections between people as paramount in issues of custody and inheritance. Children are sometimes treated as a form of property to be disposed of according to ownership rights. There have been cases of a child raised for years in mutual commitment and love with adoptive parents only to be suddenly taken away because a man who contributed one sperm wishes to claim his legal right to the child.

Thomas Murray recounts the story of Cara Clausen, who gave up her baby for adoption along with the written consent of the man identified as the father. A month later the man who had actually contributed the successful sperm discovered that Cara had a baby and demanded custody. He did not have an ongoing relationship with the mother, and he already had a son who was fourteen whom he had legally abandoned, and a daughter, twelve, whom he had never met. He had not voluntarily paid child support for either. Considering that history might cause one to wonder why he was suddenly so interested this time. The local district court did. While the judge praised the parenting of the adoptive parents and ordered an investigation as to whether the plaintiff would be a fit parent, the judge also at the same time ordered the adoptive parents to give the baby to the plaintiff. Disagreements between the courts in Iowa and Michigan extended until the two-and-a-half-year-old

---

54. Murray, *The Worth of a Child*, p. 7.

child was taken by force from the only parents she had ever known.[55] The child was treated as property, not as a human being who should have been the first concern. There is normally a personal commitment that comes with a genetic connection, especially when reinforced by carrying the child until birth. But having a genetic connection is not a property right that overrides the bonding and well-being of the child.

The Christian Tradition and Adoption

Adoption is lauded in Christian tradition. Matthew and Luke state that God chose to be incarnated in a family where he had two parents but was not genetically related to his father Joseph. When during Jesus' public ministry his mother and brothers came to see him, he used their arrival as an opportunity to proclaim that all who followed God were his family.[56] Jesus as the eldest son undoubtedly had provided for his family and even while dying on the cross was making provision for Mary,[57] yet he emphasized that to be part of his family was by choice and commitment, not genetics.[58] One is adopted into God's family.

This is not a second-best result of the fall. The opening chapters of Genesis describe Adam and Eve as receiving a world where they could choose to be with God or not. They walked with God on occasion, yet they were not allowed to eat yet from the tree of life that could grant them immortality. If they had chosen to follow God at the tree of the experiential knowledge of good and evil, they would have gained that knowledge-experience by choosing well and would have entered more fully into God's plan and relationship with them. Instead they sought to be their own gods. God still cared for them, as exemplified in the sewing of garments for them, but they were cast out of the garden, denied the tree of life, and in need of redemption. From creation they had to make a choice to become part of God's family. They chose not to be. They were not automatically God's people. That comes by invitation and response even as created.

Adoption into God's family is characteristic of God's work through the rest of the Old Testament. God is the God of Israel by covenant, not birth. God chose to work through Abraham and his progeny, yet this was not an exclusive call to be God's people but rather a call to a particular task. Melchize-

---

55. Murray, *The Worth of a Child*, pp. 41-44.
56. Matthew 12:46-50.
57. John 19:26-27.
58. Luke 8:19-21.

dek and others completely unrelated genetically to Abraham are recognized as people of God.[59] This is the theme of the book of Jonah. Even being part of the people of Israel follows commitment to know and follow God, not mere blood lines. Ruth from Moab[60] and Rahab from Jericho[61] are welcomed in, while many in the blood line are cast out.[62] The Davidic line is traced to Jesus not because it is the only one that God cares about, but as a sign to help recognize the Messiah when he comes.

On into the time of the church, children are welcomed into the church community by dedication in some churches and by infant baptism in others. It remains for the child to understand and affirm for herself that she wants to become part of the church. That decision is publicly affirmed at a later age in baptism or confirmation. In essence, God has no grandchildren. It is not sufficient to be a descendant of someone in God's family. Each individual must choose to welcome that relationship or turn away. If one does receive that welcome provided by God, one becomes part of the body of Christ on earth. All human beings in God's family are welcomed as children by adoption. One is part of the people of God by conviction, not physical descent.

The new life that begins now and extends forever includes immortality of the person, not one's genes. There is no call from the tradition to be concerned about passing on one's genes into the future. Immortality is personal, not by progeny. Genes are important to physical and mental health, but they are not the point of this life or the next. Christians are to have children to share intimate love with new people, not in a doomed quest for genetic immortality. One is to love one's neighbor as oneself, not only one's genetic relatives. The adoption of children unrelated genetically is one of the closest human recapitulations of the life-giving adoption that has welcomed people into God's family. Whether genetically related or not, the most precious heritage that human beings can pass on to their children is to know God and receive life as it is meant to be in God.

When a couple discovers a situation such as that one of them has an autosomal dominant condition yielding a 50 percent chance of severe harm for each conception, that brings to the fore the decisions discussed in chapters 5 and 8. The Christian tradition does not require a couple to have children or to be genetically related to them. The percentage of genetic overlap with parents is also not central. Substantial changes by genetic surgery that help a

---

59. Genesis 14:18-20.

60. Ruth.

61. Joshua 6:25. Note that Rahab and Ruth actually are part of the line leading to King David, Matthew 1:5.

62. E.g., Exodus 32:25-28.

child would be welcome. The key question is not duplication of parental genes but what is best for the child.

Caring for the Child

Leon Kass raises the concern that parents might abdicate responsibility for their children and the children for their parents if the genetic link is broken. Indeed the genetic connection is "an indispensable foundation for sound family life."[63] He does not explain why it is indispensable. Of course he is quite right that one has a responsibility to care for a child that one begins in sexual intimacy, yet if one of the couple that began that new life dies or flees, the child is not condemned to no possibility of "sound family life." Sharing genes may tap into sociobiological drive, but that is not the only possible grounds for committed love. Family resemblance may encourage the extension of self-love, but it is hardly determinative by itself. Is one more likely to sustain a relationship with someone because she has the same eye color or a similar nose? If such has any effect at all, it would be quickly overwhelmed by the responses and attitudes of the person.

Phenotype matters for aesthetics and initial identification. Similarities imply connection between persons. But if there is no more connection between persons than a few physical traits, they quickly fade as a point of interest. Physical appearance helps us to identify persons, but it is the person that is worth identifying. A genetic link with parents does not guarantee care or safety. Susan Smith

> feared that her two children, three-year-old Michael and fourteen-month-old Alex, stood in the way of her boyfriend's making a long-term commitment to her. In 1994 she strapped them into their child safety seats in her Mazda. Then she sent the car rolling down the embankment into the John D. Long Lake near Union, South Carolina. The two boys were drowned. . . . The safety and security of her children could not be guaranteed by the fact that they were in the care of their birth mother.[64]

Sexual intimacy should embody a commitment to each other and any resulting child. If that cannot be honored due for example to death, or simply is not, the resulting child can still be welcomed into a family. Sometimes the

---

63. Leon Kass, *Toward a More Natural Science: Biology and Human Affairs* (New York: The Free Press, 1985), p. 113.

64. Ted Peters, *For the Love of Children: Genetic Technology and the Future of the Family* (Louisville: Westminster/John Knox, 1996), p. 1.

most loving thing a young single mother can do for a child is to give that child to a family that can better raise the child. Actual family care is often more formative than genetic connection for identification and commitment. Adopted and blended families can be examples of caring concern that do not rest on common lineage.

Studies have found adoptive families and families with less than complete genetic duplication to be at least as psychologically healthy as traditional birth families. "Children conceived by ART [artificial reproductive technology] did not differ from naturally conceived children in emotions, behavior, or quality of family relations."[65] In fact, "adoptive parents and parents conceiving children through ART expressed greater warmth and emotional involvement with children, as well as greater satisfaction with parenting roles, relative to birth parents."[66] The ART families may have a genetic link through one or both parents, the adoptive families no genetic link, and most function quite well.[67]

Janet Fenton has been quoted as saying that "if adoption were made into a kind of guardianship, children . . . would know who their real parents were and that they could go back to them at any time."[68] Fenton is equating being a genetic source with being the "real parents" and assumes that such parents who already gave up the child would take the child back at a later point. If not called the "real parents," sometimes those who provided the gametes are called the "biological parents." Even this adjective is misleading. The couple that raises a child, feeding, protecting, caring, and clothing him or her for years, has contributed more to the child's biological survival than two gametes.

With the growing potential for even more separation between types of genetic sources, gestation, and parenting, it may be that the most accurate terms for each role in raising a child are "gene sources," which would include whoever provides sperm, egg, nuclear material for an egg, or the nonnuclear

---

65. Susan Golombok et al., "Families Created by the New Reproductive Technologies: Quality of Parenting and Social and Emotional Development of the Children," *Child Development* 66 (1995): 285, 295; Frank van Balen, "Child-Rearing Following in Vitro Fertilization," *Journal of Child Psychology and Psychiatry and Allied Disciplines* 37 (1996): 687, 692.

66. Nancy L. Segal, "Behavioral Aspects of Intergenerational Human Cloning: What Twins Tell Us," *Jurimetrics* 38, no. 1 (1997): 61.

67. Whether or not the use of donor gametes is ethically acceptable on other grounds is discussed in chapter 8.

68. Janet Fenton, president of Concerned United Birthparents, as quoted by Murray, *The Worth of a Child*, p. 41.

egg, "gestational mother" for the woman who may be unrelated genetically yet intimately carries the developing child to term, and "parents" for the mother and the father who give the rest of their lives in permanent commitment to raising the child. Qualifying the latter with the adjective "social" mother or father neglects their essential and extended biological contribution. Being a mother or father in the fullest sense is about living out a lifelong commitment of love, provision, and care for someone who depends on you, not genetics.

The Case of Human Cloning[69]

You have probably already met a human clone. We commonly call them identical twins. About fifteen sets are born each day in the United States.[70] In the case of identical twins, shortly after one ovum and sperm united, the resulting embryo split into two or more separately developing embryos. The individuals that are then born are distinct individuals. They share the same genetic heritage and the same womb. Usually they go on to share the same general parenting and environment, yet they still develop into distinct people. They share many physical traits and some aspects of temperament, yet one may choose to be a carpenter and the other a college professor. One may choose to become a follower of Jesus Christ and the other not. One may die in her twenties, the other reach seventy-five. Even their fingerprints will be different. Genotype is not phenotype, and genotype does not make a person or a life. To call an identical twin a "clone" is pejorative and misleading as the term is currently perceived. The connotation of clone is one who is a complete copy of an adult personality that already exists. Identical twins are not that.

A human baby with the same genetic code as an adult would not be a "clone" in that sense either. The adult that contributed the genes and the baby that received them would not be "identical twins" in that they would not share the same womb, parenting, or age. Yet they would have largely the same genetic start, although not as close as that of identical twins. Nuclear transfer does not transfer all of the genes, because mitochondrial genes are not within

69. Cloning can refer to several different processes. For example, cellular cloning consists of conditioning a cell to keep replicating indefinitely. The result is a uniform cell line useful for laboratory work. In contrast, the famous ewe Dolly came from the fusing of an adult nucleus with an ovum that had its original nucleus removed. The end result in each case of cloning is a duplication of genetic material from one organism to the next. It is this later form applied at the human level that we will address here.

70. Richard Lewontin, "The Confusion over Cloning," in *The Human Cloning Debate,* p. 127.

the nucleus. The best title might simply be "genetic twins." The National Bioethics Advisory Committee (NBAC) has used the term "delayed twins."[71] This well emphasizes the time difference in birth but may imply more contextual connection than genetic twins may have. Genetic twins could be born generations apart or in different families. The qualifier "genetic" recognizes the genetic parallel without implying the same womb or family of identical or delayed twins.

Chances are that when you first meet a child who is the genetic twin of an adult, you will not realize it, even if you know the adult who contributed the genes. Since a child typically inherits a gene from each parent for each characteristic, it is quite possible for a child to favor one parent over the other in expressed appearance or temperament. The statistical bell curve would predict a varied but relatively even mix from both parents for most children and less frequently that one child might more strongly favor one parent than the other. With all the children born, we do not find it surprising to see a child from one end or the other of the statistical curve who strongly favors one parent in appearance or temperament. This is common enough that we have stock phrases for it. She is "the spitting image" of her mother, or that boy is "a chip off the old block." The genetic twin would physically favor one parent but still grow and develop as a unique individual. Since genetic twins would probably be dramatically different in age and in quite different environments, they would probably always have significant differences in appearance.

But why not mix genetic heritage for every new person? Why repeat one set of genes? There are at least three reasons doomed to failure that are sometimes offered and four reasons that have more weight.

Inadequate Reasons

To Bring Back to Life a Child Who Has Died

This motivation is doomed to failure. Even if the same mother carries the genetic twin, the experiences in the womb and after cannot be a perfect duplicate. The child will make unique decisions and develop into a distinct person. It is a great tragedy to lose a child. The Christian confidence is that the child has survived in the loving presence and care of God. It cannot return in a new body.

71. National Bioethics Advisory Commission, *Cloning Human Beings,* vol. 1 (Rockville, Md., 1997), p. 3.

The Gene Source Seeks Genetic Immortality

If the genetic source is seeking immortality by gene duplication her effort will fail on two counts. (1) If her younger genetic twin marries and has children, the founder's genetic heritage mixes and disperses quickly at each new generation. Duplicating one's genes only delays the dispersal by one generation. One could conceivably financially endow an indefinite series of clonings of one genetic set, which could then last as long as the law and each new generation tolerated it, but what would be the point? (2) The originator of the plan is passing on what is an important but small part of who she is. The genetic twin will be a unique person in her own right. The Christian tradition promises eternal life for those who belong to God. Sustaining a particular combination of genes seems comparatively paltry.

The Gene Source Seeks Only Spare Parts

The key here is that the new baby would have all the rights and due respect of any other human being just as the second born of identical twins does now. The child would be a genetic match for tissue donation but could only do so when protected as any other child. She could not be merely spare parts. This is exemplified in the famous Ayala story. Abe and Mary Ayala were spurred on to have another child in hopes that the newborn might be a marrow donor for their daughter Anissa. Their new daughter, Marissa, happened to be the needed one-in-four sibling match for her sister's treatment of last resort. Anissa's life was saved and her little sister is loved as every other member of the family. Considering the lengths the Ayalas were willing to go to save their eldest daughter, it is not surprising that they treasure their youngest. Marissa was not harmed by donating and is reported to have a special bond with her sister whose life she saved. The fact that she was able to help is not demeaning. In the Christian tradition we are all healed to serve. To be able to help another as one can, let alone a family member, is a privilege and a universal calling. Service is not antithetical to self-worth.

Renewable fluids such as blood could be shared in both directions to the benefit of each genetic twin, but more extensive donation would require consent of the grown child as a legal adult and considerable caution. If the senior genetic twin needs a transplant due to kidney failure, it might be expected that the younger genetic twin is also vulnerable to that problem and hence not a good candidate to give up a kidney. Avoiding this problem by duplicating not a person but just tissue is a possibility.

Grotesquely cloning a headless body, as some have conjectured, would be incredibly labor intensive, expensive, and medically difficult in that the brain coordinates bodily systems. Also the time delay of normal growth and then maintenance to serve one person would be lengthy. It would also tend to dishonor the human form. We go to significant lengths to honor the human body even after a person is no longer present. Deliberately growing and maintaining recognizable partial bodies over decades would be desensitizing to the human body's importance.[72] A better alternative may be the promising work in cell and tissue lines that are genetically altered to prevent graft rejection, hence available to all.[73] Eventually organs such as kidneys may be separately grown in animals or *in vitro* for transplant to those who need them.

Reasons That Have More Weight

Someone Who Wants to Be a Parent Does Not Produce Gametes

Adoption is one of the best metaphors for how God welcomes people. The New Testament tradition is that while God cares about all the world and all the people in it, human beings are not automatically God's children. That is a gift to be received as an adoption into God's family. "To all who received him, he gave the power to become children of God."[74]

While adoption is highly honored in the Christian tradition, it is difficult to practice in the United States today. There are long waiting lists of couples who wish to adopt. "In 1984, two million infertile American couples competed for 58,000 new-born American children."[75] Many couples find that by the time it is clear that they will not be able to have children biologically, they are considered too old to adopt newborns. Older children often need homes, but foster care is designed to reunite biological family units. By the

---

72. Michael Tooley only considers how the torso is harmed in this scenario. For Tooley, since there would be no conscious individual present, there could be no harm. See "The Moral Status of Cloning Humans," in *Human Cloning,* ed. James M. Humber and Robert F. Almeder (Totowa, N.J.: Humana, 1998), p. 74.

73. National Bioethics Advisory Commission, *Cloning Human Beings: Report and Recommendations of the National Bioethics Advisory Commission* (1997), supra note 3, p. 31.

74. John 1:12-13.

75. Lee M. Silver, *Remaking Eden: How Genetic Engineering and Cloning Will Transform the American Family* (New York: Avon, 1998), p. 160.

time children have passed through years of the present overworked and byzantine system, it is often difficult for them to bond with anyone.

There are state agencies that further restrict adoption by only allowing same-race adoptions. Many children that could be placed in loving families remain without any family because of that policy. Further, some state laws make adoption a tentative affair for the adopting parents. It is difficult to give one's life unconditionally to a child as your child with the awareness that the genetic parents or state may permanently take the child away at any time. Genetic parents can use this option to threaten open-adoption parents. In adopting the child in an open system, one becomes deeply involved with and even subject to people who could not handle life well enough to raise the child themselves.

Such risks make adoption a daunting task when it is available at all. Some of the problems involved are man-made and can and should be changed in the best interests of the children. In the meantime international adoptions, as complicated as they are, have become an often-pursued alternative. Genetic twinning could offer another alternative. The resulting baby would be genetically related to both parents if the father donates the nucleus. The mother's ovum would still contribute the mitochondrial DNA[76] and her womb the bonding of pregnancy. A child would be born into a family that welcomed the child. Somatic cell nuclear transfer would make a genetically related child physically possible for almost everyone. Infertility would not be as insurmountable as it is for millions of couples now.

Avoiding Genetic Disease

If the mother carries two alleles for a dominant genetic disease, no matter which allele she randomly passes on to the baby, the baby will have the affliction. Using the nucleus from the father would again lead to a baby that is genetically related to both parents, since the mitochondrial DNA and gestation experience would come from the mother, yet not give the child the dreaded affliction. Genetic twinning the mother because of the father's genetic problem would not bring about a child genetically related to both but would still at least be fully within the family unit. Another genetic possibility would be if the mother has a mitochondrial condition. She might request cloning of her husband or herself with a donor ovum.[77] That would allow her to be genetically connected to her child without passing on the mitochondrial DNA problem.

---

76. Granted, that is just about sixty genes out of a total of fifty thousand.

77. Of course this would not be an appropriate intervention if gamete donations are always illicit. The argument to that effect is evaluated in chapter 8.

Increasing the Chance of Birth

Only about one in three conceptions survives to birth. When aiding conception outside the body, a selected zygote still has only that one in three chance of survival when placed in the womb. If the zygote is twinned once or twice, the odds that at least one will survive to birth might be increased. Since embryo failure often stems from chromosomal abnormalities, duplicating an abnormality would not increase the chance of survival. The original and its twin would come to the same end. However, if a healthy zygote is twinned, the possibility of twins or triplets at birth is raised. The human body and sleepless parents generally do better having one child at a time, but two or three at once are usually a manageable challenge. In the United States, "Mothers of Twins and Multiples" clubs are a ubiquitous source of helpful information and support. Implanting more than two or three zygotes at once often leads to the practice of selective reduction when many embryos are successful. "Selective reduction" raises all the issues of abortion and fetal status described in chapter 5.

Best Available Genetic Endowment

A couple might be aware that a particular relative hit the genetic lottery. Great-Aunt Sally was as sharp as ever at ninety years of age and remarkably had no ill health in her long and productive life. A genetic twin would not be Aunt Sally, but what a gift for the couple's child to start with a combination of genes that had already proven a clear predisposition for excellent health and multiple talents. With Aunt Sally's permission would it be right for them to pass those genes directly to their daughter to be?[78] The ethical acceptability of this option hinges not only on the ethics of cloning but also on the ethics of using donor gametes (or here, their reproductive equivalent), as discussed in chapter 8.

Concerns about Deliberate Genetic Twinning

It is not enough to say that there might be good goals such as better health achieved by genetic twinning. Are there also harms or transgressions in such a process? For example, intentional genetic twinning is likely to require conception outside the body or removal of a zygote from the body for the inter-

---

78. This kind of scenario first came to my attention in a book by Gregory E. Pence, *Who's Afraid of Human Cloning?* (Lanham: Rowman & Littlefield, 1998), p. 104.

vention. All the concerns discussed in chapters 3 and 8 about intervening in the usual process of procreation would then be raised. Three concerns more specific to genetic twinning follow.

Genetic Diversity

Genetic diversity is important to disease resistance and the richness of human life, but for genetic twinning to actually have an effect on genetic diversity millions of parents would have to choose the same few genetic sources to twin. As will be discussed in chapter 15, that is unlikely.

Psychological Harms

There are concerns about psychological effects for the newborn. In particular, does genetic twinning lead to a loss of individuality?[79] It does not seem to with identical twins. Actually genetic twins might find it a sign of parental love that his or her parents went to lengths to provide an excellent genetic start or appreciate a special bond with her or his genetic twin.[80] That common heritage and intimacy is usually highly valued among randomly occurring genetic twins as a source of emotional support, trust, and understanding. It is not perceived by twins as confusing personal identity.[81]

Common genetic heritage might be a significant aid in raising a child. Does not everyone remember some event or period in their life when they wish their parents had understood or reacted in a different way? Each life and time is different, but genetic twinning might offer significant insights to help in raising the younger twin.[82] On the other hand, the shared genetic heritage may be no more a point of special closeness or contention than it already is when a child is most closely related genetically to only two people. The younger genetic twin's life is not predetermined; genetics is not destiny.[83]

Would a genetic twin feel predestined? Only if she is mistakenly a ge-

79. Craig M. Klugman and Thomas H. Murray, "Cloning, Historical Ethics, and the NBAC," in Humber and Almeder, eds., *Human Cloning*, p. 15, and George J. Annas, "The Prospect of Human Cloning: An Opportunity for National and International Cooperation in Bioethics," in *Human Cloning*, pp. 51-64.

80. John A. Robertson, *Children of Choice: Freedom and the New Reproductive Technologies* (Princeton: Princeton University Press, 1994), pp. 168-69.

81. Segal, "Behavioral Aspects of Intergenerational Human Cloning," pp. 60-61.

82. Tooley, "The Moral Status of Cloning Humans," p. 89.

83. See, for example. Thomas J. Bouchard Jr. et al., "Sources of Human Psychological Differences: The Minnesota Study of Twins Reared Apart," *Science* 250 (1990): 223-26.

netic determinist. However, she could learn from her twin's experiences. It might be useful to see that the older twin strove mightily over well-disciplined years to run a marathon in under two hours and just could never do it. She might save time and energy for more likely success.[84]

Could parental expectations be too high? Little League fathers and stage mothers who push their children find they can only require so much. High expectations can encourage one to do one's best. Undue pressure can be debilitating. Knowing what has been achieved with a set of genes would probably increase confidence in the possibility of some achievements, but genes are not destiny. We are a complex combination of factors such as genes, environment, and our own choices. While the recipient might well appreciate being given the best available genetic start, that would not dictate what the recipient would do with it. Parents could adjust if a child did not reach proven potential in a way they had hoped. That already happens. Recognizing the shaping influence of environment, they might well take personal responsibility for their part in how they raised the child.

Would the child know too much of her future health? That is a mixed blessing already available in part by looking at parents. Genetic testing is making such prognostication more precise and readily available to all whether one has an older twin or not.

Would family relationships be confused? Active genetic twinning probably would lead to new terminology. As suggested earlier in this chapter, one might refer to the "gene source" who provided the full complement of genes for the younger genetic twin. The woman who carried the child for nine months would be the gestational mother, and the mother and the father who lovingly give their lives in permanent commitment to raising the child would be the parents. If a person is the genetic source for a twin and raises the child from pregnancy on, that person is the child's parent, not a sibling. *Sibling* and *parent* are both much richer terms than just genetic relationship. *Parent* has to do with relative age, care, and responsibility, not same, different, or partly different genes. The key in the family would be raising the child in consistent love and support, actively pursuing what is best for her or him.

Human Rights and Dignity

This may be the most compelling constraint on genetic twinning, yet it is difficult to define exactly what the concern is. UNESCO declared in 1998 that "practices which are contrary to human dignity, such as reproductive cloning

---

84. Example from Tooley, "The Moral Status of Cloning Humans," p. 84.

of human beings, shall not be permitted."[85] Other organizations have proclaimed a basic human right to a unique genetic endowment as well. Short of executing one of each pair of identical twins, this right is not sustainable. One could qualify the concern to the right to have at least no *intentional* genetic twins.

As with any legislated proclamation, motivations and reasons can vary tremendously among the participants that affirm it. One reason for the above prohibition might be that we should not violate people's future autonomy by choosing their genes for them. This concern will be addressed at length in the next chapter. In brief now it is the nature of human procreation that parental choices such as that of mate and timing affect the genetic heritage of future individuals. The question is not so much whether we will make choices that affect future generations as whether we will do so consciously and responsibly. If an intervention increases the recipient's choices rather than limiting them, his or her autonomy has not been violated.

A related objection that discerns a violation of human dignity in intentional genetic twinning might be that unless one is cloning perfection, one is deliberately endowing a fellow human being with not only good characteristics but also bad ones. In that case, the net effect might be to limit rather than increase the recipient's choices.[86] While this concern could eventually be met if techniques are developed to correct any clear disabilities in the genes being twinned, we do not have that ability now.

A concern that might also be intended here is the one raised by some and discussed in chapter 8 that all genetic heritage should be from parents covenanted together in marriage. In genetic twinning a parent, all genes would be from within the one-flesh relationship, but the proportion of genes from each parent could vary. If it is the mother's nucleus that is twinned, only the mother would be directly related to the child genetically. If the father provided the nucleus, the mother would still pass on some genes through the mitochondria in her ovum. There would then be the mutuality of genetic heritage from both parents but not a symmetrical inheritance.

In chapter 11 I proposed four standards that should be included in considering any genetic intervention. (1) Would the proposed genetic twinning be safe? There are currently techniques available that could probably clone a

85. UNESCO Thirty-first General Assembly, "Universal Declaration on the Human Genome and Human Rights," *Journal of Medicine and Philosophy* 23, no. 3 (1998): 338.

86. John Kilner raises this concern in "Human Cloning," in *The Reproduction Revolution: A Christian Appraisal of Sexuality, Reproductive Technologies, and the Family,* ed. John F. Kilner, Paige C. Cunningham, and W. David Hager (Grand Rapids: Eerdmans, 2000), p. 135.

human being, but they involve the loss of many embryos and pregnancies. Two hundred seventy-seven embryos were lost in achieving one Dolly, the first cloned mammal. As techniques improve in animal husbandry, the restraint from loss of embryos may soon fade. The standard of safety is not met now.[87] (2) Would the intervention bring about a genuine improvement for the recipient? That standard might be met in comparison to likely inheritance or nonexistence. It of course would not be if genetic twinning is inherently in some sense contrary to human dignity or basic rights. (3) Would the recipient's future be open? To maintain an open future, genes would have to be chosen that have been shown to be quite capable, and all involved would have to remember the error of genetic reductionism. (4) Would the intervention be the best use of available resources? Genetic twinning is unlikely to be justifiable for government support when compared to the cost effectiveness of meeting other priorities. That does not address, however, what a couple does with their discretionary income. In our society after-tax income can go to church tithes or betting at the race track. Would genetic twinning be the best way a couple could serve God with the substantial financial sums (at least as introduced) that it would probably require?

Chapter Summary

In this chapter we have seen some of the possible services of changing a person's genes. That of course raises basic questions of whether human beings are by nature static or should welcome change. Concerns about making people and distancing genetic links between parents and children have been considered as needed but not prohibitive cautions. Deliberate genetic twinning (cloning) has also been evaluated. It requires more accurate understanding and careful assessment than it usually receives.

---

87. National Bioethics Advisory Commission, *Cloning Human Beings*, p. 104.

CHAPTER 14

Changing Genes and the Family Line

Linda did the best she could to take care of her father. It was difficult. By his thirty-ninth birthday it had become clear that something was wrong. For the first time in his life he was bothered by increasingly serious mood swings. He had trouble concentrating. There was a twitch in his face and fingers that he attributed to nerves, but that did not explain his loss of balance. His speech slurred, and it became difficult to swallow. Just when he most needed the support of his family, he began to have trouble even recognizing them and slipped into dementia. For his last years he could not control his body nor recognize anyone who helped him. Now he was dead at the age of forty-six. The doctors had warned Linda that they knew of no way to slow the course of the disease. They also told her that she had a 50 percent chance of developing the same condition. About 150,000 people in the United States who have a parent afflicted with Huntington disease are at the same risk.

A method to cure Huntington disease does not exist yet, but if there was a treatment for Linda we would call it somatic. If the treatment affected her ova so that her descendants would not inherit the disease, the treatment would be called germline. It is commonly argued that somatic treatment is welcome but not germline. Yet if in treating Linda we could also protect her children, would that not be better? Some changes in the germline could help our descendants. Should we provide that for them when we can?

Distinguishing Somatic and Germline Intervention

There is a substantial physical difference between somatic cell intervention and intervention in the human germline. Somatic cell intervention directly involves one recipient, whereas germline changes are inheritable. This distinction is qualified by the fact that somatic cell intervention sometimes results in unintended germline effects. If a somatic intervention is early enough to affect basic body structure, it is likely to affect gametes as well. Further, it is not unusual for somatic interventions to extend a person's life long enough for him or her to have children. Genes are passed along that would not have been otherwise. That is a germline effect. Nongenetic medical interventions such as heart surgery can also affect the germline in that way.

Granted these qualifications, the distinction between somatic and germline intervention has been considered by many to be sufficiently clear to serve as a crucial ethical divide.[1] By the mid-1980s there was already considerable consensus that somatic cell therapy was acceptable.[2] For the most part it is contiguous with other therapies and had already received careful and convincing ethical analysis.[3] In contrast, germline intervention was and remains contested. On 8 June 1983 a diverse group of clergy from several faiths sent a letter to President Carter asking for a ban on human genetic engineering. J. Robert Nelson clarified afterwards that the involved clergy were not warning against somatic cell therapy. There were many examples of somatic cell therapy that most of the signers would affirm. What they wanted to discourage was modification of the human germline.[4]

Many philosophers and theologians described somatic cell therapy as

1. W. French Anderson, "Human Gene Therapy: Scientific and Ethical Considerations," *The Journal of Medicine and Philosophy* 10 (1985): 286.
2. Office of Technology Assessment, *Human Gene Therapy: A Background Paper* (Washington, D.C.: Superintendent of Documents, U.S. Printing Office, 1984), p. 47; National Institutes of Health, "Recombinant DNA Research; Proposed Actions under Guidelines," *Federal Register* 50, no. 160 (19 Aug. 1985): 33464.
3. Eve K. Nichols, *Human Gene Therapy* (Cambridge: Harvard University Press, 1988), p. 164; John C. Fletcher, "Ethical Issues in and Beyond Prospective Clinical Trials of Human Gene Therapy," *Journal of Medicine and Philosophy* 10 (1985): 293-309; W. French Anderson and John C. Fletcher, "Gene Therapy in Human Beings: When Is It Ethical to Begin?" *New England Journal of Medicine* 303, no. 22 (Nov. 1980): 1293-97; Bernard D. Davis, "Ethical and Technical Aspects of Genetic Intervention," *The New England Journal of Medicine* 285, no. 14 (30 Sept. 1971): 800.
4. J. Robert Nelson, "Genetic Science: A Menacing Marvel," *Christian Century* 100 (July 1983): 636-38.

acceptable but argued vigorously against germline intervention.[5] Government bodies emphasized the distinction as well. For example, a National Ethics Committee in France approved somatic cell therapy but recommended the complete prohibition of even research related to germline genetic therapy,[6] and the United States governing body for gene intervention protocols still emphasizes the distinction in its "Points to Consider."[7] The widespread acceptance of somatic intervention, while germline intervention is consistently questioned, reflects the influential line drawn between the two.

Why Change the Germline?

What is contested by those who reject the distinction is not the reasonable clarity of a physical difference but the moral relevance.[8] An intervention can usually be fairly characterized as primarily somatic or germline, but why does that matter ethically? Some genetic interventions to be early enough to help the individual will be early enough to affect how their gametes develop. Considerable human suffering is physiologically untreatable by therapy with only somatic effects. Some cells such as those in the central nervous system cannot be reached unless the intervention is early in development. For example, Lesch-Nyan disease, which includes aggressive self-mutilation, cannot be treated once it is in place. Replacing the disease-causing gene with a functional gene as the body first develops offers a better chance that the normal gene will express fully and cooperatively, plus have no competition with the old gene. Such could correct dominant disorders.[9] Because the intervention would be early in development, one intervention would affect all body cells including gametes, and it would not need to be repeated in the patient or his

5. Paul Ramsey, *Fabricated Man: The Ethics of Genetic Control* (New Haven: Yale University Press, 1978), p. 44.

6. National Ethics Committee, France, *Report Relative to Research Work on Human Embryos in Vitro and Use Thereof for Medical and Scientific Purposes* (Dec. 1986), p. 24. Note Comité Consultatif National d'Éthique, "Problèmes d'éthique posés par les essais chez l'homme nouveaux traitements" (Paris: Comité Consultatif National d'Éthique, 101 Rue de Tolbiac, 75654, 1984).

7. Subcommittee on Human Gene Therapy, Recombinant DNA Advisory Committee, National Institutes of Health, "Points to Consider in the Design and Submission of Human Somatic-Cell Gene Therapy Protocols," *Federal Register* 50 (1989): 33463-33467.

8. For example, R. Moseley, "Maintaining the Somatic/Germline Distinction: Some Ethical Drawbacks," *Journal of Medicine and Philosophy* 16 (1991): 641-49.

9. LeRoy Walters and Julie Gage Palmer, *The Ethics of Human Gene Therapy* (New York: Oxford University Press, 1997), pp. 72-74.

or her descendants. LeRoy Hood has led researchers at California Institute of Technology in preventing tremors in shiverer mice and extending their life span to the normal range by inserting the gene they would otherwise lack into their parents before their conception and birth.[10]

There are also conditions which are treatable without germline intervention, but only marginally so. The gene for retinoblastoma is inherited as an autosomal dominant in 50 percent of the children of a parent with the disease. Genetic diagnosis at birth can lead to heightened vigilance that tempers the threat of retinal cancer, yet later these patients tend to develop osteogenic sarcoma, another type of cancer. Early intervention with germline effects would with one step eliminate more than one kind of cancer the patients would probably otherwise face.[11] For those patients there would be less suffering, financial cost, and risk.

The often-voiced concern about changing genes with potential germline effects is that it may affect some descendants if it is passed on and expressed. This would compound good and bad results. However, harmful results need not be allowed to continue and interventions could be incremental in number and degree until proven as with any other medical innovation. If the intervention has a damaging side effect, it could then be redone to modify or eliminate it. If the effect is positive, many people could benefit from one intervention. That is markedly more efficient than intervening in each generation.

Spotting PKU early and changing diet has kept many people from devastating mental retardation. On the other hand, sufferers find the bland and expensive diet hard to keep. Also many have survived to child-bearing age, hence increasing the number of children afflicted with the disease. A single germline intervention could eliminate a person's need for lifelong treatment and the risk of passing the condition on to children. Germline intervention requires fewer interventions with attendant costs and problems than lifelong treatment and retreatment in every generation. If it would be appropriate to treat one person for deliverance from Huntington disease, why not treat that one person in such a way as to deliver future children as well? One can argue that "the benefits of forever eliminating diseases such as spina bifida, anencephaly, hemophilia and muscular dystrophy would seem to make germ-cell gene therapy a moral obligation."[12] Prudentially, more people are affected by a germline intervention. In that sense the stakes are higher, but the proce-

10. Walters and Palmer, *The Ethics of Human Gene Therapy*, p. 61.
11. Walters and Palmer, *The Ethics of Human Gene Therapy*, p. 79.
12. Art Caplan, "An Improved Future?" *Scientific American* (Sept. 1995): 143.

dure should not be done even for the first patient until its success and safety are substantially assured. Once it is clear that the procedure is helpful, provision for descendants is an advantage.

It has also been argued that zygote selection will almost always be superior to direct genetic alteration.[13] Why pursue direct germline intervention rather than zygote selection or selective abortion? There are at least three reasons.

1. Germline intervention values the zygote, embryo, or fetus, which is treated rather than discarded. Selection methods usually lead to the immediate or eventual destruction of what is not chosen for implantation. It would be more respectful and welcoming of developing life if it is treated rather than destroyed.

2. Germline intervention can enhance beyond what occurs in a particular family line. Two homozygous parents both afflicted with cystic fibrosis could not provide a healthy zygote to select. "Much more commonly, two prospective parents will both carry alleles causing milder forms of disease that do not typically prevent people from reaching adulthood or having children. Diabetes, heart disease, obesity, myopia, asthma, a predisposition to some cancers, and many other conditions that adversely affect the functioning of a human organ, tissue, or physiological system are examples. And preemptive cures for all could be achieved by genetic engineering."[14]

3. Germline intervention could potentially enhance beyond what occurs in anyone's family line.

Since germline intervention affects future descendants, it raises the question of what concern we should have for people in the future.

Concern for Future Generations

For the Christian tradition described in chapter 3, we are responsible to do the best we can with what have. We have received the mandate to sustain, restore, and improve our physical world and ourselves. Does that include taking into account people of the near and distant future?

Ethical systems usually include some degree of concern for the welfare

---

13. Gerd Richter and Matthew D. Bacchetta, "Interventions in the Human Genome: Some Moral and Ethical Considerations," *Journal of Medicine and Philosophy* 23, no. 3 (1998): 314-15; Marc Lappe, *Ethical and Scientific Issues Posed by Human Uses of Molecular Genetics* (New York: Academy of Sciences, 1976), pp. 634-36.

14. Lee M. Silver, *Remaking Eden: How Genetic Engineering and Cloning Will Transform the American Family* (New York: Avon, 1998), p. 268.

of people, but does that include people in the future? Traditional theories of obligation have a basic problem in this case.[15] Future persons cannot make contracts or promises. The historic paradigm of *obligation* has three requirements: a specifiable service is required of one person, two parties are involved, one to provide the service and one to receive it, and a prior transaction has created the promise.[16] One who does not exist cannot fulfill the criterion of making a promise, so the usual description of obligation cannot apply.

However, the term *obligation* may be broader than that. Obligations toward those who are not able to speak for themselves but who are recognized persons, such as infants, can be as clear as obligations toward those who are competent. The obligations may be even more clear due to the recipient's need for special protection. *Having* claims does not require being able to *make* claims.[17] Claims can exist without mutual agreement. In many cases the obligation of one human being to another is extensive whether claimed or not. A requirement that comes with position such as that of a parent may be called a "duty" but still exemplifies this broader sense of obligation.[18] One may have obligations to people who have not made a reciprocal promise.

While obligations to children who have not entered an agreement are relatively familiar, obligations specifically to those who do not exist yet have not been as carefully addressed. Can obligations extend to unnamed future human beings? Yes, some obligations may fall to unspecified persons.[19] One may have an obligation to build adequate brakes in a car even if one does not know who will eventually drive it, and that eventual purchaser has a right to sound brakes even if he was not born when the car was manufactured.[20] One could say that people in the future should have clean air. If so, whoever now makes choices that affect air quality should consider that obligation. Even those who do not recognize specifically "obligations" to future human beings often argue for taking future needs into account.

15. J. Brenton Stearns, "Ecology and the Indefinite Unborn," *The Monist* 56 (1972): 613.

16. R. B. Brandt, "The Concepts of Obligation and Duty," *Mind* 73, no. 291 (July 1965): 387. Other examples include Martin P. Golding, "Obligations to Future Generations," *The Monist* 56 (1972): 85-99, and several articles in *Responsibilities to Future Generations: Environmental Ethics,* ed. Ernest Partridge (Buffalo: Prometheus, 1981).

17. Carol A. Tauer, "Does Human Gene Therapy Raise New Ethical Questions?" *Human Gene Therapy* 1 (1990): 414.

18. Brandt, "Concepts of Obligation," p. 387.

19. Galen K. Pletcher, "The Rights of Future Generations," in *Responsibilities to Future Generations,* p. 168.

20. Pletcher, "The Rights of Future Generations," p. 170.

Among those future needs are those of our children, but what do we owe *their* children? Led by powerful commitments and motivations such as love and hope, people often make tremendous efforts on behalf of their own children. That intervention for their children has effects for the children of their children. John Passmore argues that one should act deliberately to benefit the descendants of one's children.[21] However, he emphasizes how limited that beneficence could be. Human ignorance is great, capacity to change the future limited, and unintended effects more influential than intended ones. Yet Passmore argues that we do cherish people such as our children and the institutions that are important to us. If one cares for other people, one will also care for what happens to them after one's own death. Concern from personal love extends into the future. According to Passmore the extension depends on the usual, although not always present, commitment of parents to the happiness of their children. Witness the extensive efforts that go into trust funds and life insurance policies. One's children will probably be most happy if their children are happy, as those children are likely to be most happy if their children are happy. Passmore calls the resulting connections "a chain of love" from the present on into future generations. The progression continues, making a chain of love that if not directly broken still does gradually diminish over time. Passmore suggests in this light that the best service for future generations is to create the best possible world now. However, this generation should be willing to forego some enjoyments to better secure the needs of the near future, when we are able to project a higher degree of probability that the effort will be substantially beneficial. Love for people we do know and care for leads to concern and effort toward their future and beyond.

Passmore's chain of love calls for concern for one's descendants. Is there a further case to care for those who are not closely and directly related? For Jonathan Glover, one's obligation is to whoever follows.[22] Glover argues from the principle of equality that the worth of each individual calls for equal consideration regardless of where or when that person lives. One should be concerned to aid and not harm others "even if one does not know their names." He cites the analogy of a bus with many passengers getting on and off. It would not be acceptable to leave a time bomb on the bus simply because one does not know who will be on board when the bomb explodes. One's place in time makes no more difference than one's place geographically. "The tempo-

---

21. John Passmore, "Conservation," in *Responsibilities to Future Generations,* p. 54. For his complete argument see John Passmore, *Man's Responsibility for Nature* (New York: Scribner, 1974).

22. Jonathan Glover, *What Sort of People Should There Be?* (New York: Penguin, 1984), pp. 143-44.

ral location of future people and our comparative ignorance of their interests do not justify failing to treat their interests on a par with those of present people."[23] Harms should be avoided and recognized goods should be pursued for future generations.

For Daniel Callahan, to exclude any human beings, present or future, from our moral community invites abuses such as that of slavery or other oppression. He grants that "to state that we have moral obligations to the community of all human beings introduces its own problems. One of them turns on the practical impossibility of effectively discharging obligations to all human beings."[24] The problem is compounded if concern for future generations of human beings is included. Yet whenever human beings may live, they are still human beings. As human beings they warrant consideration if our actions can affect them.

Callahan then goes on to emphasize that our actions will affect future human beings. The very existence of future generations depends on the present generation. The present generation has a responsibility to future people due to their biological dependency and their need as fellow human beings. Callahan argues as well that this biological link incurs a further obligation. As we have received from the past, so we have an obligation to pass on to the future. He labels this obligation with the Japanese term *on*.[25] One repays the care received from one's parents by taking equal or better care of one's own children. With no exact correspondence in the English language, the term carries an idea of both gratitude and justice in passing on what the present has received in trust. It parallels the mandate to do our part to sustain what we have received.

Thomas Sieger Derr sees concern for future generations as common to the worldviews of the Western religious traditions of Judaism, Christianity, and Islam.[26] Each refers to an idea of covenant, as in the case of Abraham, where individual choices have consequences for descendants as God interacts with children of the covenant on through the generations. Emphasis is placed on each generation fulfilling and carrying on that covenant. In the Christian tradition this is exemplified in the prominence of infant dedication or baptism. There the community welcomes newborns into the church to be raised as part of that community. The congregation promises to teach and encour-

23. Gregory Kavka, "The Futurity Problem," in *Obligations to Future Generations,* ed. R. I. Sikora and Brian Barry (Philadelphia: Temple University Press, 1978), p. 201.

24. Daniel Callahan, "What Obligations Do We Have to Future Generations?" in *Responsibilities to Future Generations,* p. 76.

25. Callahan, "What Obligations Do We Have to Future Generations?" p. 77.

26. Thomas Sieger Derr, "The Obligations to the Future," in *Responsibilities to Future Generations,* pp. 41-42.

age the children with the expectation that someday as adults they will publicly affirm their own commitment in baptism or confirmation and so seek to raise their children.

The Western traditions also usually describe history in a linear sense. Granting the laments in Ecclesiastes that complain of endless empty repetition,[27] history is usually described not as a repeated cycle but as having a beginning in creation, a consistent working of God within it, and a definite culmination followed by transformation. The future does not merely repeat the past but can change and develop in substantially new ways. With that potential can come responsibility to contribute to positive change.

In Judaism and particularly in the Christian tradition such responsibility is often summarized as love for one's neighbor as oneself. Donald MacKay advocates that one should benefit one's neighbor, including neighbors in the future, with whatever tools are available.[28] In Luke 10, where the command to love one's neighbor is affirmed, the question is immediately raised as to who is included in the category of neighbor. The response is the story of the Good Samaritan, which culminates in the conclusion that one's neighbor is whomever one is able to help. Neighbor-love would then extend to future generations to the degree one is able to benefit them effectively. To love one's neighbor means to seek the best for others as one is able, whoever the other may be racially, culturally, geographically, or temporally. Such intervention for MacKay does not lead to salvation or perfection, yet human beings are responsible to God to improve life for one another rather than drift in complacency.[29] This is the model of Christ held up to follow both in the Gospels and in the apostolic letters.[30]

For MacKay one should be motivated not only by love of neighbor but also by "the fear of the Lord." Sins of omission are as serious as sins of commission, sloth as dangerous as pride. The steward who buried his talent rather than multiplying it was rebuked for his inaction. Knowledge and neighbor-love bring responsibility. Human beings will be held accountable for what they have achieved compared with what they could have done for the service of others and the glory of God. For MacKay, planning and action on behalf of future human beings is not arrogant but a duty for the responsible steward.[31] Care for future generations is part of the mandate to do what we can to sustain, restore, and improve ourselves and our world.

27. Ecclesiastes 1:1-10.
28. Donald M. MacKay, *Human Science and Human Dignity* (Downers Grove, Ill.: InterVarsity Press, 1979), p. 60.
29. MacKay, *Human Science and Human Dignity,* p. 79.
30. For example, John 13:3-15; Philippians 2:4-7.
31. MacKay, *Human Science and Human Dignity,* p. 58.

Reasons for considering the needs of future generations have included love for one's own children, the worth of all human beings, membership in the moral community of humanity, love of neighbor, and fear of God. However, none of these have been argued as an unqualified absolute. What else may counterbalance these claims or be distinctive about applying them to the future?

Three Often-Voiced Constraints

1. Do We Have a Right to Make Choices
Affecting Future Human Beings?

Part of the difficulty of action or restraint on behalf of future generations is that members of society are making choices that affect future generations but cannot consult the people of those generations. To choose wisely on their behalf parallels the role of a parent making formative decisions for a child, but this would not be an instance of rightfully rejected ethical paternalism. "Paternalism may be defined as a refusal to accept or to acquiesce in another's wishes, choices, and actions for that person's own benefit."[32] One can act on behalf of future generations, but it is not possible to override the expressed wishes, choices, or actions of people who have not yet made any.

Since they do not yet exist, to what degree can there still be concern for their autonomy?[33] In chapter 4 we noted theological and philosophical reasons for respecting persons. That respect should also be for whoever lives in the future. Out of respect for persons, whoever they may come to be in particular, they should have choices rather than be predestined to someone else's design. Although it is not possible to honor the autonomy of future individuals by consulting with them as we act, it is possible to be concerned about their autonomy as an end state. Current choices should avoid limiting the level of autonomy they will one day possess.

It is not enough to hope for ratification of our actions.[34] A later approval is problematic in that the intervention cannot be undone and the re-

32. James F. Childress, *Who Should Decide? Paternalism in Health Care* (New York: Oxford University Press, 1982), p. 13.

33. "Autonomy simply means that a person acts freely and rationally out of her own life plan, however ill-defined." Childress, *Who Should Decide?* p. 60.

34. Childress, *Who Should Decide?* p. 93, and Alan Soble, "Deception and Informed Consent in Research," in *Bioethics,* ed. Thomas A. Shannon, rev. ed. (Ramsey, N.J.: Paulist, 1981), p. 364.

cipient may be substantially influenced by the received choices. Aldous Huxley referred to an extreme form of this problem in *Brave New World*. "That is the secret of happiness and virtue — liking what you've got to do. All conditioning aims at that: making people like their unescapable social destiny."[35] In Huxley's brave new world all choices for the next generation were made and set by the controllers. People were shaped to their role rather than shaping roles and environment to the needs and desires of people.

Such a concentration of choice in the hands of a comparative few who choose the conditioning could limit the self-determination of future generations. Does one generation have a right to make choices of such influence for future generations? The European discussion has at times responded with an emphatic no. In an appeal to the French *patrimonie* or the German *Erbgut*, the broad collective environment of human beings must remain just as received. Mauron and Thévoz give the example that one cannot tear down a Gothic chapel for one's own convenience.[36] We should not in any way change our given heritage.

Yet in an important sense the question of right to influence is inapplicable. "The human autonomy we are required to respect is not an absolute individual sovereignty. No one has created himself."[37] We do make formative choices that then shape who we become, but we start with a long list of givens bestowed on us by those who precede us. Past generations have made countless choices for the good and ill of the present generation. This generation's choices will unavoidably shape the world the next generation enters.[38]

The choice is not whether this generation will shape the next or not, but rather to what degree and in what direction. Where we build our homes and cities shapes the environment that is passed on. Medical intervention that enables people with genetically based myopia, diabetes, retinoblastoma, and other diseases or disabilities to survive and bear more children spreads those genetic propensities and diseases through the population. The present generation could refuse to restrain or act deliberately on behalf of future generations, but it cannot escape its influence, nor the fact that by avoiding conscious intervention a different heritage is established from what could have been. Some risks are avoided and others are retained. When one generation

35. Aldous Huxley, *Brave New World* (New York: Harper and Row, 1969), p. 10.
36. Alex Mauron and Jean-Marie Thévoz, "Germ-Line Engineering: A Few European Voices," *The Journal of Medicine and Philosophy* 16 (1991): 654-55.
37. Paul J. M. Van Tongeren, "Ethical Manipulations: An Ethical Evaluation of the Debate Surrounding Genetic Engineering," *Human Gene Therapy* 2 (1991): 74.
38. Willard Gaylin, *Adam and Eve and Pinocchio: On Being and Becoming Human* (New York: Viking Penguin, 1990), pp. 258-59.

builds beautiful Arcadian neighborhoods far from the smells and noise of manufacturing, most of the next generation has to drive to work.

Are there ways to protect the autonomy of future human beings? If our shaping genetic heritage is incremental, no one generation would so change perception and experience as to determine all who follow. Over time small initial changes can lead to vast divergences as described in chaos theory, but each ongoing overlapping generation would have the opportunity to adjust before long-range implications became set. Intervention could increase choice rather than narrow it. Future generations might then be even more able to adapt to their unique environment and perspective. The current generation would not need to master the impossible task of predicting and balancing all the preferences of future generations.

Also reversibility is a major concern for implementing change.[39] Future generations should not have to suffer indefinitely an earlier mistake. If choices are incremental and reversible, future generations could restore a pattern that had been deleted or changed. It might be argued that some genetic heritage such as Tay-Sachs disease has little chance of being helpful in any scenario. Since we are finite beings considering a distant future, there might be other changes that seem desirable now that would not be appreciated later. Vigilant caution is in order. Out of autonomy concerns the future should not be predestined to one narrow vision.[40] Chapter 15 will describe a process that welcomes ongoing diversity.

2. Do We Really Know What Will Help Future Human Beings?

It can be argued that one's place in time should make a difference in considering the needs of others precisely because as one goes further into the future the circumstances and needs of future generations become harder to predict. The increasing uncertainty makes the weight of such concerns of less import. One cannot have an obligation to benefit remote future generations when one does not know what will benefit them.[41] Since we do not know all that the future holds, we might not know what would be a desirable genetic endowment for future generations.

39. James F. Childress, *Priorities in Biomedical Ethics* (Philadelphia: Westminster, 1981), p. 110.
40. Robert Nozick, *Anarchy, State, and Utopia* (New York: Basic Books, 1974), pp. 313-14.
41. Martin P. Golding, "Ethical Issues in Biological Engineering," *UCLA Law Review* 15, no. 267 (Feb. 1968): 457.

This is especially true for remote human generations. Capacities that receive widespread acclamation now may not in the future. Even widely lauded and flexible capacities such as intelligence are controversial in their definition and measurement. Brigitte Berger writes that current IQ tests measure only a narrow band of intellectual capability that reflects a "modern consciousness" of high abstraction.[42] Daniel Boorstin observed from writing the book *The Discoverers* that there was no prototype of one who greatly contributes. Those who have made important discoveries have ranged from mystics such as Parcelus to establishment figures such as Harvey. Charles Frankel notes in particular the tendency of people under the different circumstances of various decades to emphasize different values.[43] Choices of any given generation reflect more their temporary circumstances than future desires and needs.

Others have responded that while one does not know completely what will positively benefit future human beings, to a considerable distance in time one has a good idea what will harm them. The starting point would be to relieve burdens and in the process, as in the Hippocratic tradition of *primum non nocere,* first do no harm. Thomas Szasz has written skeptically about such a commitment in that according to Szasz often one person cannot be helped without hurting another.[44] He cites an example of prolonging the life of a patient who harms others, or correctly diagnosing a woman as psychotic to protect her husband and then seeing her lose her freedom to involuntary commitment. While one cannot predict all the effects of one's actions, that fact does not lead to the conclusion that all choices are equally desirable, nor that random choice would be as positive in its net effect as deliberately selected choices. Szasz is right that life is complex, but he also appears to be assuming that life is a zero-sum game with losers always in direct proportion to winners. Life may not always be a zero-sum game, and even if it is in some cases, justice might still come into play as to who might appropriately bear which burdens.

Faced with these human limitations one could argue that the wisest course is not to intervene at all.[45] Intervention should not take place where

42. Brigitte Berger, "A New Interpretation of the I.Q. Controversy," *Public Interest* 50 (Winter 1978): 29-44.

43. Charles Frankel, "The Specter of Eugenics," *Commentary* 57, no. 3 (1974): 31.

44. Thomas Szasz, "Ethics and Genetics: Medicine as Moral Agency," *Genetic Engineering: Its Applications and Limitations,* Proceedings of the Symposium held in Davos, 10-12 Oct. 1974, p. 114.

45. Leon Kass, "New Beginnings in Life," in *The New Genetics and the Future of Man,* ed. Michael P. Hamilton (Grand Rapids: Eerdmans, 1972), p. 62.

human beings lack the knowledge to proceed. The argument at this point, however, appears to be not that such intervention is of itself immoral but rather that it would be immoral to act imprudently. Acting with insufficient knowledge would be immoral, but if human beings gained sufficient knowledge so that the intervention would not be imprudent, it would not then be immoral on that count. Unless one is convinced that human beings will never discern *any* correction or improvement as is constantly attempted in health care, this concern calls more for caution than for complete and permanent prohibition of intervention.

While it can be difficult to know exactly what will always be most beneficial to future generations or how to balance competing concerns, there is enough likely continuity to have a good idea at least of what would be likely to harm them. There is more ethical responsibility than merely the avoidance of harm, but that is at least a minimal place to start. While we do not know the future situation and ideals, passing on capable and well-functioning bodies to future generations is likely to be helpful to them.

3. Are Not the Needs of the Present Already All-Consuming Without Adding Concern about Future Human Beings?

Even if widely perceived as beneficial, important for equality of opportunity, and promising eventual cost advantage, the initial costs of extensive intervention would be high. How might the competing claims between needs of the present generation and future generations be justly balanced? Would amelioration of current evils always be of the highest priority so that any effort on behalf of future generations would be postponed indefinitely?[46]

John Rawls suggests a method for discerning fair warrants for choosing resource use. One is to imagine deciding generational duties without knowing which generation one will be in. The intent of deciding behind this "veil of ignorance" is simply to lead people to count people in other generations as of equal concern with themselves. Each other person counts as much as oneself in such a calculation because by the rules of the thought experiment one does not know which one *is* oneself. By such criteria reasonable people might choose to expect each generation to invest in some improvement for the future as long as it is at minimal cost to their generation. These savings would include that each generation would without sacrificing its own welfare set aside some resources and pass on information and culture to start the next

---

46. Golding, "Ethical Issues in Biological Engineering," pp. 458-59, and Glover, *What Sort of People Should There Be?* p. 140.

generation off a little better than it did.[47] Each generation would be expected to contribute "justified savings" which while of minimal cost to each generation would add to an accelerating cumulative benefit. From such a policy every generation would benefit but the first.[48] If the first generation's sacrifice is minimal, it may not be too much to ask.

Such a rubric might be one way to distribute justly between generations the costs and benefits of genetic intervention. Rawls applies the standard specifically to genetic endowment. "It is also in the interest of each to have greater natural assets. This enables him to pursue a preferred plan of life. In the original position, then, the parties want to insure for their descendants the best genetic endowment (assuming their own to be fixed). The pursuit of reasonable policies in this regard is something that earlier generations owe to later ones."[49] Of course risk, surety of benefit, and other considerations would still need attention.

Chapter Summary

There is a substantial physical difference between somatic cell intervention and intervention in the human germline. Somatic cell intervention directly involves one recipient, whereas germline changes are inheritable. The widespread acceptance of somatic intervention, while germline intervention is consistently questioned, reflects the influential line drawn between the two. What is contested by those who reject the distinction is not the reasonable clarity of the distinction but its relevance. It is often quite clear whether an intervention is primarily somatic or germline, but why is that an ethically important distinction? Some genetic interventions, to be early enough to help the individual, will be early enough to affect how gametes develop. There are also conditions which are treatable without germline intervention, but only marginally so. The concern about intervention with potential germline effects is that it has the potential to affect some descendants. If the intervention has a damaging side effect it could be redone to modify it or eliminate it. If the effect is positive, many people could benefit on into the future from one intervention. More people are affected by a germline intervention. In that sense the stakes are higher, but the procedure should not be done even for the

47. John Rawls, *A Theory of Justice* (Cambridge: Harvard University Press, 1971), pp. 284-93.

48. Ronald M. Green, "Intergenerational Distributive Justice," in *Responsibilities to Future Generations*, p. 95.

49. Rawls, *A Theory of Justice*, p. 108.

first patient until its success and safety are substantially assured. Once it is clear that the procedure is helpful, provision for descendants is an advantage.

There are multiple reasons to be concerned about our descendants. The reasons considered in this chapter included love for one's own children, the worth of all human beings, membership in the moral community of humanity, love of neighbor, and fear of God. One standard objection to such concern is that making choices for future generations violates their autonomy. Actually, choices that shape future generations are unavoidable. What we can do out of respect for future persons is make changes incrementally and reversibly that are likely to increase future choice. A second concern was that our knowledge of future generations and their contexts is too limited to choose for them. While our knowledge is limited, we can safely expect that clear harms will not be beneficial to near future generations. Also some goods are so basic, such as a well-functioning body, that it is reasonable to expect them to be appreciated. A further concern was how to balance needs of future generations with the overwhelming ones of the present. Just distribution between generations may call for justified savings which while of minimal cost to each generation are of great cumulative benefit. In fact the cumulative benefit would more than cover the contributions of each generation except the first one that made minimal sacrifices to begin the process.

The genetic heritage we pass on to future generations should be a considered part of our current reflection as we make choices that will deeply affect our children and theirs. Appealing again to the framework explained in chapter 3, we should at least sustain what has been entrusted to us in order to pass it on to them. Genetic heritage should not be worse for our presence. We should also restore what we can. This healing work is part of our reflection of God's gracious redemption. To sustain and restore ourselves and the rest of the physical world could be enough to absorb our attention for some time, yet a third interest is warranted. That is improvement. Improvements are appropriate as the opportunities for them are clear. Such required clarity would recognize the immense interdependence of ourselves and the physical world, yet that it may not already be ideal. The elimination of smallpox from the globe was an appropriate alteration of our environment. Wiping out Tay-Sachs, Huntington, or Alzheimer's disease from our genetic heritage would be as well.

CHAPTER 15

Changing Genes and Community

Coercion and Eugenics

Can we protect human freedom and diversity while pursuing genetic intervention? It was not long ago that the eugenics movement was party to some of the most horrific abuses of the last century.[1] Many minority groups worry that any genetic measurement or change will be used against them.[2] A genetic program that sets human development toward one ideal would be at risk of such abuse. Pursuing one ideal would narrow the diversity of possibilities and values carried on. While group consensus may have the advantage of eliminating some mistakes, it can also lead to "group think" that is less careful and more group aggrandizing. Since human beings tend to self-interest, one prevailing ideal would probably place the self-interest of one group over that of others.

Daniel Boorstin suggests that this is an evident observation from even the most cursory reading of history.[3] There are multiple examples from re-

1. For example, Benno Muller-Hill, *Todliche Wissenschaft: Die Aussonderung von Juden, Zigeurnern und Geisteskranken 1933-1945* (Lethal Science: The Exclusion of Jews, Gypsies, and the Mentally Ill 1933-1945) (Rowohlt: Reinbek, 1984).

2. Herbert Nickens notes that already today a substantial portion of African-American adults in the United States blames a racist conspiracy for high rates of drug abuse, AIDS, crime, broken families, and teen pregnancy in the black community. "The Genome Project and Health Services for Minority Populations," in *The Human Genome Project and the Future of Health Care*, ed. Thomas H. Murray, Mark A. Rothstein, and Robert F. Murray Jr. (Bloomington: Indiana University Press, 1996), p. 59.

3. Daniel Boorstin, *The Washington Post*, 5 Aug. 1984, p. C3.

cent history of groups using genetic concern as a weapon against other groups. Paul Ramsey has summarized this often-cited concern:[4] "the culmination or abuse of eugenics in the ghastly Nazi experiments would seem to be sufficient to silence forever proposals for genetic control."[5] This argument is so pervasive in discussion about genetic intervention, and appropriately so, that it warrants extended analysis. In recent history a highly educated culture applied genetic concern in a horrific way. What can we learn from that experience?

Arguing from historical analogy, such as Ramsey's above reference to the Nazi experience, is fraught with difficulties. It depends on establishing the historical detail needed for an accurate comparison and to show that the case is close enough to draw an effective parallel.[6] With the Nazi movement as the prime exhibit, the history of human attempts to improve human genetic endowment, often called *eugenics,* has been abhorrent in its interference with the choice of marriage partners, sterilization programs, and deadly racism. On the one hand, the parallel of proposed genetic intervention with eugenic sterilization or choice of marriage partners is not directly applicable, in that genetic intervention as now contemplated does not require those means. In fact, freedom in the choice of mates and whether to have children could be increased, since the extensive elimination of inheritable diseases would allow many people to have children who would not otherwise have been able to have them. On the other hand, there may be a relevant link between genetic concern and racism. At least such an argument is cited so frequently that it warrants a particularly thorough appraisal. The first question then to test the analogy is to what degree genetic concern has been racist.

---

4. Examples of others who have cited this concern would include Martin P. Golding, "Ethical Issues in Biological Engineering," *UCLA Law Review* 15 (Feb. 1968): 448-50; Hans Schwartz, "Theological Implications of Modern Biogenetics," *Zygon* 5 (Sept. 1970): 264; Arno G. Motulsky, "Government Responsibilities in Genetic Diseases," in *Genetics and the Law II,* ed. Aubrey Milunsky and George J. Annas (New York: Plenum, 1980), p. 238; World Council of Churches, Church and Society, *Manipulating Life: Ethical Issues in Genetic Engineering* (Geneva: World Council of Churches, 1982), p. 9; and W. French Anderson, "Genetics and Human Malleability," *Hastings Center Report* 20 (Jan./Feb. 1990): 24.

5. Paul Ramsey, *Fabricated Man: The Ethics of Genetic Control* (New Haven: Yale University Press, 1978), p. 1.

6. The *Hastings Center Report* has organized two discussions of this problem specifically as it applies to the analogy of Nazi practices with current choices: "Biomedical Ethics and the Shadow of Nazism: A Conference on the Proper Use of the Nazi Analogy in Ethical Debate," *Hastings Center Report* 6 (Aug. 1976), special supplement, and "Contested Terrain: The Nazi Analogy in Bioethics," *Hastings Center Report* 18 (Aug./Sept. 1988): 29-33.

Racism

Eugenics was first coined as a study and term by Sir Francis Galton in his *Inquiries into Human Faculty and Its Development*. Galton's interest was sparked by the work of his cousin Charles Darwin on natural selection. Darwin warned that helping "the weak" human beings to survive and propagate was "highly injurious to the race of man."[7] Galton responded with a program to purposefully encourage the positive evolution of human beings through marriage choice.[8] He defined that positive evolution as follows:

> We would include among our standards of eugenic value sound physical health and good physique, intelligence, and moral qualities which make for social cohesion. The latter would comprise courage (but not aggressiveness), serenity or contentment, and cooperativeness. We would also here include the quality described above as genophilia (love of children).[9]

It is noteworthy that eugenics did not start with race as the ideal nor with race listed as the epitome of the ideal.[10] According to Mark Haller in his study of how eugenic ideas began in the United States, eugenics "began as a scientific reform in an age of reform," which was pursued by the more liberal leadership.[11] It did, however, begin to be used to explain poverty in terms of bad inheritance. Studies of family lines such as the "Jukes," "the Tribe of Ishmael," and "the Kallikaks" were published as proof that no environmental reforms could salvage some family lines.[12] From finding some family lines incorrigible it was a small step to rejecting the wider families of particular races.

Legislation that claimed the justification of eugenics was enacted to re-

---

7. Charles Darwin, *The Descent of Man and Selection in Relation to Sex*, 2nd ed. (New York: D. Appleton, 1922), p. 136.

8. C. P. Blacker, *Eugenics: Galton and After* (Cambridge: Harvard University Press, 1952), pp. 107-8.

9. Blacker, *Eugenics*, p. 289.

10. As typical of his period, Galton did begin to articulate also an imperialist motivation for eugenics. Part of the purpose for eugenics became "to give the more suitable races or strains of blood a better chance of prevailing speedily over the less suitable than they otherwise would have had." Quoted by Golding, "Ethical Issues," p. 464.

11. Mark A. Haller, *Eugenics: Hereditarian Attitudes in American Thought* (New Brunswick: Rutgers University Press, 1963), p. 5. Diane Paul traces this pattern across Europe in "Eugenics and the Left," *Journal of the History of Ideas* 45, no. 4 (1984): 567-90. Gunnar Broberg and Mattias Tyden found this connection specifically in Sweden in *Eugenics and the Welfare State: Sterilization Policy in Denmark, Sweden, Norway and Finland* (East Lansing, Mich.: Michigan State University Press, 1996).

12. Haller, *Eugenics*, pp. 106-7.

strict immigration from certain countries. A number of prominent citizens connected the quality of the American character with the propagation of "superior races," namely, northern European whites. Francis A. Walker, the president of MIT and director of the 1870 census, warned that massive immigration of inferior stock was overwhelming the "native" Anglo-Saxon stock.[13] Henry Cabot Lodge wrote in 1891 that "immigration of people of those races which contributed to the settlement and development of the United States is declining in comparison with that of races far removed in thought and speech and blood from the men who have made this country what it is."[14]

In the spring of 1894 several young Harvard University graduates started the Immigration Restriction League and in 1895 Henry Cabot Lodge introduced the Immigration Restriction Law, which limited immigration by race and was passed with strong support by the United States Congress. It was vetoed by President Grover Cleveland, but the immigration restriction movement continued to gain strength as well as the attention of the eugenics movement.[15]

E. A. Ross, a leader in the eugenics movement, raised the alarm that as a result of the continuing immigration of inferior races, there would be a diminution of stature, a depreciation of morality, an increase in gross fecundity, a considerable lowering of the level of average natural ability, and a falling off in the frequency of good looks in the American people.[16] The Second International Eugenics Congress, which met in New York City in 1921, laid heavy emphasis on racial issues, reflecting the concerns of the wider American culture at the time. The racially based Immigration Restriction Bill passed again in the United States Congress in 1925 and this time was signed into law. Across the Atlantic, the eugenics movement also reflected the racist tack of the Second International Eugenics Congress and the wider culture. In England the emphasis was not so much on limiting immigration as on subjugating the lesser races in colonies for the lesser races' own benefit.[17]

At about the same time Nazism began to gain power in Europe and pursued a racial policy combining a drive for racial purity with a quest for racial dominance that then elevated both to a level of unrestrained terror. It has become a paradigm for the abuse of power and has become intertwined with

13. Haller, *Eugenics,* p. 139.

14. Haller, *Eugenics,* p. 56.

15. Kenneth Ludmerer, *Genetics and American Society: A Historical Appraisal* (Baltimore: Johns Hopkins University Press, 1972), p. 84.

16. Paul Bowman Popenoe and Roswell Hill Johnson, *Applied Eugenics* (New York: Macmillan, 1918), p. 301.

17. Geoffrey Searle, *Eugenics and Politics in Britain 1900-1914* (Leyden: Noordhoff International, 1976), pp. 35, 42-43, 74; also Haller, *Eugenics,* pp. 13-14.

eugenics in the perception of many. "No other historical experience has the place in our ethical discourse as the Nazi one. It is as though in a relativist and pluralist society this is our single absolute evil."[18]

While being held in the Landsberg am Lech prison (1923-24), Adolf Hitler wrote in *Mein Kampf* his vision for the future. "Everything we admire on this earth today — science and art, industry and invention — is the creative product of but a few peoples, perhaps originally of one race. Upon them the subsistence of this whole civilization depends. If they are destroyed, the beauty of this earth will be buried with them."[19] For Hitler, "the triumphant advance of the best race" is "the sine qua non of all human progress." The world is a struggle where "it is necessary and just for the best and strongest man to be victor."[20] According to Hitler that strong man on whom civilization depends is the Aryan race.

> So it is no accident that the first civilizations arose where the Aryan, encountering lower races, subjugated them and made them do his will. They were the first technical tools to serve a dawning civilization. . . . Thus the road which the Aryan must travel was clearly marked. As a conqueror he subjugated the inferior peoples, and regulated their practical activity under his orders, according to his will, and for his own purposes. But in thus setting them to a useful if a hard task, he not only spared the lives of the conquered, but gave them a fate which perhaps was actually better than their previous so-called "freedom."[21]

For Hitler, these were not idle concepts to be left in the abstract. From a 1933 speech:

> Implementation of the fundamental political concept of race, which has been reawakened by National-Socialism and expressed in the phrase "blood and soil," implies the most far reaching revolutionary transformation that has ever taken place. The fundamental necessity for consolidation of the racial foundation of our people, which is implicit in these words . . . governs all the aims of National-Socialism both externally and internally.[22]

18. Peter Steinfels, "Biomedical Ethics and the Shadow of Nazism," *Hastings Center Report* 6 (Aug. 1976), special supplement, p. 1.

19. Adolf Hitler, *Mein Kampf*, trans. Ludwig Lore (New York: Stackpole, 1939), p. 281.

20. Hitler, *Mein Kampf*, p. 281.

21. Hitler, *Mein Kampf*, p. 287.

22. Werner Maser, *Hitler's Mein Kampf: An Analysis* (London: Faber and Faber, 1970), pp. 133-34.

The "purity of the volk" was to be and indeed became the center point of Hitler's policies. *Disease* was defined to include racial judgments.[23] Eugenics was to purify the health of the people of Germany as individuals and as a society.[24] "It will be the first task of the People's State to make race the center of the life of the community."[25] In eugenics Hitler found a "scientific" cloak and dagger for his destruction of "international Jewry." When Germany came under his power,

> the eugenics movement became inextricably interwoven with the Nazi regime. Hitler's Minister of the Interior, Wilhelm Frick, proclaimed, "the fate of race-hygiene, of the Third Reich and the German people will in the future be indissolubly bound together." Prominent eugenicists became Nazi officials, and Hitler filled his government with other men who at least sympathized with the eugenics program. Many of the private organizations concerned with eugenics education were reorganized as government agencies, the most prominent of which was the Kaiser Wilhelm Institute for Anthropology, Human Genetics, and Eugenics, directed by Fischer.[26]

The eugenics movement became a racist movement in Nazi Germany, as it did largely in the American movement for immigration restriction and in the British eugenicists' support of colonialism.

Using the Analogy

Are such racist tendencies somehow inherent to the idea of eugenics? Rainer Hohlfeld charges that they are. "All concepts of a positive eugenics are therefore constant expressions of the class and race thinking of a ruling elite."[27] Amitai Etzioni continues the concern, "Even before genetic engineering developed very far, the mere question of how it might be used would invite a resurgence of racist ideologies and conflicting racist camps, each advocating its version of the desired breed."[28]

23. William E. Seidelman, "Mengele Medicus," *The Milbank Quarterly* 66, no. 2 (1988): 223.

24. Ludmerer, *Genetics and American Society*, p. 116.

25. Hitler, *Mein Kampf*, pp. 338-39.

26. Ludmerer, *Genetics and American Society*, pp. 115-16.

27. My translation of "Alle Konzepte einer positiven Eugenik sind daher stets Ausdruck eines Klassen und Rassendenkens einer herrschenden Elite." Rainer Hohlfeld, "Jenseits von Freiheit und Wurde: Kritische Anmerkungen zur gezielten genetischen Beeinflussung des Menschen," *Reformatio* 32 (May 1983): 220.

28. Amitai Etzioni, "Biomedical Ethics and the Shadow of Nazism," *Hastings Center Report* 6 (Aug. 1976), special supplement, p. 14.

Here Etzioni has pointed, perhaps unintentionally, to a key distinction between eugenics and racism. As Cynthia Cohen has emphasized, for the Nazi analogy to retain its power, it must be used with precision where it most accurately applies.[29] It is accurate that eugenics like racism depends on the valuation of some genetic endowments as more desirable than others. For example, it would be preferable to be born with healthy eyes than with genetically caused blindness.

Theresia Degener argues that disability is actually a neutral condition, rendered difficult only by society's discrimination. Society should adapt to whatever one's abilities are.[30] Granting that society should welcome and adapt to its members, whatever their abilities, it would still be to the individual's advantage and society's benefit to increase each one's physical options. Those who are blind should be welcomed. If the genetic option is available not to be blind, it would be helpful for the individual and society. Pope Pius XII has written that "the fundamental tendency of genetics and eugenics is to influence the transmission of hereditary factors in order to promote what is good and eliminate what is injurious. This fundamental tendency is irreproachable from the moral viewpoint."[31] If one is attempting to improve in some sense the genetic endowment of newborns, one must have some goal as to what a better or ideal genetic endowment would be. Without such a goal there would be no direction or point to genetic intervention. It is in the choice of the ideal that racism has often entered eugenics.

If "normal" health is the ideal, eugenics will not be racist unless normal health is defined as the distinct characteristics of one particular race. If intelligence is the ideal, eugenic attempts will not be racist unless "intelligence" is defined in an inherently racist manner, such as through a culturally biased test. It is only if the ideal is defined by the characteristics of a particular race that eugenics will be racist.

While the term *eugenics* has been used to refer to coercive breeding and racist policies, its literal definition is simply "good birth." In that sense eugenics is the attempt to improve the genetic endowment of newborns. The ideal goal of eugenics is from outside eugenics and is what determines the

29. Cynthia B. Cohen, "Contested Terrain: The Nazi Analogy in Bioethics," *Hastings Center Report* 18 (Aug./Sept. 1988): 33, also in more detail in " 'Quality of Life' and the Analogy with the Nazis," pp. 113-35.

30. Theresia Degener, "Female Self-Determination Between Feminist Claims and 'Voluntary' Eugenics, Between 'Rights' and Ethics," *Issues in Reproductive and Genetic Engineering* 3 (1990): 94, 98.

31. Pope Pius XII, "Moral Aspects of Genetics," 7 Sept. 1953, in *The Human Body: Papal Teachings,* ed. The Monks of Solesmes (Boston: St. Paul Editions, 1979), p. 256.

racist or nonracist intent of the particular eugenic program. Hitler could use the eugenics movement not because eugenics is inherently racist but because the German eugenists were willing to adopt his Aryan race model as the ideal genome.

The harmful use of a technology does not automatically prove that such use is inherent to the technology. Josef Mengele, the Nazi "doctor of death," claimed to be doing medical research as he tortured prisoners. While his acts are abhorrent and justly condemned, they are not of themselves an argument that all medicine and medical research is sadistic, but rather they call for careful safeguards against abuse such as rules of informed consent and institutional review boards for human-subjects research as discussed in chapter 4. Nazi racism, in the name of a healthy genetic endowment, does not prove that seeking a healthy genetic endowment is pursuing an inherently racist ideal.

However, the analogy's most powerful warning may be the one described by Gary Crum:

> I believe we should strive to see Nazis as individual persons such as ourselves; persons whose rationales and actions were sometimes despicable, but not always so. We should make an effort to stop using the experience of Nazism as a metaphor for "The Cosmic Evil" and instead try to read it like a warning label on a bottle under our own kitchen sinks.[32]

Many people who strove to be moral were drawn to the Nazi programs. The Roman Catholic response was resistance to the destruction of mentally or physically handicapped people, yet considerable compromise with Nazism at other points has been described by Donald J. Dietrich.[33] Milton Himmelfarb has argued that many in the theologically leftist branch of German Protestantism were among those who supported Nazism.[34] It is a chilling warning to take care in what physical goals are chosen. Societies have repeatedly made race part of their ideal. It is not logically necessary, but is it socially avoidable? A possible defense might be the dissemination of choice to the point where it could not be used as a weapon of one racial group against another. I will propose such a process later in this chapter.

---

32. Gary E. Crum, "Contested Terrain: The Nazi Analogy in Bioethics: Commentary," *Hastings Center Report* 18 (Aug./Sept. 1988): 31.

33. Donald J. Dietrich, *Catholic Citizens in the Third Reich: Psycho-Social Principles and Moral Reasoning* (New York: Transaction Books, 1988).

34. Milton Himmelfarb, *Hastings Center Report* 6 (Aug. 1976), special supplement, p. 11. Paul Weindling found eugenics advocates within the political left that was working against Nazism: *Health, Race and German Politics Between National Unification and Nazism 1870-1945* (Cambridge: Cambridge University Press, 1989).

Welcoming Diversity

Encouraging a variety of uses of genetic intervention is important not only for defense against group abuses but for human flourishing. Genetic intervention directly addresses the physical form that is so central to being human. Deciding what is appropriate is then related to the most basic questions of the purpose and place of human life. Our physical nature is foundational to our very existence. It should only be changed with great care, yet it is probably not now the best expression of what human beings can physically be. Physical change could be for the better. While genetic intervention should be incremental in an area of such implications, it would best also take place in the light of considered long-range goals lest small steps culminate in unwanted results. Since long-range goals tend to reflect deeply held values and worldviews, a working consensus is even more difficult to obtain than immediate cooperative choices. A detailed social consensus on what human beings should be is unlikely.

"The United States, like many other societies, is morally pluralistic: no one set of beliefs about how it is good or fitting for human beings to live their lives prevails in American society. Although some quite general beliefs about human good are widely shared in American society, many beliefs about human good are widely, deeply, and persistently disputed."[35] By pluralistic I do not mean secular. A secular worldview is only one of many competitors for shaping our perception and commitment. Our society in all the variety of its different conceptions of the end and purpose of being human, of justice and autonomy, of benefits and harms,[36] is likely to continue to respond to genetic intervention in a plethora of different ways.

While the conception of what is most human varies dramatically, there may be, however, at least four characteristics of human beings that are widely affirmed and could structure a response to the question of direction. Those four traits of human beings are that we are finite, fallible, self-concerned, and diverse. First, human beings are finite in that they do not have access to all information, nor could they comprehend it if they did. As finite beings, people tend to have varied incomplete sets of knowledge that lead to different choices. One set of knowledge may lead toward different solutions from that of another or partially overlapping set.

35. Michael J. Perry, *Love and Power: The Role of Religion and Morality in American Politics* (New York: Oxford University Press, 1991), p. 8.
36. James F. Childress, *Who Should Decide? Paternalism in Health Care* (New York: Oxford University Press, 1982), p. 48.

Second, human beings are fallible in that even when all applicable information is available, they may still make mistakes in understanding and judgment. Fallibility multiplies the diversity of choices from fact sets by further varying the responses to any one set. Even if there were one clear choice that followed from a shared set of information and values, a variety of choices would probably be made. Human fallibility is further compounded by the human tendency to misjudge the degree of one's finiteness when pursuing a desired goal.[37]

Third, it is widely held that human beings tend primarily to be self-concerned. In the Christian tradition this is seen as a corollary of sin. Yet even apart from sin, each person tends to be more concerned with his or her own personal welfare than that of others. While there are many instances of exception and tempering, such as in the care of offspring, the tendency is prevalent. Choices made on behalf of others bear the risk of being made more for the one who is choosing than for those who will receive the intervention. That is as much from the impinging proximity of one's own needs and desires as from any intentional priority. The self-interest of one often calls for a choice different from the self-interest of another. Rarely do the self-interested choices of all completely coincide. Even choices that are harmful for most people usually benefit someone. Since there are many different individuals and interest groups, choices of direction will vary further.

Fourth, people are diverse. There is a great degree of variation in values and the weight given to each value from one person or group to another. People often have different end goals as to what is desirable. Faced with the same information, logic, and joint interests, they may still weigh them differently for different ends. Considering these characteristics of human beings, complete consensus on direction for intervention in human genetics is unlikely. A probable lack of consensus in this case is not a pessimistic expectation. Consensus might be undesirable even if it could be obtained.

In light of the above human characteristics, it may be that complete consensus concerning the best direction for genetic intervention would not be helpful. One enforced ideal too easily lends itself to the loss of enriching diversity and even worse to the type of abuses seen in the recent Nazi past. A wide latitude for different choices is preferable. What is both needed and possible for implementation in the national community is a process of widely disseminated choice within the broad bounds of limited societal consensus, not a unanimous and enforced ideal. Such a process of decision is described below.

37. James B. Nelson, *Human Medicine: Ethical Perspectives on New Medical Issues* (Minneapolis: Augsburg, 1973), p. 94.

Who Decides?

If detailed consensus on the direction of genetic intervention is unlikely as a community and undesirable, how could choices best be made? Rather than one set conclusion, a process of who decides may be our best community response. There are a number of proposed arbiters.

Chance

Currently genetic heritage appears to be primarily a matter of chance. Some theological perspectives would argue that it is God who sets genetic endowment through apparent chance and so the process should not be interfered with, while others say that God could exercise providence through human choice. An advantage to chance is the diversity of combinations it presents. Any intentional and effective influence on human genetics would be likely, by definition, to lessen the range closer to those endowments deemed desirable.[38] The exception to such narrowing would be if choice is widely disseminated and those who choose, choose substantially differently.

Robert Sinsheimer argues that human genetic endowment is not best left to chance. By intervening human beings could enlarge collective freedom and concurrent responsibility.[39] He goes on to argue that much of human progress has been the consequence of human effort to reduce the role of chance in the chance of hunger or cold, the chance of attack, the chance of plague. By limiting chance human beings have actually increased their choice and freedom.[40]

Future Generations

Directional choices could be left to future generations, but this generation cannot avoid the choice of intervening or not. If the choice is not to intervene,

38. Jonathan Glover, *What Sort of People Should There Be?* (New York: Penguin, 1984), p. 47.
39. Robert L. Sinsheimer, "Genetic Intervention and Values: Are All Men Created Equal?" in *Modifying Man: Implications and Ethics,* ed. Craig Ellison (Washington, D.C.: University Press of America, 1978), pp. 122-23.
40. Sinsheimer, "Genetic Intervention," pp. 122-23. Roger L. Shinn states this point as well in "The Ethics of Genetic Engineering," in *The Implications of the Chemical-Biological Revolution* (North Dakota State University, 1967), p. 16.

future generations would begin with a different set of givens than if this generation began intervention. Such reticence would probably reduce future choices by not enhancing abilities, hence opportunities. If advantages of intervention are ever to be realized, some present generation must begin on behalf of the next. Always leaving genetic intervention to future generations would lead to infinite deferment. Instead genetic intervention could emphasize the increase of capacity to expand future generation choices and be incremental to lessen any unexpected harm.

Research Scientists

If genetic intervention is begun within a given generation, the involved decisions could be left to scientists, physicians, an appointed panel, parents, or a legislative body. Genetic scientists would have the best grasp of the involved technical information and possibilities. They would be the experts on what was materially feasible.

For E. O. Wilson this expertise in empirical knowledge, particularly of the theory of evolution, is the only sure and worthy guide for human choice. Wilson argues that "like everyone else, philosophers measure their personal emotional responses to various alternatives as though consulting a hidden oracle."[41] The irony according to Wilson is that a person's emotional-ethical responses or "oracle" are part of the survival kit bequeathed by the natural selection of evolution to aid survival and propagation. The "authority" of emotions in ethical matters is that in general they help their bearer make choices that will help that bearer or near relatives propagate, hence spreading the genes for that emotional-ethical response. Since scientists, indeed sociobiologists, best know the mechanism of such evolving thought processes, they would be the appropriate ones to guide future selection.

Empirical knowledge of our biological nature does reveal the limits of what physical change is currently feasible, but it is more difficult to show that it offers guidance for what is desirable. It demarcates what is physically possible, but not clearly what should be sought. The perspectives of geneticists and other scientists would probably not be representative of the concerns of the population at large, and they would be as subject to the vicissitudes of self-interest as any other group. Institutional finance, personal prestige, and fear of liability could gain inordinate influence.

41. Edward O. Wilson, *On Human Nature* (Toronto: Bantam, 1982), p. 6.

Physicians

Fowler, Juengst, and Zimmerman advocate extending the patient-centered ethic of clinical therapy to genetic intervention decisions.[42] The medical model is already in place and effective. Genetic intervention would begin as help for the presenting patient to overcome a reproductive health problem of possibly conceiving children with genetic disease.[43] Decisions would be made by parents with physician guidance. As characteristic of clinical medicine, the focus would be on compassion for those immediately present with little if any concern for long-run implications.[44] Questions of what genetic intervention is appropriate would be "ultimately questions for professional conscience and vision, not public policy."[45] The apparent assumption is that only fellow professionals can fully understand and guide the practice of the profession, and hence set the appropriate standards. When faced with new therapies, "If pediatricians do not respond to these dilemmas thoughtfully, carefully, and forcefully, other decision makers, whose decisions may reflect political, social, or market forces, may then respond in ways that do not reflect the best interests of children."[46]

Their concern is perceptive and laudable for all involved in such choices, including physicians. Traditionally physicians have had complete control of medical decisions because they purport to have solely the patient's best interests in mind. Fortunately, this may often be substantially the case. Unfortunately, medical expertise does not guarantee it. Physicians are as subject to the vicissitudes of humanity as any other human beings. Troyen Brennan writes to his fellow physicians that they need to take care to follow "a truly ethical stance consistent with just doctoring and not merely, as it appears they have in the past, turn these ethical propositions to their own advantage."[47] Beneficence, particularly if practiced unilaterally, can

42. Gregory Fowler, Eric T. Juengst, and Burke K. Zimmerman, "Germ-Line Gene Therapy and the Clinical Ethos of Medical Genetics," *Theoretical Medicine* 10, no. 2 (June 1989): 151-65.

43. Eric T. Juengst, "The NIH 'Points to Consider' and the Limits of Human Gene Therapy," *Human Gene Therapy* 1 (1990): 430.

44. Arthur Zucker and David Patriquin, "Moral Issues Arising from Genetics," *Listening: Journal of Religion and Culture* 22, no. 1 (1987): 65-85.

45. Eric T. Juengst, "The NIH 'Points to Consider' and the Limits of Human Gene Therapy," *Human Gene Therapy* 1 (1990): 431.

46. John Lantos, Mark Siegler, and Leona Cuttler, "Ethical Issues in Growth Hormone Therapy," *Journal of the American Medical Association* 261, no. 7 (1989): 1024.

47. Troyen A. Brennan, *Just Doctoring: Medical Ethics in the Liberal State* (Berkeley and Los Angeles: University of California Press, 1991), p. 238.

mask self-interest. "Motivating reasons can diverge from justifying reasons."[48]

One can see advantages to the current clinical model such as the working out of these issues in the privacy of the doctor-patient relationship. Choices could be made personally with minimum outside interference, yet professional expertise would guide the process. Problems such as the general public's tendency "to underestimate familiar risks and overestimate risks that are unfamiliar, hard to understand, invisible, involuntary, and/or potentially catastrophic," would be tempered by knowledgeable and professional counsel.[49] However, all the limits of current medical practice might be carried over such as lack of access or accountability.[50]

Expert Panel

A wider based group of experts could be assembled to include along with genetic scientists other leaders such as clergy, ethicists, and political scientists to deliberate and choose. Such a body would have the advantage of sustained, careful, and more comprehensive reflection on the involved questions. It would still, however, lack the breadth of society, for no one committee can hold representatives of all. This would be true even if the committee was as broadly based as the Public Policy Advisory Committee proposed by Jeremy Rifkin and his Foundation on Economic Trends.[51] He has argued that such an expert advisory committee should include one or more experts in each of the following fields: protecting medical care and insurance consumers, workplace discrimination, women's rights, rights of minorities, disabled rights, and legal rights to privacy and civil liberties. While such a committee would presumably care for people's rights, it could not possibly represent all perspectives and concerns. No committee could.

Even with such a broadly based committee, committee work could lend itself to self-interested manipulation and would, if it could choose for all, potentially eliminate some traits that are highly valued by minority groups or

48. Childress, *Who Should Decide? Paternalism in Health Care*, pp. 43-44.

49. W. French Anderson, "Human Gene Therapy: Why Draw a Line?" *Journal of Medicine and Philosophy* 14, no. 6 (1989): 691.

50. For an example of a modified structure for clinical medicine, see Brennan, *Just Doctoring*.

51. Foundation on Economic Trends, "Proposed Amendment to the National Institutes of Health Guidelines for Research Involving Recombinant DNA Molecules to Establish a Public Policy Advisory Committee," *Human Gene Therapy* 2 (1991): 133.

subsets of the groups that do have representation. To the degree such an expert committee gave specific recommendations, it would centralize the thrust of intervention. Observing the allotment of kidney dialysis by a Seattle committee, Sanders and Dukeminier wrote that "the Pacific Northwest is no place for a Henry David Thoreau with bad kidneys."[52] Given centralized power to choose, even a committee of minorities threatens diversity. Such a council could serve as a long-range advisor but not well as the sole locus of decision.[53] No matter how erudite the committee it could not match the best efforts and innovations of millions of people all trying to do the best for themselves and their children. Insights can come from countless unheralded corners.

Parents

Societal diversity adds not only valued variety to life, but also greater societal adaptability and room for minority perspectives.[54] Garland Allen has written that science and technology both in research and application have always been controlled by whichever class was in power.[55] By disseminating choice, minority perspectives are protected as required by respect for those who hold them and as potential precursors for views that could be someday adopted by the society at large. Disseminating choice lessens the chance of unified movement but also maximizes maintained options for the future. Diversity would probably be served best by dispersing intervention choices to the widest level, the choices of persons for themselves and parents for their children. There is precedent for the latter in that parents already make most decisions on behalf of their children.

On the one hand the same freedom, however, that is advocated to protect parents from coercion may be turned to unbridled choices of too much intervention. "If the changes that will be achieved are sufficiently desirable so

52. David Sanders and Jesse Dukeminier Jr., "Medical Advance and Legal Lag: Hemodialysis and Kidney Transplantation," *UCLA Law Review* 15 (1968): 378.

53. For a parallel with past bioethics commissions established by the federal government see John C. Fletcher and Franklin G. Miller, "The Promise and Perils of Public Bioethics," in *The Ethics of Research Involving Human Subjects: Facing the Twenty-first Century*, ed. Harold Y. Vanderpool (Frederick, Md.: University Publishing Group, 1996), pp. 155-84.

54. Richard A. McCormick, *The Critical Calling: Reflections on Moral Dilemmas Since Vatican II* (Washington, D.C.: Georgetown University Press, 1989), p. 268.

55. Garland Allen, "Genetics, Eugenics, and Class Struggle," *Symposium on the History and Teaching of Genetics: Thirteenth International Congress of Genetics* 79 (June 1975): 29.

that they not only avoid social harms but also facilitate individual success in a high-technology, post industrial society, then the genetic changes will be desired by individuals for their children."[56] In the American tradition of personal freedom, genetic intervention might be pushed more aggressively by autonomous individual parents than by the authoritarianism often feared. On the other hand, parental objection to needed intervention has been raised as a potential problem. Genetic intervention could be expected for some cases as efficacious medical treatment is required in some cases for children now.[57] Conflicts between parental choice and what society comes to regard as a decent minimum could develop. Parental choice may be limited to be sure that the child is protected from an intervention or lack of intervention known to be overwhelmingly harmful.[58] Such legislated constraints would raise concerns of past eugenic coercion.[59]

Other problems with parental choice might include intentional or unintentional usurpation of parental choice by the counselors who present the possible options, a lack of understanding and foresight when faced with complicated and technical choices,[60] or self-centered choices by the parents. Generally our society hopes that parents have the best interest of their children in mind. Would there be a role for government provision or restraint?

Legislatures

Four arguments need to be considered for the role of government. First, William Vukowich advocates that legislatures alone should make the involved choices for more effective unity of effort. Legislation should be passed that sets standard intervention for each child.

> The selection of desirable and undesirable traits can be left to the legislature. Indeed, legislatures have enacted negative eugenic laws in the past. Al-

---

56. H. Tristram Engelhardt Jr., "Human Nature Revisited," in *Ethics, Politics, and Human Nature*, ed. Ellen Frankel Paul, Fred D. Miller Jr., and Jeffrey Paul (Oxford: Basil Blackwell, 1991), p. 189.

57. Hardy Jones, "Genetic Endowment and Obligations to Future Generations," in *Responsibilities to Future Generations*, ed. Ernest Partridge (Buffalo, N.Y.: Prometheus, 1981), p. 250.

58. World Council of Churches, Church and Society, *Manipulating Life: Ethical Issues in Genetic Engineering* (Geneva: World Council of Churches, 1982), p. 8.

59. Juengst, "The NIH 'Points to Consider,'" p. 430.

60. Bo Lindell, "Ethical and Social Issues in Risk Management," in *Faith and Science in an Unjust World*, ed. Roger L. Shinn (Philadelphia: Fortress, 1980), p. 126.

though legislative selection of genetic traits could open the door for oppressive practices, traditional constitutional limitations should insure against abuses. Legislative rather than parental choice would provide greater unity of effort: it would be a more effective means of diminishing detrimental genes and propagating superior ones than parental choices that would vary from couple to couple.[61]

Utopias generally have been optimistic about government leadership. Plato's guardians and Well's samurai would be examples. In contrast, Justice Brandeis has written that "experience should teach us to be most on our guard to protect liberty when the government's purposes are beneficial. Men born to freedom are naturally alert to repel invasion of their liberty by evil-minded rulers. The greater dangers to liberty lurk in insidious encroachment by men of zeal, well-meaning but without understanding."[62] Joseph Fletcher counsels that descriptions of dystopias such as that of *Brave New World* involve totalitarian regimes abusing human engineering, but this does not mean that human engineering necessarily leads to totalitarianism.[63]

The legislature as sole decider could grant an efficiency of unified vision and development,[64] and may be less finite than one set of parents, but it may be more fallible than all parents acting separately. At least the mistakes of parental decisions would be spread out, for potentially less possible maximum gain of unified development but also less chance of unmitigated disaster of everyone enhanced in a deleterious way. Usually acting by majority decision, a legislature as sole decider would probably limit diversity and be more easily manipulated to one group's self-interest. "Nations, like individuals, are endlessly tempted to claim that they are more moral than they are."[65]

Second, for Margery Shaw, the legislature has a duty to protect public health. "It should be incumbent upon the law to control the spread of genes causing severe deleterious effect just as disabling pathogenic bacteria and vi-

---

61. William Vukowich, "The Dawning of the Brave New World — Legal, Ethical, and Social Issues of Genetics," *University of Illinois Law Forum* 2 (1971): 202.

62. Justice Louis Brandeis, "Olmstead vs. U.S." 277 U.S. 479 (1928).

63. Joseph Fletcher, "New Beginnings in Life: A Theologian's Response," in *The New Genetics and the Future of Man*, ed. Michael Hamilton (Grand Rapids: Eerdmans, 1972), p. 84; Aldous Huxley, *Brave New World* (New York: Harper and Row, 1969).

64. John Naisbet argues the importance of such focused vision for any effective development in *Megatrends: Ten New Directions Transforming Our Lives* (New York: Warner, 1984), pp. 98-99.

65. Kenneth Thompson, *Morality and Foreign Policy* (Baton Rouge: Louisiana State University Press, 1980), p. xi.

ruses are controlled."[66] This is not only for the sake of the treated individual but also because of the effect on all the other people in the community. Fewer individuals incompetent to care for themselves and hence in need of protection mean more talent and more productivity for one another. Such a legislated requirement would lead into direct conflict with institutions and individuals that do not accept genetic intervention.

Third, others expect a right to intervention that would probably require public provision and enforcement. "It seems quite certain that with further advances in genetics our concept of human rights, and our concern with the quality of life, will be enriched with a new right: that of being born without the handicap of a readily preventable serious genetic defect."[67] To date, many children are born with genetic conditions that cause pain and disability. When nothing can be done about it, that is unfortunate, not unjust.[68] However, if the means is available to greatly increase the chance that a child will be born with a healthy genetic endowment, and we do not do that for them, that omission may indeed be unjust and uncaring. There might be a "genetic decent minimum" that is widely recognized and should be provided as helpful to any human being.[69]

The feminists Goerlich, Kronnich, and Degener each argue that individuals and society should adapt to given genetic endowments, not genetic endowments to individual or social desires,[70] yet many genetic disabilities would reduce choice in any society. "Huntington's is, above all, a disease of endless replication, reducing the wonderful multiplicity of human lives to a dreary, deadening sameness, repeating over and over again the same awful saga."[71] Disease lessens diversity by decreasing opportunity for different

---

66. Margery Shaw, "Conditional Prospective Rights of the Fetus," *Journal of Legal Medicine* 5 (1984): 63-116, as quoted by Neil A. Holtzman, "Recombinant DNA Technology, Genetic Tests, and Public Policy," *American Journal of Human Genetics* 42 (1988): 628.

67. Bernard D. Davis, "Ethical and Technical Aspects of Genetic Intervention," *The New England Journal of Medicine* 285, no. 14 (1971): 800.

68. Leonard Fleck writes of this shift in "Justice, Rights, and Alzheimer Disease Genetics," in *Genetic Testing for Alzheimer Disease: Ethical and Clinical Issues*, ed. Stephen G. Post and Peter J. Whitehouse (Baltimore: Johns Hopkins University Press, 1998), p. 202.

69. Allen Buchanan, Dan W. Brock, Norman Daniels, and Daniel Wikler, *From Chance to Choice: Genetics and Justice* (Cambridge: Cambridge University Press, 2000), pp. 81-82, 174.

70. Annette Goerlich and Margaret Krannich, "The Gene Politics of the European Community," *Issues in Reproductive and Genetic Engineering* 2 (1989): 214; Degener, "Female Self-Determination," pp. 94, 98.

71. Alice Wexler, *Mapping Fate: A Memoir of Family, Risk, and Genetic Research* (New York: Random House, 1995), p. xxv.

choices. There is little variety among the lifestyle choices of corpses. Alive and valued people who suffer genetic disabilities make important choices in how they leverage what they can do, yet they do not have as many choices as people who have more capable bodies. There may be consensus that government should reduce the most horrific genetic harms for the sake of each person.

Walters and Palmer argue that it is preferable that genetic disease be eliminated by voluntary genetic screening and therapy. They are aware that "if a voluntary program has been tried and has failed because of public inertia or unreasonable resistance, a mandatory program might seem to be morally justifiable in this case, even to a civil libertarian, as a reasonable means to a highly desirable end."[72] Walters and Palmer are thinking about an attempt to rid a population of cystic fibrosis, a laudable goal. Such an effort has parallels in the eradication of smallpox. On the other hand, the phrase "unreasonable resistance" is fraught with dangerous precedence. Walters and Palmer argue then that such a campaign may well be offered and would probably be widely accepted as current screening programs, but it should not be required by any state. Such would violate a basic human right.[73] Yet we do mandate immunization for the sake of the child, even against parental wishes. This difficult balance will be discussed further under the workable model proposal.

The fourth community concern that is likely to draw legislative interest is that of long-term mass effects. What might be advantageous for individuals could have a negative impact when widely practiced. Parents would not necessarily be trying to change society, just trying to give their own children the best possible start. Individuals and society already work sometimes successfully, sometimes not, at building communities that encompass substantial differences in levels of ability. Depending on how it is implemented, genetic intervention could lessen the distance between people's opportunities or increase them. Issues of direction and distribution would be crucial. Lee Silver has projected that if only an elite group has access to the technology, eventually two classes of human beings could eventually differentiate as separate species. They might be called the "naturals" and the gene enriched or "GenRich."[74]

Actual divergence would require population isolation. As long as there is intermarriage, genes introduced to the genrich would disperse to the naturals. The development of separate species would probably require an isolated

72. LeRoy Walters and Julie Gage Palmer, *The Ethics of Human Gene Therapy* (New York: Oxford University Press, 1997), p. 87.

73. Walters and Palmer, *The Ethics of Human Gene Therapy*, p. 88.

74. Lee M. Silver, *Remaking Eden: How Genetic Engineering and Cloning Will Transform the American Family* (New York: Avon, 1998), p. 4.

colony, such as on Mars, spurred by selection for the Martian environment. Long before there could be differentiation into separate species, government will be challenged on its commitment to equality of opportunity.

One requirement that government could well introduce from the start would be assurance of safety, as it already does for other medical interventions. This would include insisting on incremental steps. Our history has been to use new methods and products before we know their full impact. That has included untoward effects from aerosols harming the ozone layer to DDT weakening eagle eggs.

A Workable Model

A workable model that allows choice, yet remains thoughtful and accountable, might be that of combining the best contribution of each decision group with accountability to the others. The process might call for genetic scientists to tell what is technically possible, expert panels to offer long-range integrated advice, chance to set a starting point, legislatures to set a choice perimeter, physicians to enable thoughtful choices, parents to choose with caring diversity, and future generations to reverse or augment the incremental changes as appropriate to them.

The resulting society could develop considerable variety from one group to another. Robert Nozick has called such a process a utopia of utopias.[75] A process would be in place, not a set ideal, for people to make their own choices within limits so universally felt by society that they would be required. The balance would be constantly tested by practical choices, changing circumstances, and contributing judgments. An extensive literature has already developed addressing the integration of moral and religious belief, politics and law, in our pluralistic society.[76]

75. Robert Nozick, *Anarchy, State, and Utopia* (New York: Basic Books, 1974), pp. 297-332.

76. Robert Audi and Nicholas Wolterstorff, *Religion in the Public Square: The Place of Religious Convictions in Political Debate* (Lanham: Rowman & Littlefield, 1997); Ronald F. Thiemann, *Religion in Public Life: A Dilemma for Democracy* (Washington, D.C.: Georgetown University Press, 1996); Stephen L. Carter, *The Culture of Disbelief: How American Law and Politics Trivializes Religious Devotion* (New York: Doubleday, 1994); Michael J. Perry, *Love and Power: The Role of Religion and Morality in American Politics* (New York: Oxford University Press, 1991) and *Morality, Politics, and Law* (New York: Oxford University Press, 1988). For a more sociological emphasis, see James Davison Hunter, *Culture Wars: The Struggle to Define America* (New York: Basic Books, 1991).

Our current system of medical care for minors might be a case in point. Parents have considerable latitude in where and how they seek medical care for their children, yet the law requires a societally perceived basic minimum of certain essential care.[77] Parents are held responsible to meet that minimum. When parental choice disregards the child's needs, society, often clumsily, can intervene as minimally as necessary. A degree of genetic intervention could someday be required as the law now forces vaccination or more dramatically a blood transfusion for an infant that would die without it. When the social consensus of a basic minimum conflicts with what the parents think is best, the clash is a tragic one, as seen in the deaths of children followed by criminal prosecution of well-meaning parents.[78] The balance of competing interests would require constant testing and renegotiation.

Another case is that of education. Parents are given considerable latitude to shape a child through education as they believe is best for the child, yet the government holds parents accountable for requirements the society overwhelmingly considers necessary. Our social consensus requires a minimum of a certain amount of education. To function in our society all children should receive the opportunity to learn basic skills such as how to read and write. How parents achieve that goal, whether through home schooling, public schools, professional tutoring, boarding schools, or other means, is open to varying degrees.

People could advocate that genetic intervention be prohibited or limited to certain types of intervention. If by persuasion such reservation became part of the broad societal consensus, it could be enforced by society. Allowing considerable latitude for such decisions does not assume ethical relativism. Hopefully the test of publicity and limits where there is broad societal consensus would constrain rank abuse while allowing the diversity of response that would be most helpful. Consensus adjustment on appropriate limits would likely occur over time.

Would human beings thereby have the wisdom to choose well in an informed, accountable process? The stakes would be high and implementation would best be in increments. As Paul Ramsey warns,

---

77. There are related issues of how involved parties exercise their role. For example, government can enter through numerous means such as taxation or forgiveness of taxes, criminal law, providing information or facilities. Note James F. Childress, *Priorities in Biomedical Ethics* (Philadelphia: Westminster, 1981), p. 101.

78. For an explanation and defense of the beliefs and practice of followers of Mary Baker Eddy in this regard see *Christian Science: A Sourcebook of Contemporary Materials* (Boston: The Christian Science Publishing Society, 1990).

Mankind has not evidenced much wisdom in the control and redirection of his environment. It would seem unreasonable to believe that by adding to his environmental follies one or another of these grand designs for reconstructing himself, man would then show sudden increase in wisdom. If genetic policy-making were not miraculously improved over public policy-making in environmental and political matters, then access to the Tree of Life (meaning genetic management of future generations) could cause grave damage.[79]

The best community goal would be one of improvement, not perfection, in a structured but adjustable process, not the application of a unanimous and enforced ideal. So constructed the process could respect the diversity of parents and recipients within the minimal constraints of broad societal consensus.

Chapter Summary

The past history of eugenics warns us about current use of genetic intervention. The pattern of coercion, racism, and narrowed choices was abhorrent. That, however, does not preclude freedom and improvement in genetic heritage today. Diversity and group protection could be served by widely disseminating the power to choose genetic intervention to those who receive it and to parents on behalf of their children. The best community goal would be one of improvement, not perfection, in a structured but adjustable process, not the application of a unanimous and enforced ideal. So constructed the process could respect the diversity of parents and recipients within the minimal constraints of broad societal consensus.

79. Ramsey, *Fabricated Man*, p. 96.

A Concluding Perspective
from the Christian Tradition

Genes shape us powerfully. With the developing techniques of genetic intervention we can increasingly understand and shape them. How should we use this new capability? We began to address that question with three chapters that set important context. The first sought to defuse a common assumption that science and the Christian tradition are engaged in a fight to the death. On the contrary, they have much in common. With that explained we could overview the genetic role in human form and behavior. In chapter 2 we recognized that technology shapes us and that if we make the effort to do so, we may shape it. However, directing technology requires answering the basic question of where we want to go. What is our purpose? In chapter 3 a response to that question was described from the Christian tradition. By God's grace and calling human beings are to sustain, restore, and improve ourselves and our world.

With the above in place we walked through four levels of human genetic intervention from least to most transforming, which is also the order from techniques currently in widest use to applications that are more distant. Those types of intervention were genetic research, genetic testing, genetically designed pharmaceuticals, and genetic surgery that directly alters a person's genes. At each level a chapter was devoted first to the involved choices and implications for individuals, then a chapter for families, and finally a chapter on community concerns. Every chapter ended with a brief summation. For this concluding perspective I will not repeat those findings but rather highlight two themes that have run throughout the study. They are that (1) genes are an *important* part of human life, but just as significantly that (2) genes are *only* a part of human life.

344

Genes Are an Important Part of Human Life

Matter matters. So much of being human is physical. We are almost always conscious of our experiences of hunger, thirst, sleep, posture, pleasure, heat, and cold. Soren Kierkegaard wrote that the curse of the philosopher is always having to turn aside from the sublime world of thought in order to sneeze. In the words of Genesis, we are dust and to dust we shall return. In the meantime we are physical beings. Adam, the first human being in Genesis, is described as being made from the dust yet uniquely in-breathed with God's Spirit. Human beings have a special calling yet are of this earth. It is difficult to pray if one is physically exhausted. By use we build physical patterns of memory and response into our brains, which are then manifested as our character. The physical side of who we are is deeply interrelated with all that we are. Being in part physical beings is not a bad thing. God created both this material world and our physical form and declared it "good." The physical is not our ultimate concern, but we should care about it because it is part of our God-given character, place, and stewardship.

Genetics can play an important role in the suffering and comforts, capacities and infirmities, of our physical life. Genetic tests that trigger effective treatment for hemochromatosis or warn one in time to plan for dementia can be a boon. Genetic pharmaceuticals such as Humulin that sustain a body afflicted with diabetes and genetic surgery that may someday increase capacity to fight disease or improve one's memory are examples of genuine service genetic intervention can or may eventually offer. As physical beings we can be deeply affected and greatly helped by genetic intervention.

However, the lasting import of our physical bodies is not so much in what they become as in what kind of persons we become as we live and work with and through our physical bodies. Our bodies are where we have the most choice. They are the limited spheres of our greatest influence. We have significant choices in what we place in our mouths and what we speak with them, in what we hold onto with our hands or drop, in what we choose to see with our eyes or turn away from, in where we direct our feet to take us and where we stay. Since this is where most of our decisions are, this is where we have the greatest opportunity to learn to be the kind of people we are meant to be. This is where we can develop patterns of self-discipline and priorities of service. This is where we can learn to pursue what matters. As creative creatures made in God's image, called to be like Christ, and motivated by love for God and one another, we are to carry on the God-given pattern of creation, redemption, and transformation in our physical world. That includes the larger environment and our bodies. We are to sustain our physical bodies, which for our focus here prominently includes our genes. We should restore our bodies when they are damaged, and we should improve them as we can to better serve God and our neighbors.

Our task of growing and improving here is progressively more complicated by simply needing to sustain and restore our physical form. Our physical life inevitably, eventually, falls apart. How fitting that as the task of living through our bodies becomes more difficult, we are hopefully developing the needed discipline and skills to handle it. It is who we are, not just our current physical form, that has the potential by God's grace to go on beyond this life, dwarfing its fragile and limited starting point. In this school for souls God is creatively sustaining, restoring, and improving the people of God as individuals and as a community. Part of that process is for us to respectfully and joyfully join in, reflecting God's image as individuals and community. That goal is achieved in part by how we sustain, restore, and improve our physical world, including our bodies. The gift of "Sabbath rest" is not an excuse for passive indolence.[1] Whether because of the damage of sin or because the good creation can still grow, or both, there are aspects of the physical world entrusted to us that could be better. We are called by God's grace to care for and develop ourselves and our world, including our genes.

Genes Are Only a Part of Human Life

Genes Are Only Part of Our Physical Form

Even if genetic intervention is perfected and widely implemented, there will be limits to what it can physically accomplish. It can be a boon to the cure of disease and increase of physical capacity and all the opportunities that may produce. Most of its eventual benefits are probably not even conceived of as yet. However, there would still be handicaps, accidents, and disease unresponsive to genetic intervention. Correcting the genetic code cannot cure all physical ills and a maximum of successful genetic intervention would not meet all genuine physical concerns. There would still be a need to treat nongenetic diseases, to protect human genes from environmental damage, and to provide people with opportunities to develop the capabilities that they have.[2]

1. Michael Banner extols the importance of living all of life in Sabbath rest in *Christian Ethics and Contemporary Moral Problems* (Cambridge: Cambridge University Press, 1999), pp. 223-24.
2. World Council of Churches, Church and Society, *Manipulating Life: Ethical Issues in Genetic Engineering* (Geneva: World Council of Churches, 1982), p. 9.

A Concluding Perspective from the Christian Tradition

Our Physical Form Is Only Part of Human Life

Genes are not all there is to physical life, and the physical is not all there is to life as a whole. Our God-given bodies should be appreciated in their own right, but whether they last for seventy, one hundred, or someday two hundred years, they are wonders dwarfed by what lasts. What goes on past this physical life is our person. That is our will, priorities, loves, character, discipline, and joys, not our physique. Our physical form is essential here, but instrumental for while we are here. This is where we temporarily live and choose as we become who we are meant to be. Part of that process is learning to give our lives in care for others. The physical world is a temporary stage, a place to choose and learn. It is also a stage in the process of what God has in store for his children. Human destiny is outside and beyond this physical world,[3] which has temporary written all over it. Even our sun will burn out eventually, probably exploding to engulf our planet in flame before leaving anything left of it alone in the dark. Our own lives flicker with no guarantee from day to day and with a maximum length that looks shorter as we approach it.

It is difficult for most of us to remember life in the womb, although it is probable that all readers of this work have spent considerable time in one. It was dark and warm. All the food and oxygen that we needed was provided, although we had no idea of the source of that sustaining cord that conveyed it. It was a puzzling place. We were growing legs that only made the close quarters more cramped. Eyes had nothing to see. Lungs were ill suited for our fluid environment. A voice was available but there was no one to talk to. Then things took a nasty turn. We were flipped upside down and squeezed hard between the walls. The following hours were probably painful for everybody involved. Then suddenly there were colors for the eyes and the freedom to swing legs and arms. We each took a deep breath and our voices were heard for the first time. And we met face-to-face the one who had been sustaining us all along.

The classic Christian tradition sees this life compared to the next much as life in the womb was to this one. The brief transition of death leads to a life more colorful and free than this one. God's children will meet face-to-face the One who has sustained them all along and feel more at home than ever before.

But one should not be born prematurely. There are things in the womb and this life that are best done in each. There is growth and development that

3. Charles E. Curran, "Theology and Genetics: A Multi-faceted Dialogue," *Journal of Ecumenical Studies* 7 (Winter 1970): 75-76.

takes place uniquely in each. For a time human beings are material and dependent on the material, but only for a time. The womb matters. The physical world matters. Both deserve appreciation for their Maker and how they are used to make us. Neither is the point or the end. They are places that give opportunities for becoming. Learning to rightly sustain, restore, and improve our physical world, including our bodies, is part of that process.

Saint Basil wrote almost two millennia ago that "whatever requires an undue amount of thought or trouble or involves a large expenditure of effort and causes our whole life to revolve, as it were, around solicitude for the flesh must be avoided by Christians."[4] Richard McCormick concurs from the Roman Catholic tradition that "excessive concern for the temporal is at some point neglect of the eternal."[5] Since human beings are physical beings, our physical nature and condition cannot be well ignored,[6] yet the body is not the purpose of human life. Augustine writes that people are to use *(uti)* the world as we enjoy *(frui)* God.[7] Too often people reverse the designed order and try to use God to enjoy the world. As essential as the physical is, it is not all-encompassing. The physical world and health in it is a good, but it is not God. We are all at best only temporarily able-bodied. Life goes on for most without physical perfection. Jesus' own priorities were that he forgave sin first, healing the inner human being and only then the outside body. When Jesus declared that his followers should be perfect-complete, he said nothing of their physical bodies.[8] The physical is important to sustaining our current existence. It is worth sustaining, restoring, and improving, but it is only a part of who we are.

While genetic intervention could offer substantial changes for the better in individual and community lives, the most significant choices of life will not be enhanced or set right by any kind or degree of physical intervention.[9] The most important choices in life are not primarily physical ones. Some hope that the use of genetic intervention over time and enhanced by culture

---

4. From "The Long Rules" as quoted by H. Tristram Engelhardt Jr., "Genetic Enhancement and Theosis: Two Models of Therapy," *Christian Bioethics* 5, no. 2 (1999): 198.

5. Richard A. McCormick, "Theology and Bioethics," *Hastings Center Report* 19 (Mar./Apr. 1989): 10.

6. See Mary Midgeley for a philosophical analysis of this in *Beast and Man: The Roots of Human Nature* (New York: Meridian, 1978), p. 310, or for a theological perspective, C. S. Lewis, *Screwtape Letters* (Grand Rapids: Baker, 1943), pp. 24-25.

7. George Forell, *History of Christian Ethics* (Minneapolis: Augsburg, 1982), p. 169.

8. Matthew 5:48.

9. Hessel Bouma III, Douglas Diekema, Edward Langerak, Theodore Rottman, and Allen Verhey, *Christian Faith, Health, and Medical Practice* (Grand Rapids: Eerdmans, 1989), p. 267.

will eventually lead to earthly utopia.[10] Others have observed that in human experience solutions consistently lead to further problems and questions.[11] This has been reflected in most modern utopian literature, which describes worlds of indeterminate process, hopefully progressing, rather than making claims of an achieved final standard.[12] Yet many religious and nonreligious traditions are quite hopeful that despite severe setbacks good can and will increase. Whatever one's eschatology, it is not genetic intervention that will bring heaven or a utopia.[13] Published hopes that genetic intervention might deliver us from homelessness, alcoholism, criminality, divorce, and more are expecting more than physical change can provide by itself.[14]

Genetic intervention, however, could make a portion of life better for people. Some increases in one kind of capability would be mutually exclusive with other enhancements. The physique of a weight lifter is not suitable for a long-distance runner. Yet probably all could benefit from greater resistance to cancer or a better ability to remember what one wants to remember. Allen Verhey has suggested a balance of vision and realism, hope and prudence, that knows both the common grace of God and the intransigence of human pride and sloth.[15] Genetic intervention in human beings at best can increase the physical capacity of human beings. What human beings would choose to do with that increased capability remains in question. We do well to keep in mind not only the boon of the proposed changes but also the limitations of

10. Robert Sinsheimer, as quoted by Leon Kass, "New Beginnings in Life," in *The New Genetics and the Future of Man*, ed. Michael P. Hamilton (Grand Rapids: Eerdmans, 1972), p. 59.

11. Roger L. Shinn, "The Ethics of Genetic Engineering," in *The Implications of the Chemical-Biological Revolution* (North Dakota State University, 1967), p. 22; Curran, "Theology and Genetics," p. 81.

12. Elizabeth Hansot, *Perfection and Progress: Two Modes of Utopian Thought* (Cambridge: MIT Press, 1974), p. 13. An example of the former might include Plato's *Republic*, while of the latter Marge Piercy's *Woman on the Edge of Time* (New York: Fawcett Crest, 1976) or B. F. Skinner's *Walden II* (New York: Macmillan, 1948).

13. John Passmore, *The Perfectibility of Man* (New York: Charles Scribner's Sons, 1970).

14. Neil A. Holtzman, "Policy Implications of Genetic Technologies," *International Journal of Technology Assessment* 10, no. 4 (1994): 570-71; Robert N. Proctor, "Genomics and Eugenics: How Fair Is the Comparison?" in *Gene Mapping: Using Law and Ethics as Guides*, ed. George J. Annas and Sherman Elias (New York: Oxford University Press, 1992), pp. 76-93.

15. Allen Verhey, "The Morality of Genetic Engineering," *Christian Scholar's Review* 14, no. 2 (1985): 133. John Stott has described such a course as refusing to be deceived by utopic dreams but also refusing to give up in hopelessness: *The Year 2000* (Downers Grove, Ill.: InterVarsity Press, 1983).

what genetic change can do for physical health and capacity, as well as of what physical change can achieve for life in general.

So What Place for Human Genetic Intervention?

We should sustain, restore, and improve our bodies genetically so that we can better pursue what matters. Our bodies do not have to be perfect to well serve God. Yet more capable bodies are more capable. Having a body genetically freed from consuming pain frees us to focus on more important things. Having a body genetically enabled to a greater capacity may give us more ability to pursue what matters most. If genetic intervention could someday increase one's powers of perception and understanding of human emotion, that could be used for more effective counseling or to run a more lucrative con game.[16] Increasing our ability does not automatically direct those newfound possibilities to good ends.

Genetic technology, as other technologies, at least makes nature less demanding. Central heating guided by a thermostat frees us from devoting a major portion of the day to chopping and hauling wood. Central heating offers a gift of time that can be used to watch television or to listen to Scripture, enjoy a friend, learn a skill, make a table, teach a child, or create a work of art. With the modification of ourselves and our environment, there is less external requirement for conscious choice and self-discipline. It used to be that if one lazed away the day, one was cold that night. Now one can be just as comfortable anyway. The newfound freedom from meeting natural needs gives greater choice. With more choices there is more responsibility. Jesus said, "To whom much has been given, much shall be required."[17] We are responsible for what we do with what we have. The freedom that used to be available only to the wealthy, who could use servants to insulate themselves from daily tasks, is now available to most of us in technologically rich countries. We can go on to new challenges or not challenge ourselves at all.

Genetics does not so much make us automatically better as it can make us more capable. Genetic intervention, like many technologies, frees us from some constraints and increases our abilities and choices. Pursued as an end in itself it is at best a distraction, and when all-consuming, idolatry. If all we

16. Allen Buchanan, Dan W. Brock, Norman Daniels, and Daniel Wikler, *From Chance to Choice: Genetics and Justice* (Cambridge: Cambridge University Press, 2000), pp. 179-81.

17. Luke 12:48.

manage to do is relieve physical suffering and to control our physical world in the finest degree, our potential will be wasted.[18] Such an effort is worthwhile as a means but is not the ultimate point. The gain is not substantial if the newly available time and energy are squandered. Genetics can free and empower us in some ways, but for what? Retirement golf communities frittering away years of hard-won growth and insight on self-entertainment? Having a genetically honed body is potentially helpful to more worthy goals, an instrumental good, not an intrinsic one. It can even be harmful if the pursuit or use of its extended life and capacities either distracts us from what most matters or so insulates us from challenges for a time that we fail to realize our most important needs in time.

Genetically healing or increasing our physical capacity gives one more opportunity to do the things that matter. As in the parable of the talents, we are to multiply what we have in order to better serve.[19] Having great memory is just a parlor trick by itself. Using an expanded ease in memory to learn a new language so that one can encounter, enjoy, and serve people one could not communicate with before, is a worthy use. An improved immune system that frees one from cancer or the common cold so that one can more comfortably play bridge or hit the slot machines misses the point. Freedom from disease and increase of capacity so that one can better worship, care, serve, discover, *live*, is the point. The womb was essential to our presence and development, but it would not be a fully human life in all that it was meant to be if our lives stopped there. The physical world, including our genetics, is the place where we choose, learn, and grow now. It does not have to be our end. It can be by God's grace our beginning.

18. Gerald P. McKenney calls the single-minded effort to relieve suffering and expand human choice "the Baconian Project." He critiques it in *To Relieve the Human Condition: Bioethics, Technology, and the Body* (Albany: State University of New York Press, 1997).

19. Matthew 25:14-30.

Author Index

Adolfsson, Rolf, 256n.
Alberman, E., 200n.
Alexander, D., 201n.
Allen, David B., 267n.
Allen, Garland, 336
Almeder, Robert F., 299n., 302n.
Alper, Joseph S., 212n.
American College of Medical Genetics, 105n.
American Society of Human Genetics, 175
Anderson, Norman, 285n.
Anderson, V. Elving, 27, 41n., 42n., 203n., 211, 236n., 261n.
Anderson, W. French, 7n., 233, 251, 307n., 323n., 335n.
Annas, George J., 10n., 134, 302n., 323n., 349n.
Aquinas, Thomas, 83
Aristotle, 121
Arnold, Matthew, 188
Aronowitz, Robert A., 243n.
Arras, John D., 127
Asch, Adrienne, 145n., 203n.
Audi, Robert, 16n., 341n.
Augustine, 83, 85, 348
Ayer, A. J., 30n.

Bacchetta, Matthew D., 310n.

Bacon, Francis, 29n., 71n.
Bailey, J. Michael, 47-48
Balen, Frank van, 181n., 295n.
Banner, Michael, 187n., 346n.
Barbour, Ian, 36, 55n.
Barr, Patricia A., 151n.
Bartholome, William G., 109n., 110n.
Beauchamp, Tom L., 8n., 97n., 100-101, 253n.
Beck-Gernsheim, Elisabeth, 13, 163, 198n., 247n., 259n.
Beckwith, Francis J., 111n., 112n., 124n.
Beckwith, Jon, 11n., 212n.
Beecher, Henry, 96n.
Behe, Michael J., 30n.
Benjamin, Martin, 236n., 265n., 267n.
Berger, Brigette, 318
Berger, Edward M., 243
Bernhardt, B. A., 161n.
Berry, R. J., 25-27, 45n., 46n., 55n., 125
Binstock, Robert H., 211n.
Bird, Thomas D., 141n.
Birkner, John H., 12
Blacker, C. P., 324n.
Blackmun, Harry, 112
Blaese, Michael, 7n.
Boethius, 78n.
Boivin, Michael J., 257n.
Bonnicksen, Andrea L., 121n.

352

Subject Index

Abortion: to avoid disability, 202; to avoid disease, 183-84, 222-24; Christian scripture 113-16; legal right to, 199; status of the embryo, 122-26; status of the fetus, 113-23; status of the zygote, 126-34; techniques of, 131-32, 199-200

Adoption: to avoid disease, 179; Christian affirmation, 292-96, 299; difficulties in, 179, 299-300; of zygotes 179-80

Alzheimer's disease: diagnosis, 154-55; familial type (FAD), 140-41; perceiving risk, 160; test availability, 150-51

Attitudes: commodification, 143; illustrated by genetic discoveries, 43-51; shaped by intervening, 195-97; 288-90

Autonomy: definition of, 98; protecting future, 304, 315-17; respect for, 97-98

Behavior: genetic influence on, 39-43, 46-49, 255-58, 277, 345; influenced by pharmaceuticals, 255-58

Cancer: BRCA1 testing, 153-54, 157, 159-60, 171-72; patent on a method of inducing cancer, 142-43; predictive tests, 153-54; source of genetic research funds, 4; therapy, 276; TP53, 197

Children: deciding on their behalf, 108-

10, 254, 336-37; desire to have, 176-77; obligation to future generations, 310-13; obligation to have, 178-79; open future, 259-61; pro-creation of 176-204; property claims, 291-92; seeking best physical base, 286, 301; sharing genes with parents, 180-85, 291-96; testing for adult onset disease, 174-76

Christian resources: church community, 224-27; counseling test results, 168-69; importance for bioethics, 3, 15-17; policy statements, 141-44, 225-26, 277-78, 307; purpose of life, 64-90, 344-51

Cloning: concerns, 301-5; definition of, 296; genetic twins, 296-97; inadequate reasons to clone, 297-99; psychological effects, 302-3; reasons to clone that have more weight, 299-301; spare parts, 298-99

Commodification, 143, 288-90

Community: autonomy, 98; church support, 224-27; church teaching 141-44, 225-26, 277-78, 307, 346, 348-51; connections from genetics, 46, 138; directing genetic intervention, 341-43; diversity, 330-32; employment, 216-19; equality of opportunity, 266-71; goals of medicine, 263-66; group consent, 139-41; insurance, 206-16; patents,

MATTHEW

A Devotional Commentary